Collapsing Gracefully: Making a Built Environment that is Fit for the Future

Emilio Garcia · Brenda Vale · Robert Vale

Collapsing Gracefully: Making a Built Environment that is Fit for the Future

Springer

Emilio Garcia
School of Architecture and Planning
University of Auckland
Auckland, New Zealand

Brenda Vale
School of Architecture
Victoria University of Wellington
Wellington, New Zealand

Robert Vale
Wellington, New Zealand

ISBN 978-3-030-77785-2   ISBN 978-3-030-77783-8 (eBook)
https://doi.org/10.1007/978-3-030-77783-8

© The Editor(s) (if applicable) and The Author(s), under exclusive license to Springer Nature Switzerland AG 2021

This work is subject to copyright. All rights are solely and exclusively licensed by the Publisher, whether the whole or part of the material is concerned, specifically the rights of translation, reprinting, reuse of illustrations, recitation, broadcasting, reproduction on microfilms or in any other physical way, and transmission or information storage and retrieval, electronic adaptation, computer software, or by similar or dissimilar methodology now known or hereafter developed.

The use of general descriptive names, registered names, trademarks, service marks, etc. in this publication does not imply, even in the absence of a specific statement, that such names are exempt from the relevant protective laws and regulations and therefore free for general use.

The publisher, the authors and the editors are safe to assume that the advice and information in this book are believed to be true and accurate at the date of publication. Neither the publisher nor the authors or the editors give a warranty, expressed or implied, with respect to the material contained herein or for any errors or omissions that may have been made. The publisher remains neutral with regard to jurisdictional claims in published maps and institutional affiliations.

This Springer imprint is published by the registered company Springer Nature Switzerland AG
The registered company address is: Gewerbestrasse 11, 6330 Cham, Switzerland

Preface

This book examines what has caused the societal collapse in the past and applies this to the present, in the face of the latest impacts of climate change at the poles, the need to reduce 70% of our carbon emissions in 11 years, and the growing disproportional environmental impact between a rich minority and the rest of the world. It is also the first time in history that the human habitat is clearly identified with urban landscapes and the concentration of people in cities. This increases the dependence of cities on rural areas to obtain a continuous supply of food and ecosystems services. Moreover, it exposes millions of people living in coastal cities to the threat of sea-level rise. Regardless of these facts, cities keep on developing and growing, investing more energy and resources in their built environments without accounting for the social and environmental costs of doing this. For these reasons, this book focuses on the built environment. With this focus, the aim is to consider what needs to happen to the built environment now to avoid sudden, enforced change in the future. This is not a book about the end of the world, hopeless apocalyptic scenarios, or the struggles of ancient societies to persist. This is a book about understanding critical changes in the context of social and environmental crises and how this could be instrumental in taking future decisions about our habitat.

The book is about applying what has been learned about the societal collapse in the past to the present. Reading it, the aim is to make all involved in making decisions about the built environment—politicians, economists, engineers, planners, designers, educators—think differently about it in order to cope with a very uncertain future, given how long the built environment lasts.

Auckland, New Zealand	Emilio Garcia
Wellington, New Zealand	Brenda Vale
Wellington, New Zealand	Robert Vale

Contents

1	**What Do We Mean by Collapse?**	1
	Introduction	1
	Types of Collapse	2
	The Faith in Economic Growth	3
	The Faith in Technological Development	7
	A Plan B: Collapsing Gracefully	12
	References	14
2	**Current Ideas for Future Built Environments**	17
	Introduction	17
	Smart Cities (Even When It Is a Dumb Idea)	18
	Buildings All at Sea (a Good Place from Which to Watch Tsunamis)	21
	Living in Space (Because We've Made the Earth Uninhabitable)	24
	Grand "Sustainable" Buildings (for the Rich)	26
	The Sustainability of "Sustainable" Houses (also for the Rich)	30
	Investigating the Technical Aspects of a Building That Is Claimed to Be "Sustainable"	30
	Urban Design	31
	Politically Correct Solutions (Even Though They Are Not Correct)	33
	Refugee Camps	33
	Design for Refugees; the IKEA Better Shelter	35
	Housing Refugees	36
	The Solution Must Be Digital and Employ Robots (Even When the Analogical Works Fine and We Have a Lot of People)	37
	Climate Change Solutions Must Sound "Scientifically Plausible" (Even Though They Are not Feasible or Even Necessary)	39
	The Role of Design	41
	Conclusion	42
	References	42

3	**What Can We Learn from the Collapse of Societies in the Past?**	49
	Introduction	49
	Collapse and Survival	50
	Theories Behind the Collapse of Civilisations	50
	Complexity and Societal Hierarchy	55
	More Recent Views of Collapse	57
	Resilience and Collapse	60
	Collapse Theories and the Built Environment	63
	Proposed Solutions Post-collapse	64
	What Can Be Learned from Collapse Theories	65
	References	65
4	**The Modern Built Environment and Its Relationship to Collapse**	69
	Introduction	69
	Perceptions of Collapse in the Built Environment	70
	Reciprocity Between Habitat and Culture	71
	Built Environments, Ecosystems and Collapse	75
	Dealing with Collapse Through a Better Understanding of Sustainability and Resilience	76
	Engineering Resilience and Collapse in the Built Environment	77
	The Ecological Resilience Approach and Collapse	79
	The Importance of Scales in Resilience: Panarchy	83
	Panarchy and the Scales of Collapse in the Built Environment	84
	The Links Between Sustainability, Resilience, Collapse and the Built Environment	86
	Similarities and Differences Between the Understanding of Collapse in History, the Built Environment, Sustainability and Resilience	90
	Why Are These Issues Relevant for Designers?	92
	Some Final Thoughts About the Collapse	93
	References	94
5	**Technology and Collapse**	99
	Part 1: Technology and Complexity	99
	Invention	99
	What Is Technology?	101
	Energy Return On Investment (EROI)	105
	Prefabrication	107
	Technology and Complexity	109
	Progress	112
	Part 2: A Case Study of Technology and Climate Change	113
	Introduction	113
	Risk	114
	Flood Prevention	118
	Living with the Effects of Coastal Erosion and Flooding	121

	Living with Regular Inundation	123
	Pipe Dreams	128
	Flooding and Collapse	130
	Conclusion	133
	References	134
6	**Inequality, Collapse and the Built Environment**	**143**
	Part 1: The Problem of Inequality	143
	Introduction	143
	The Theoretical Link Between Inequality and Collapse	144
	Inequality and Collapse of Ancient Societies	145
	Development and Inequality in Ancient Civilisations	145
	Past Inequality and the Role of the Built Environment	147
	Are Crises and Collapse Levellers of Inequality?	148
	Inequality in Contemporary Built Environments: Clustering Processes	149
	Inequality, Gentrification and Segregation	150
	Real Estate and Gentrification	151
	Other Characteristics of Inequality in the Built Environment	153
	Conclusions	155
	Part 2: Inequality in the Urban Landscape of New Zealand: From the Country to the Plot	156
	Introduction	156
	Inequalities and New Zealand	156
	Maori	159
	New Zealand Housing	159
	The Built Environment as a Vehicle for Accumulating Wealth and Increasing Inequalities	161
	Inequalities in the Built Environment at the Country Scale	165
	Inequality at City Scale: Auckland	167
	Auckland Topography	167
	Renting	168
	Gentrification	169
	Inequality at Neighbourhood Scale: Gentrification in the Suburb of Glen Innes	171
	Economic Impact of Gentrification	174
	Collapse and Gentrification	179
	Inequality at the Plot Scale: A Divided Garage City	179
	Conclusion	181
	References	183
7	**Growth and Collapse**	**189**
	Introduction	189
	Growth	191
	Economic Growth	193
	Growth and Buildings	194

		Urban Growth Patterns	195
		How Its Citizens Perceive the City	197
	Design of the Built Environment		198
	Growth and Collapse		199
	How Should We Grow the Modern Built Environment?		201
	References		203
8	**Growth and Resources**		207
	Introduction		207
	City Living		208
		High-Rise Buildings	209
	Ecological Footprint and GHG Emissions		213
		GHG Emissions and Density	216
	Food and Urban Settlement		217
	EF and Urban Settlement		218
		Urban Cuba	221
	Density and Collapse		223
		Marginal Returns, Urban Complexity and Collapse	224
	References		226
9	**Epidemics, Pandemics and Collapse**		231
	Introduction		231
	Epidemics and the Collapse of a Civilisation: The Case of Tenochtitlan		232
	Pandemics and Collapse		234
		The Impact of the Black Death	235
		The Impact of the Black Death on the Built Environment	236
		Containment Within the Built Environment: Quarantine	237
		Isolation from Urban Landscapes: Lazarettos	238
		Eyam: The Plague Village	238
		What Can Be Learned from Previous Pandemics?	239
	Covid-19		240
		Modern Communication	240
		COVID-19 and the Economy	243
		Environmental Impact of the Pandemic	247
	The Built Environment		248
		COVID-19 and Cities	251
		COVID-19 and Housing	252
	The Resilience of the Built Environment to Changes Induced by COVID-19		254
	Conclusion		257
	References		258

10	**The Architecture of Wealth**	263
	Introduction	263
	Expression of Wealth	264
	Measuring Built Assets and Wealth	265
	Dwellings as Investment	265
	The 2008 Global Financial Crisis and Near Collapse	266
	The Landlord	268
	Flipping	269
	My House—My Castle	270
	Large Houses	274
	Plot Size	276
	Manufacturing and Commerce	278
	The Quest for the Tallest Building	279
	"Green" Buildings as Investment	281
	Empty Buildings	282
	Partial Occupancy	282
	Second Homes	283
	Ghost Cities	285
	Cities of the Dead	286
	Architects and Wealth: Do Architects Only Work for the Wealthy?	289
	Monuments, Wealth and Collapse	291
	References	292
11	**What Should We Do?**	299
	Introduction	299
	The Cost of Climate Change	299
	Paying to Fix It	300
	Who Should Pay to Fix It?	302
	The Cost of Giving up Fossil Fuels	303
	Time to Eat the Rich?	304
	The Built Environment	304
	Cities	304
	Neighbourhoods	305
	Plots and Buildings	306
	What Should Designers Do?	307
	Final Thoughts	308
	References	308

List of Figures

Fig. 1.1	Global GDP growth (per capita) and depletion of biocapacity (per capita). *Data sources* *Global Footprint Network (2019), ** WID.world (2019a)	4
Fig. 1.2	Global carbon dioxide emissions. *Data source* Boden et al. (2017)	6
Fig. 1.3	Global technological development in transistors per chip. *Data source* Rupp (2018)	8
Fig. 1.4	Global energy consumption. *Data source* BP (2020)	10
Fig. 1.5	Income inequality. The chart shows that in 2016 10% of the world population accounted for 52% of global income while the bottom 50% lived with less than 10% of global income and the middle 40% represented 38% of the income share. *Data source* WID.world (2019b)	11
Fig. 2.1	Carbon footprint of cities (per capita). Cities in blue are the 10 "smartest cities" and cities in green are the 10 "stupid cities". The horizontal scale ranks the 20 cities according to their carbon footprints per capita, with 1 being the most sustainable and 20 the least sustainable (adapted from Moran et al., 2018 except for Reykjavik, Douala, Cairo and Nairobi where national averages from 2016 were used from DataBank, 2020)	20
Fig. 2.2	Floating school, Lagos (adapted from https://www.dezeen.com/2014/03/25/makoko-floating-school-nigeria-nle/)	22
Fig. 2.3	Lunar Habitation (adapted from https://www.fosterandpartners.com/projects/lunar-habitation/#gallery)	25
Fig. 2.4	CopenHill Energy Plant (adapted from http://emag.directindustry.com/copenhill-a-waste-to-energy-plant-with-a-ski-slope/)	27
Fig. 2.5	Meridian First Light House (adapted from https://www.firstlightstudio.co.nz/the-meridian-first-light-house)	32

Fig. 2.6	ZEB Multi-Comfort House (adapted from https://www.archdaily.com/773383/zeb-pilot-house-pilot-project-snohetta)	34
Fig. 2.7	The IKEA Better Shelter (adapted from https://bettershelter.org/)	36
Fig. 2.8	3D printed house, Austin, Texas (adapted from https://singularityhub.com/2018/03/18/this-3d-printed-house-goes-up-in-a-day-for-under-10000/)	38
Fig. 3.1	Law of diminishing returns (adapted from https://bohemianeconomics.wordpress.com/2018/11/01/a-criticism-of-diminishing-returns/)	51
Fig. 3.2	Pagoda of Fugong Temple and Home Insurance Building (adapted from https://en.wikipedia.org/wiki/Pagoda_of_Fogong_Temple and http://architectuul.com/architecture/home-insurance-building)	53
Fig. 4.1	The Pyramid of the Sun, Teotihuacan, Mexico City. At the top is the Pyramid of the Sun. It is the oldest and the largest building in the complex (200 AD). The profile of the Pyramid of the Sun seems to copy the profile of the mountains. Architectural ornaments are humble (adapted from author's photograph)	73
Fig. 4.2	Details of complex carvings in the Temple of Quetzalcoatl (Feathered Serpent) that were added to the building during the high period of development of Teotihuacan (AD 350–650) (adapted from author's photograph)	74
Fig. 4.3	Collapse curves (adapted from Tainter, 1988:125)	78
Fig. 4.4	The adaptive cycle (adapted from https://www.resalliance.org/adaptive-cycle)	80
Fig. 4.5	Shape of the conservation curve in the adaptive cycle (see Fig. 4.4) and area covered by the Roman empire (adapted from Tainter, 1988:125)	81
Fig. 4.6	Growth in the population of Rome (the dashed line indicates the change in scale) (adapted from https://romabyrachel.weebly.com/the-timeline.html)	82
Fig. 5.1	Weavers' cottages, Wardle, UK (adapted from https://en.wikipedia.org/wiki/Weavers%27_cottage)	103
Fig. 5.2	1771 Cromford mill, water-powered spinning (adapted from https://historystack.com/Cromford)	104
Fig. 5.3	Kapiti Coast map. Hatched areas indicate major settlement (adapted from https://www.gns.cri.nz/Home/Our-Science/Land-and-Marine-Geoscience/Regional-Geology/Urban-Geological-Mapping2/Kapiti-Coast)	115
Fig. 5.4	Kapiti coast section (adapted from Nolan, 2017:11)	115

List of Figures

Fig. 5.5	Tai O stilt houses (adapted from https://en.wikipedia.org/wiki/Wikipedia:Featured_picture_candidates/Tai_O#/media/File:1_tai_o_hong_kong_2013.jpg)	119
Fig. 5.6	Happisburgh in 2001 (left) and 2014 (right). The shaded buildings are the same in both images. The groin (left) has gone to be replaced by a different sea defence (right) (adapted from Mike Page Aerial Photography) (https://www.pri.org/stories/2018-04-05/british-village-crumbles-sea-family-holds-home-cant-be-saved)	121
Fig. 5.7	The receding coastline at Dunwich: The dashed road layout shows what has been lost: The large central square was the marketplace (adapted from https://flickeringlamps.com/2016/06/12/the-last-ruins-of-dunwich-suffolks-lost-medieval-town/)	122
Fig. 5.8	Tonle Sap lake; the lake is black and the shaded area is the flood plain (adapted from https://en.wikipedia.org/wiki/Tonl%C3%A9_Sap_Biosphere_Reserve)	125
Fig. 5.9	House on stilts at Tonle Sap lake (adapted from author's photograph)	125
Fig. 5.10	Floating house on Tonle Sap lake (adapted from author's photograph)	126
Fig. 5.11	Cow in Rotterdam harbour (adapted from https://apnews.com/article/9d1f901a48b04843a06052d652b1050)	131
Fig. 6.1	Gated community of Alphaville meets Carapicuiba, Sao Paulo (adapted from Chicca, 2013:180)	154
Fig. 6.2	Gini indices for New Zealand—IRD is Inland Revenue Department, NZOYB is New Zealand Official Year Book—an official digest of statistics (adapted from Creedy et al., 2017:14)	157
Fig. 6.3	The share of the top 10% of income in New Zealand (Based on data from WIID, 2019)	158
Fig. 6.4	State house numbers (adapted from Schrader, 2012b)	160
Fig. 6.5	Home ownership in NZ (adapted from Pool and Du Plessis, n.d.)	161
Fig. 6.6	Housing prices in New Zealand from 1990 to 2018 (based on data from RBNZ and CoreLogic, 2020)	162
Fig. 6.7	Changes in household density, population and dwelling numbers (based on data from Stats NZ, 2020a)	163
Fig. 6.8	Housing ownership in New Zealand (1991–2019) (based on data from Stats NZ, 2020b)	163
Fig. 6.9	Housing and wealth (based on data from RBNZ et al., 2020)	164

Fig. 6.10	Deprivation map of Auckland. Deprived areas are more frequently found to the South and are particularly clustered in the South East (adapted from https://ehinz.ac.nz/indicators/population-vulnerability/socioeconomic-deprivation-profile/#nzdep-for-2018-nzdep2018)	168
Fig. 6.11	State houses in Glen Innes (adapted from author's photograph)	171
Fig. 6.12	Master plan of the new housing developments in Glen Innes (Tamaki Regeneration) (adapted from https://tamakiakl.co.nz/development/glen-innes)	172
Fig. 6.13	New development in Glen Innes East (adapted from author's photograph)	173
Fig. 6.14	Incomes within Glen Innes. Real values according to the period of each census (based on data from Stats NZ, 2013)	174
Fig. 6.15	State house in Glen Innes West (adapted from author's photograph)	175
Fig. 6.16	New housing units in Glen Innes West (adapted from author's photograph)	175
Fig. 6.17	State houses in Glen Innes East (adapted from author's photograph)	176
Fig. 6.18	New housing development in Glen Innes East (adapted from author's photograph)	176
Fig. 6.19	Changes in population (based on data from Stats NZ, 2013)	177
Fig. 6.20	Population changes (based on data from Stats NZ, 2013)	178
Fig. 6.21	New housing development in Glen Innes. In order to get access to houses located deep in the plot internal streets become wider and longer increasing the impervious surfaces (adapted from author's photograph)	178
Fig. 7.1	Areas covered by three empires over time (adapted from Taagepera, 1979)	190
Fig. 7.2	Built environment divisions	202
Fig. 8.1	Percentage change in US population with rural population declining and urban increasing (based on United States Census Bureau, n.d.)	208
Fig. 8.2	Russian balcony extension (adapted from https://weirdrussia.com/2015/08/27/balconies-in-russia/)	211
Fig. 8.3	Burj Khalifa, 2010, in the Dubai skyline (adapted from https://en.wikipedia.org/wiki/Burj_Khalifa)	226
Fig. 9.1	Distribution of confirmed cases and deaths (by thousands) across development levels. Rank levels of HDI are organised from 1 to 6, namely, from more developed to less developed. (Based on data collected until April 2020). *Human Development Index (2019). **COVID-19 Dashboard (2020)	244

List of Figures xvii

Fig. 9.2	Deaths per million as of 23 June 2020 against the population density of selected countries (based on Our World in Data, 2020a, b, c)	249
Fig. 9.3	Number of COVID-19 cases and population density (adapted from https://blogs.worldbank.org/sustainab lecities/urban-density-not-enemy-coronavirus-fight-evi dence-china)	251
Fig. 9.4	Change in time spent at home. 11 April 2020 (Our World in Data, 2020b)	252
Fig. 9.5	Change in time spent at home. 20 June 2020 (Our World in Data, 2020c)	253
Fig. 9.6	Reduction in complexity (city scale)	254
Fig. 9.7	Increment in complexity (plot scale)	255
Fig. 9.8	Increment in complexity (building scale)	255
Fig. 10.1	The 1761 stone pineapple sits on a seven-sided drum in Dunmore Park, forming a summerhouse above the walls that formed part of the walled garden (adapted from author's photograph)	264
Fig. 10.2	Cottages at Bournville (adapted from https://upload.wik imedia.org/wikipedia/commons/0/09/Bournville._Cot tages_in_Linden_Road_%28front_view%29.jpg)	269
Fig. 10.3	Example of Huachafo architecture (adapted from https://www.archdaily.mx/mx/786643/que-es-lo-huachafo-en-la-arquitectura)	272
Fig. 10.4	Cholets (adapted from http://arquitecturahuachafa.blo gspot.com/)	272
Fig. 10.5	House in the "New French Style" near Hanoi (adapted from Herbelin, 2013)	273
Fig. 10.6	A McMansion in Wellington, NZ (adapted from author's photograph)	274
Fig. 10.7	Three terraced houses, Northampton, UK: in the left hand one the door opens to the street, the centre has a bay window, and the right hand one has an area with railings to give light to a basement room (adapted from author's photograph)	277
Fig. 10.8	House with garage and front garden, Hawkes Bay New Zealand (adapted from author's photograph)	277
Fig. 10.9	Old Government Buildings, New Zealand (adapted from https://en.wikipedia.org/wiki/Old_Government_B uildings,_Wellington#/media/File:Old_Government_B uildings_-_whole.JPG)	279
Fig. 10.10	Cemetery in Missions de Sierra Gorda (Querétaro, Mexico) before the celebrations for Dia de Muertos (Day of the Dead) (adapted from author's photograph)	287

Fig. 10.11	"Ofrenda" made with flowers in the main square of the town. Sierra Gorda, Queretaro, Mexico (adapted from author's photograph)	288
Fig. 10.12	Celebration during the Day of the Dead in Mixquic, Mexico (adapted from author's photograph)	289

List of Tables

Table 4.1	Views of collapse	91
Table 5.1	Costs in 2016 of producing one kilogram of milk in the Netherlands (European Milk Board, 2016:6–8)	132
Table 8.1	Food as a proportion of bottom-up EFs (food, energy, transport, housing and consumer goods) calculated on the same basis (adapted from Chicca et al., 2018:149)	217
Table 8.2	Countries with near fair share EFs	219
Table 8.3	EFs, wealth and human development index	220
Table 9.1	European Union, COVID-19 and built area per capita	250
Table 10.1	Average per person dwelling area, GDP and GNI for selected countries	276

Chapter 1
What Do We Mean by Collapse?

Pride comes before a fall
Proverb

Introduction

The United Nations Refugee Agency website (UNHCR, 2020) details stories of people from Syria whose lives were turned upside down by the war. Many fled as refugees. For these people life, as they had known it, no longer existed. A way of life had collapsed and new lives had to be formed from the remnants. In contrast, for many people, and especially those living in wealthier societies, life seems stable and far from any possible collapse. For many people, life has never been better. We have a secure water supply, we have shelter, and through modern herbicides, pesticides and fertilisers we have much more control over the production of food than many previous human generations. We also have the benefits of modern medicine. We are a very mobile society, no longer living within the limits of how far we can walk in a day. Now that same day can see you moving from one continent to another. At the touch of a switch, we can light up the night, something undreamed of for many people a century ago. A modern developed society is dependent on electricity as predicted in an article from 1928: "It should not be regarded merely as a new form of light and heat; electricity provides a complete revolution in method" (Dale, 1928). Electricity underpins all modern communication systems. Electricity supply can fail and for most people power cuts are the nearest they come to experiencing the collapse of something they have come to rely on, albeit the collapse is only temporary.

All these benefits of modern society have come about through the exploitation of technological developments. The exploitation has occurred because the resources are there to provide and power the necessary hardware. At the same time, as discussed below, there have always been those who question if this supply of resources is inexhaustible and if it is possible, given human population growth, to provide the level of development seen in wealthy countries to all the people in the world, those in the less wealthy countries and also to the poorer members of the wealthy societies.

Many of the issues involved in the collapse of societies in the past (see Chap. 3), such as environmental problems, pollution, lack of resources, inequality and lethal pandemics, are relevant to wealthy modern societies. Modern citizens feel themselves far from the possibility of collapse, so, no doubt, did the citizens of the mighty Roman Empire. The purpose of this book is to look at aspects of collapse, both in general terms but also particularly how they might relate to the built environment. This first chapter introduces the environmental issues facing modern humanity. Initially, however, it may be useful to think about types of collapse.

Types of Collapse

Arnold Bennett describes a scene early on in his 1908 novel, *The Old Wives' Tale* (Bennett, 1908), in which the 15-year-old daughter Sophia tries on her mother's new crinoline skirt and subsequently falls over in a mass of silk and hoops somewhat buoyed up by the voluminous garment. This could be construed as a picture of collapsing gracefully and she is soon put back on her feet by her sister.

Sophia's graceful collapse can be compared with the death of her father John Baines in the same book a few pages further on and 2 years later. John Baines had suffered a stroke many years back and was confined to bed. Left unattended, he collapsed by slipping out of bed and asphyxiated on the floor (Bennett, 1908: Book 1, Chap. IV, Part III). He was found with his "tongue protruded between the black, swollen, mucous lips". Unlike Sophia in the crinoline, for John, there was no soft landing when he fell out of bed. This suggests that the collapse can be relatively graceful or exactly the opposite.

What this description does not define is the meaning of collapse. Both characters in *The Old Wives' Tale* suffer a collapse but Sophia recovers from it while John does not. Although we can describe what happens to both of them as a collapse, in one case the collapse is catastrophic and fatal and in the other the collapse is only mildly inconvenient and even somewhat comical, so clearly collapse can have several meanings.

Collapse can also happen in different contexts. Without entering into a deeper discussion about the definition here (see Chap. 3 for the detailed definition), collapse happens in specific contexts, situations and environments that are part of the process of collapse. Falling in a mass of silk as a young lady is very different from falling out of a bed after a stroke when you are older. The context and the environment where collapse happens could play a role in making it more or less graceful.

The assumption of this book is that the built environment, as the cultural landscape and habitat of a society where the process of collapse occurs, plays a role that deserves to be studied to avoid an ungraceful landing. Therefore, this book sets out to examine what collapse means for the built environment, not just for the societies that create it and inhabit it. Past societies that have disappeared, such as the Romans and the

Incas, have left behind built environments that themselves may have been part of the reasons behind such societal collapses. The question we wish to explore here is what type of built environment do we need to create now so that we can avoid collapse or, at the very least, collapse gracefully, given that the built environment tends to last a long time. We have to remember that much of the built environment that we use today was built by previous generations. We manage to live quite happily in what our ancestors built, even though they did not have the benefits of computers, mobile phones, space travel or fast food.

The Faith in Economic Growth

There are reasons why nearly a quarter of the way through the twenty-first century humanity should be worrying about collapse. In 1970, a group called the Club of Rome asked researchers at MIT to use the newly available power of computer modelling to model the future of humanity.

> The Club of Rome is an organisation of individuals who share a common concern for the future of humanity and strive to make a difference. Our members are notable scientists, economists, businessmen and businesswomen, high level civil servants and former heads of state from around the world (Club of Rome, 2018).

This work resulted in a book published in 1972 called *The Limits to Growth* (Meadows et al., 1972). The model compared the interaction between resources, food per capita, industrial output per capita, population and pollution for several different scenarios and concluded that whatever assumptions are made about these five factors "The basic behavior mode of the world system is exponential growth of population and capital, followed by collapse" (Meadows et al., 1972:142). Accepting that any model is a simplification of a complicated situation, this study showed the dangers of allowing exponential growth in a finite system, concluding that "Every day of continued exponential growth brings the world system closer to the ultimate limits to that growth. A decision to do nothing is a decision to increase the risk of collapse" (Meadows et al., 1972:183).

This work was, not unexpectedly, heavily criticised, not least by economists who of course cannot possibly consider the idea of "limits to growth" as that would run counter to their fundamental faith that growth without end is not only possible but necessary and desirable. The critics claimed that the MIT study had failed to factor in the effect of changes to and innovations in technology and the ability to substitute "…man-made factors of production (capital) for natural resources…" (Stiglitz, 1974). Schumacher (1973:99–102), who was against the modern economic ideals, was also a critic. He criticised the MIT group by proposing that the calculations done were redundant since the conclusions could be derived from the assumption that infinite material growth is not possible in a finite world. Moreover, he highlighted that it is hard to estimate the resource availability in the world and even more difficult to understand the impact that the "inventiveness of industry" can have on future availability and exploitation of resources (Fig. 1.1).

Notwithstanding the criticisms, *The Limits to Growth* did suggest that humanity might need to investigate its behaviour in order to be sure that current patterns of living would not lead to the collapse it predicted.

The seemingly irrational commitment of economists to endless growth in a finite world may indeed be the reason why as a society we do not seem to take seriously the idea that a collapse might be possible or even likely, in spite of evidence to the contrary. For example, in 2008 Graham Turner, a senior research scientist at the Australian government research organisation CSIRO, wondered to what extent the modelling carried out in 1970 for *The Limits to Growth* had been accurate, so he compared what had really happened in the 30 years since 1970 with the predictions of the *Limits to Growth* modelling. He found that since 1970 reality had very closely followed the path suggested by the modelling of a "business-as-usual" scenario in *The Limits to Growth* leading him to the rather shocking conclusion that "global collapse" was likely "before the middle of this century", i.e. before 2050 (Turner, 2008:37). Clearly, Turner's idea of a collapse is not the same as falling down in a crinoline. We will discuss the possible meanings of collapse and in particular what we mean by it in this book, in more detail, but at this point, it is enough to say that

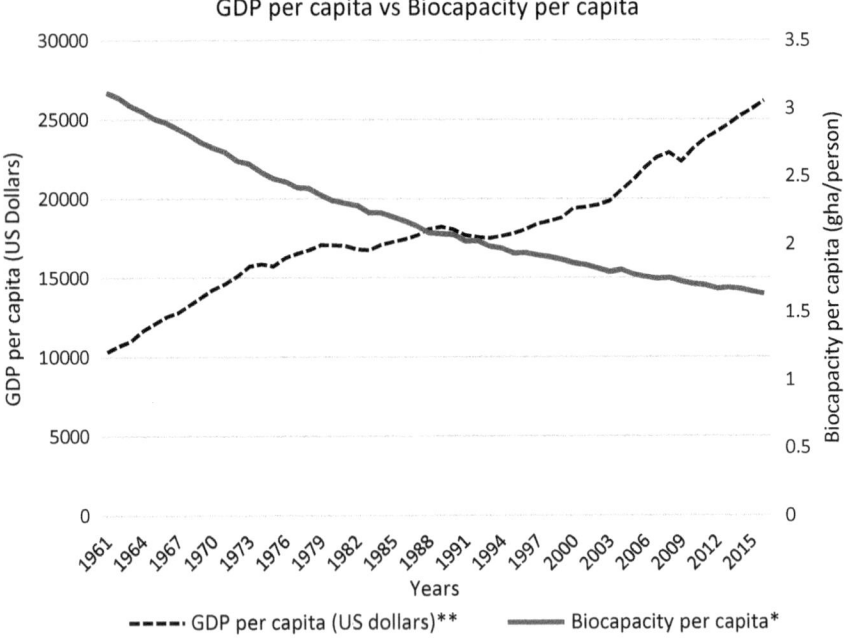

Fig. 1.1 Global GDP growth (per capita) and depletion of biocapacity (per capita). *Data sources* *Global Footprint Network (2019), ** WID.world (2019a)

we will be thinking more about Turner's idea of "global collapse" than the gentle collapse from falling over in a crinoline, buoyed up by its hoops and skirts.

If global collapse is due around 2050 as stated by Graham Turner and by *The Limits to Growth* before him, that is not very far off in time. As we write this, 2050 is about as far ahead of us as 1990 is behind us. As it happens, 1990 is the reference date for the Kyoto Protocol, the global agreement for reducing the greenhouse gas (GHG) emissions that are causing climate change.

> During the first commitment period, 37 industrialized countries and the European Community committed to reduce GHG emissions to an average of five percent against 1990 levels. During the second commitment period, Parties committed to reduce GHG emissions by at least 18 percent below 1990 levels in the eight-year period from 2013 to 2020 (UNFCCC, 2018a).

So maybe it would be a good idea to see how we have done in the 30 years from 1990 to now in order to get an idea of how well we might do in the 30 years from now until 2050 in order to try to avert collapse or at least to try to collapse gracefully.

One way to get a handle on this might be to see how things have changed since 1990 in terms of the five factors considered by *The Limits to Growth* modelling, which were pollution, population, food per capita, industrial output per capita and resources. Starting with one form of pollution, in 1750 global carbon (not CO_2) emissions were 3 million tonnes. This figure rose to 6,074 million tonnes in 1990, and in 2014, emissions were 9,855 million tonnes (Boden et al., 2017). German researchers have concluded that although the Kyoto Protocol, which applies only to its signatories, not to the whole world, may have led to reduced emissions in some of its signatory countries, this has been achieved by the signatories exporting carbon-intensive production to non-signatory countries. Overall, the Kyoto Protocol has had either no effect or may even have increased global emissions (Aichele & Felbermayr, 2011). Not all nations signed up to the Kyoto Protocol and even though some did it does not seem to have made any difference since global carbon emissions have increased by over 60% between 1990, the reference year for the Kyoto Protocol, and 2014, the last year for which there are accurate figures. Until the COVID-19 pandemic in 2020 closed down many activities, emissions had not fallen since 2014 (Mooney & Dennis, 2019).

If carbon emissions are harmful to the climate as suggested by global agreements such as the Kyoto Protocol (UNFCCC, 2018b) and the more recent Paris Climate Agreement (United Nations, 2015), so far human society has failed to acknowledge this harm because we have not done anything to reduce the emissions.

One reason for the increase in emissions might be because of population growth, even if emissions per person stayed the same, more people will mean more emissions. The world population in 1990 was 5,327,231,061 and in 2014 it was 7,295,290,765 (Worldometers, 2020). This is an increase in the population of 37%. If we had managed to keep to the same level of emissions per person, we could have expected a similar rise in emissions as the rise in population, but between 1990 and 2014 emissions rose by 60%. Emissions have risen by quite a lot more than population growth which suggests a problem ahead (Fig. 1.2). Using the 2014 population, the

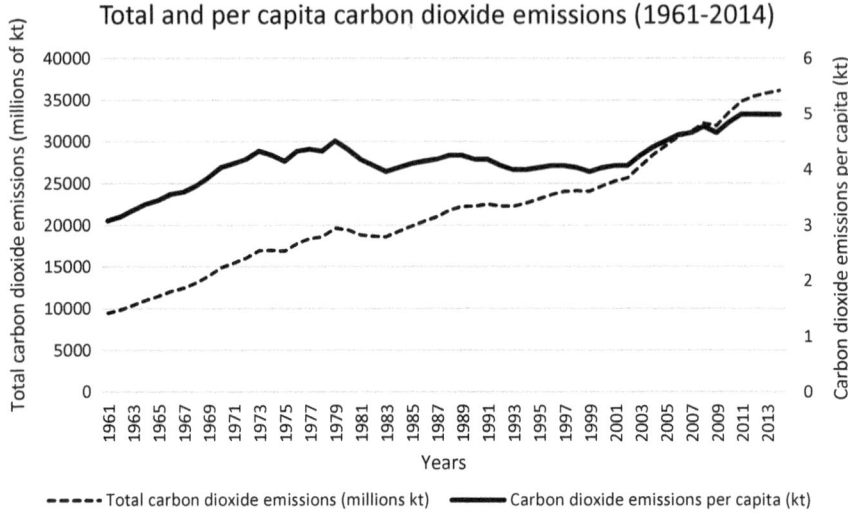

Fig. 1.2 Global carbon dioxide emissions. *Data source* Boden et al. (2017)

carbon emissions per person have risen from 1.1 tonnes in 1990 to 1.3 tonnes in 2014. Carbon dioxide is invisible, making emissions quite hard to visualise, but given that dry wood is about 50% carbon (Ecometrica, 2011) each person on earth is now responsible for throwing away the carbon equivalent of two and a half tonnes of firewood every year. This global average figure represents more energy than some households use in a year (Chicca et al., 2018:201).

Turning to a more visible form of pollution, a recent report from the UK's Government Office for Science states "Around 70 per cent of all the litter in the oceans is made of plastic". The report goes on to make the shocking statement "Globally, production of plastics exceeds 300 million tonnes per annum and it is likely that a similar quantity of plastics will be produced in the next eight years as was produced in the whole of the twentieth century" (Thompson, 2017:4). It is more than likely that quite a lot of this very durable plastic will end up in the sea. In spite of the durability of plastic, we tend to use it for ephemeral purposes—more than half of the plastic used in North America and Western Europe is used for packaging (Gourmelon, 2015)—and then we throw it away, the problem being that there really is no "away" to throw it into.

In terms of the *Limits to Growth* factor of food per capita, the value (comparable in dollar terms) of agricultural production in 1990 was US$1,431 billion, and in 2016 the figure was US$2,629 billion (FAO, 2017:88). This is an increase of 84%. On the very crude assumption that the value of production represents the amount of food produced, more food is being produced than the increase in population, meaning there could be less malnutrition. This hypothesis is supported by the figures, as the World Bank shows that whereas in 2000 14.8% of the world's population was undernourished, by 2015 the percentage had fallen to 10.7% (The World Bank

Group, 2018a), so no collapse there. On the other hand, the increased value of food production also represents higher value products, such as more meat and dairy. The United Nations Food and Agriculture Organisation predicts an increasing proportion of the world's protein input coming from meat in all countries including those which are already classed as "developed" (OECD/FAO, 2015:34).

The problem with this is that meat uses a lot of grain for its production, with grain for feeding livestock expected to be the main part of cereal use by 2024 (OECD/FAO, 2015:30). Feeding grain to livestock uses a lot more land to provide a given amount of calories or protein than feeding grain to people. It is not just the quantity of food but the type of food that has an impact. A vegetarian diet with dairy products and eggs uses less land area than one based on meat (Pimental & Pimental, 2003). The move to more meat (and dairy) may be a problem in another way since according to a recent study, humans already represent 36% of the weight of all mammals on the Earth and their farm animals are an additional 60%. Only 4% of the total biomass of mammals on Earth is wild animals, including everything from elephants and tigers to rats and mice (Bar-On et al., 2018). As the number of people and the number of farm animals continue to grow, the number of wild animals will decline further until it will no longer be a question of the elephant in the room because there will be no elephants.

Finally, and unsurprisingly, as Barry Commoner stated in the first of his four principles of ecology "everything is connected to everything else" (Commoner, 1971), modern "efficient" agriculture and food production are enormous users of energy. As far back as 2003, the production of food in the United States used not only half the country's total land area, leaving less space for the buffalo to roam, but also 80% of the fresh water and a surprising 17% of the total fossil fuel energy (Pimental & Pimental, 2003). What is often not mentioned is that this means that "when the oil runs out" there will be no food. Global proved oil reserves are currently enough to meet 50.2 years of consumption at the 2017 rate, or less if demand increases (BP, 2018:13). This may not be a problem as the Deputy Director of the United Nations FAO says that there are only 60 more harvests left to the world because of soil degradation (Arsenault, 2014).

It might not matter if the oil to grow crops runs out because there might not be any soil left to grow them in. Collapse, what collapse?

The Faith in Technological Development

The next factor used by *The Limits to Growth* to predict collapse was industrial output per capita. We saw earlier when considering pollution that "a similar quantity of plastics will be produced in the next eight years as was produced in the whole of the twentieth century" but of course one person's pollution is another person's production. Producing all that waste plastic has made a profit for someone, so must that make it an acceptable thing to do (Fig. 1.3)?

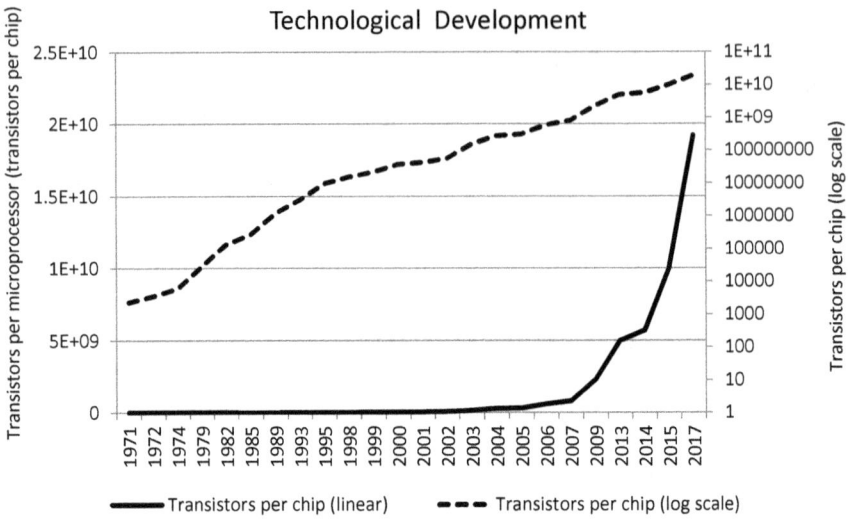

Fig. 1.3 Global technological development in transistors per chip. *Data source* Rupp (2018)

Waste from the materials produced by industry is high on several levels. Hawken et al. (1999:81) pointed out over 20 years ago in their book *Natural Capitalism* that "only one percent of the total North American materials flow ends up in, and is still being used within, products six months after their sale". This one percent is not just because things are being thrown away but also because it often takes a lot of discarded material to make the desired material. For example, the manufacture of a single gold wedding ring results in the creation of about 18 tonnes of mining waste (Farrell et al., 2004:3). As Schumacher noted (1973:97), "The most striking thing about modern industry is that it requires so much and accomplishes so little".

The OECD (2018) defines industrial production as "…the output of industrial establishments and covers sectors such as mining, manufacturing, electricity, gas and steam and air-conditioning". Since 1994 (the farthest date in the past for which figures are given) it has increased by 92% (based on data from The World Bank Group, 2018b) compared with the population which has risen by less than half this amount in the same time. The clear meaning of the figures is that there are not just more people on the Earth, there are more people with more stuff. This might work in a "circular economy" where the resources in a discarded product are used to make a new one, but the world does not work like that. Having more stuff means making more waste. Municipal solid waste, the technical term for what the garbage man collects, has risen in the United States from 2.7 lb (1.2 kg) per person per day in 1960 to 4.4 lb (2.0 kg) per person per day in 2013. In 1990, the date we are using here for comparisons, it was over 4.6 lb (2.1 kg) (EPA, 2016), so at least the figure has come down slightly but there has been a big increase from the apparently less wasteful era of the 1960s.

It is not just in the USA that waste is increasing, it is a global phenomenon. A report published by the World Bank puts it very clearly "MSW generation levels are expected to double by 2025. The higher the income level and rate of urbanization, the greater the amount of solid waste produced" (Hoornweg & Bhada-Tata, 2012:8). In Chap. 2, a building dedicated to the incineration of waste in Oslo is discussed to illustrate how this problem is being approached by a "superstar architect".

Technological developments also produce waste. The rise in ownership of electronic goods has also led to a rise in electronic goods that are no longer wanted or E-waste. In 2016, some 44.7 million metric tonnes (Mt) of E-waste were produced, and this is predicted to rise to 46 Mt in 2016 and 52.2 Mt in 2021, which means E-waste is growing at a yearly rate of 3–4% (Baldé et al., 2017:38). In 2019 the iron, copper and gold in this waste together with other lesser components had a value of around US$ 57 billion but only 17.4% was recycled (Forti et al., 2020:14–15). We may think we are buying smartphones but we are not smart enough to recover the valuable materials in them when we throw them away.

It seems the situation is that the Earth has more people emitting more carbon dioxide and producing more waste than ever before. But we have more food and more stuff, so perhaps there is no problem. Assuming that the carbon dioxide does not change the climate to the point where much of the planet becomes uninhabitable and assuming that the waste does not choke us, the problem may lie in resources, the last of the *Limits to Growth* categories to be considered here. Do we have enough of the materials we need to allow us to carry on growing? A report published in 2001 (Tilton, 2001) suggests that there is little likelihood of mineral resources "running out" in spite of their being finite, because as a resource becomes harder to get either its price increases, which makes it cost-effective to extract it from sources that are more expensive to exploit, or ways are found to substitute it with something else. It can be argued that humanity is unlikely to run out of materials soon.

Turning to one particular category of resources, it seems somewhat perverse that humanity has chosen to build a society with a growing population powered by increasing consumption of the finite or non-renewable sources of energy, petroleum, natural gas, coal and uranium. But perhaps this finite nature is more conceptual than real, are the non-renewable fuels really finite in practice? A German source is helpful in this respect.

> In relation to the conditions in 2007, conventional crude oil will be available worldwide for 42 years, natural gas for 61 years, black coal for 129 years and lignite for 286 years. In relation to the reserves in 2005, uranium has a reserve-to-production ratio of 70 years. However, this represents a snapshot and assumes that the consumption, based on the existing reserves, is continued at the current level in the future. What it does not take into account is that advancements in energy-saving technology and substitution successes reduce the consumption while the discovery of new deposits, as a result of improved exploration technologies, can increase the reserves (Kraftwerk Forschung, undated).

From that point of view, there seems to be something to worry about, if oil will be used up in 42 years from 2007, assuming consumption "at the current level", that means by no later than 2049, with increased consumption between 2007 and now suggesting the date might come rather sooner. Something else will have to replace it

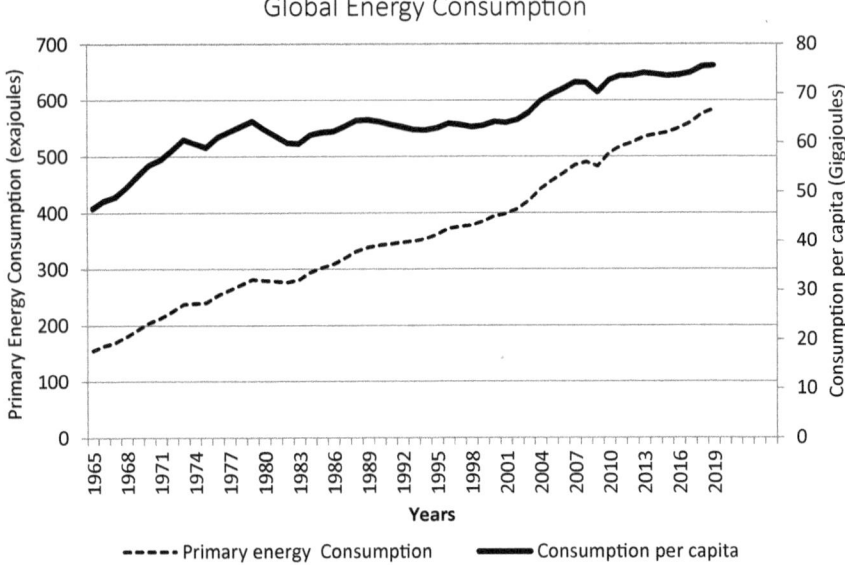

Fig. 1.4 Global energy consumption. *Data source* BP (2020)

and that will mean that the use of other fuels will increase and their life will be accordingly shorter. However, as the human population grows so also does the demand for energy (Fig. 1.4). This demand is also increased as economic development raises the standard of living.

The recent introduction of new techniques has increased the availability of non-renewable energy sources. The use of hydraulic fracturing has increased oil production in the USA from 5.5 million barrels per day in 2000 to 9 million in 2015 and half of oil production is now obtained using this technique (EIA, 2016). Another unconventional source of non-renewable energy is the vast deposits of oil bound up in the sand, which is a considerable part of some nations' oil reserves. "Of the 170 billion barrels of Canadian oil that can be recovered economically with today's technology, 164 billion barrels are located in the oil sands" (CAPP, 2018). This all sounds good, resources are being found to add to those that are already known, extending the life of finite reserves. However, a factor that needs to be taken into account is not only the financial cost but the energy cost, known as the energy return over (energy) invested or EROI (see further discussion in Chap. 5). If it takes 100 kWh of energy input to produce 100 kWh of oil from an unconventional source, that source is not a net producer of energy. It appears, now that the "easy" oil, gas and coal have been found, that not only is the EROI of all non-renewable energy sources declining over time, meaning that more energy has to be put in to get a unit of energy out, but the EROI of oil sands is considerably lower than that for conventional oil. Unfortunately, the EROIs of the various renewable sources of energy such as solar and wind are also not very favourable. It will require a huge input of non-renewable energy to build the

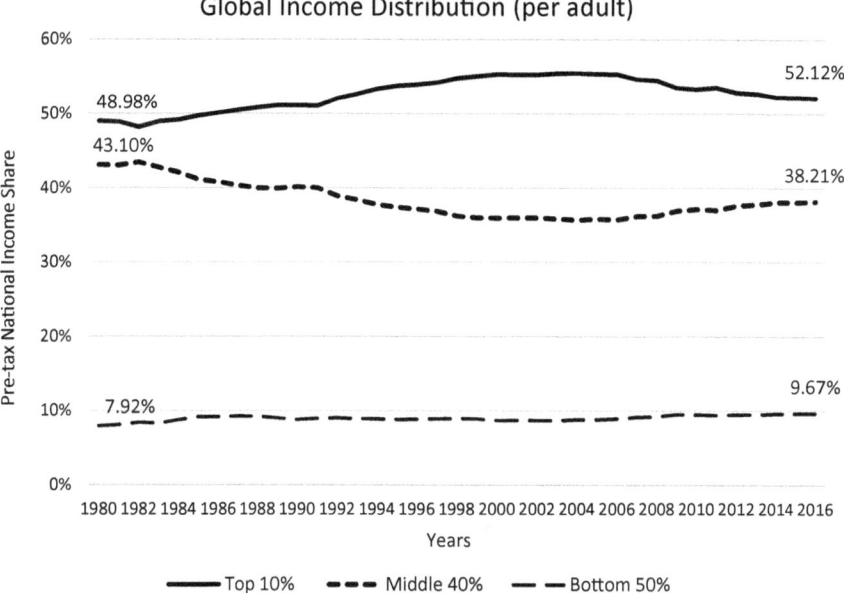

Fig. 1.5 Income inequality. The chart shows that in 2016 10% of the world population accounted for 52% of global income while the bottom 50% lived with less than 10% of global income and the middle 40% represented 38% of the income share. *Data source* WID.world (2019b)

equipment that would allow the world to operate on renewable energy (Hall et al., 2014), but at least the energy sources once built would be carbon-free. If unconventional fossil fuel resources are exploited, like Canada's oil sands, then the emissions produced by obtaining them have to be added to the emissions produced by burning them, a double burden.

To the issues of growth in energy demand, and rising levels of waste and GHG production needs to be added the rise in inequality (Fig. 1.5). It seems that as development occurs living standards rise, and demand for energy and higher cost foods such as meat and dairy also rise—but only for some. Development does not make every member of a society equally rich.

The distribution of global income between 1980 and 2016 (Fig. 1.5) shows that "inequality decreased between the bottom and middle of the income distribution and increased between the middle and the top" (Piketty, 2020:25). Since Fig. 1.5 illustrates a "distribution" of income, it means that if someone gets a big slice of the pie the rest will get smaller slices. This fact shows that the global economic growth (see Fig. 1.1) has not benefited all groups equally but has contributed to accentuating the gaps between the rich, the poor and the middle class, with few rich people getting half of the pie and half of the world sharing a thin slice. High income inequality is also linked with contrasting lifestyles and consumption levels. The top 1% of the world population produces more than nine times the global average emission per capita, which accounts for more than the bottom half of the world population emissions combined (Piketty, 2020:666). If inequality keeps on growing, it might lead to greater emissions for one population sector and scarcity of resources in the other. Since inequality can contribute to climate change issues and cannot be solved by economic and technological growth, it should be considered a big threat to our civilisation.

A Plan B: Collapsing Gracefully

It seems possible that there may be enough reserves of non-renewable energy to allow society to continue to grow, at least until the fuels became too expensive to extract, in either money or energy terms. However, it may not be wise to use these carbon-based fuels. If all the world's known fossil fuels were burned, a recent estimate is that the global mean temperature would increase by between 6.4 and 9.5 °C, but in the Arctic, the increase would be between 14.7 and 19.5 °C, so all the ice would melt. In addition, there would probably be more than four times the present rainfall (Tokarska et al., 2016). This might not be very beneficial for the continuation of human society. Chapter 5 deals with the challenges that flooding and sea-level rise pose to the urban habitat of societies.

To look at this question of changing the climate and the effects it might have, such as sea-level rise, the UN's Intergovernmental Panel on Climate Change (IPCC) was set up in 1988 by the United Nations Environment Programme (UNEP) and the World Meteorological Organization (WMO) with a membership of 195 separate nations. Its purpose is to provide government policymakers with regular scientific assessments of all aspects of climate change. The aim of these assessments is to provide scientific information to governments so that they can develop policies related to climate change.

> The IPCC assesses the thousands of scientific papers published each year to tell policymakers what we know and don't know about the risks related to climate change. The IPCC identifies where there is agreement in the scientific community, where there are differences of opinion, and where further research is needed. It does not conduct its own research (IPCC, 2018a:3–4).

Thirty years ago, when the IPCC was founded, there were just over 350 parts per million (ppm) of carbon dioxide in the atmosphere (NOAA, 2018), while NASA quotes a figure of 409 ppm for August 2018 (NASA, 2018). These figures are from the continuous records taken at Mauna Loa in Hawaii since 1958. When these records began the CO_2 level was 315 ppm, so it rose by 35 ppm in the first 30 years and by nearly 60 ppm in the next 30 years following the foundation of the IPCC, so it appears that so far since its inception the IPCC has been able to do little to reduce the likelihood of climate change.

At the time of writing this, in October 2018, the IPCC had just issued a report with the lengthy title of *Global Warming of 1.5 °C, an IPCC special report on the impacts of global warming of 1.5 °C above pre-industrial levels and related global greenhouse gas emission pathways, in the context of strengthening the global response to the threat of climate change, sustainable development, and efforts to eradicate poverty.* The press release accompanying the report stated "Limiting global warming to 1.5 °C would require rapid, far reaching and unprecedented changes in all aspects of society…With clear benefits to people and natural ecosystems, limiting global warming to 1.5 °C compared to 2 °C could go hand in hand with ensuring a more sustainable and equitable society" (IPCC, 2018.a). An accompanying document states "Actions that can reduce emissions include, for example, phasing out coal in the energy sector, increasing the amount of energy produced from renewable sources, electrifying transport, and reducing the 'carbon footprint' of the food we consume" (IPCC, 2018b:8). Because the IPCC's work is based on inputs from a large group of independent scientists from all over the world, their reports have to represent something on which all the participants can agree, so the inbuilt tendency will be for these reports to be quite conservative.

Responses from the world's elected politicians to the IPCC report were almost instantaneous. As for phasing out coal, the Australian Deputy Prime Minister, Michael McCormack, immediately asserted that Australia would carry on using coal for electricity generation rather than moving to renewable energy (Karp, 2018), while the UK's Minister of State for Energy and Clean Growth, Claire Perry, who was responsible for "carbon budgets", "international climate change" and "climate science" (GOV.UK, 2018) said a week later that it was not the government's job to advise people to eat a diet that would reduce emissions (Harrabin, 2018).

From the evidence, collapse seems quite likely on more than one front (food, energy, climate…) and democracy has provided society with leaders whose only idea seems to be to lead us over the cliff. As a result, maybe we should follow the idea of the seventeenth-century French mathematician and philosopher Blaise Pascal. In his celebrated "wager" he proposed that since it was not possible to prove or disprove the existence of God the safer bet would be to assume that there was a God and live your life accordingly. When you died, if there was no God you would not have lost out as you would have had a good life. On the other hand, if you lived a vice-ridden and godless life and then found on death that there was indeed a God, you would be condemned to eternal damnation (Pascal, 1932).

Following Pascal's wager, if we are going to collapse, might it not be a good idea to have a "plan B" for collapsing gracefully, just in case the politicians, and the voters, have got it wrong? Perhaps the next step is for those dealing with the design of the built environment to consider what a Plan B built environment might be like. The aim of this book is to examine this and come up with suggestions for what needs to happen now to collapse gracefully in the future. Chapter 2 begins this investigation by looking at current built environment initiatives to deal with the problems outlined in this chapter.

References

Aichele, R., & Felbermayr, G. (2011). *Kyoto and the carbon footprint of nations: Ifo Working Paper No. 103*. June. Munich, IFO Institute for Economic Research.

Arsenault, C. (2014). *Only 60 years of farming left if soil degradation continues*. 6 Dec Reuters. Retrieved October 9, 2018, from https://www.reuters.com/article/us-food-soil-farming-idUSKCN0JJ1R920141205.

Bar-On, Y. M., Phillips, R., & Milo, R. (2018, May). The biomass distribution on earth. *National Academy of Sciences*. 201711842. Retrieved October 2, 2018, from http://www.pnas.org/content/early/2018/05/15/1711842115

Baldé, C. P., Forti, V., Gray, V., Kuehr, R., & Stegmann, P. (2017). *The global e-waste monitor—2017*. United Nations University (UNU), International Telecommunication Union (ITU) & International Solid Waste Association (ISWA).

Bennett, A. (1908). *The old wives' tale. Book 1, Chapter II Part II* Retrieved October 1, 2018, from Project Gutenberg at http://www.gutenberg.org/files/5247/5247-h/5247-h.htm#link2HCH0001.

Boden, T., Marland, G., & Andres, R. (2017). *Global, regional, and national fossil-fuel CO_2 emissions*. Oak Ridge, Tenn, USA, Carbon Dioxide Information Analysis Center, Oak Ridge National Laboratory, U.S. Department of Energy. Retrieved September 6, 2020, from https://cdiac.ess-dive.lbl.gov/ftp/ndp030/global.1751_2014.ems.

BP. (2018). *BP statistical review of world energy, 2018*. BP.

BP. (2020). *Bp statistical review of world energy*. Retrieved September 6, 2020, from https://www.bp.com/en/global/corporate/energy-economics/statistical-review-of-world-energy.html.

CAPP (Canada's Oil and Natural Gas Producers). (2018). *Canada's oil sands*. Retrieved October 11, 2018, from https://www.canadasoilsands.ca/en/explore-topics/our-energy-future.

Chicca, F., Vale, B., & Vale, R. (2018). *Everyday lifestyles and sustainability: The environmental impact of doing the same things differently*. Routledge.

Club of Rome. (2018). *About us*. Retrieved May 28, 2018, from https://www.clubofrome.org/about-us/.

Commoner, B. (1971). *The closing circle: Man, nature and technology*. Knopf.

Dale, V. (1928). Electricity and the home: Modern marvels of heat and light. *The Graphic, 122*(3069), iv, vi.

Ecometrica. (2011). *A one tonne carbon tree*. Retrieved June 13, 2018, from https://ecometrica.com/assets/one_tonne_carbon_tree_discussion_paper_3.pdf.

EIA. (2016). *Oil production in the United States (2000–2015)*. March 15. Retrieved October 11, 2018, from https://www.eia.gov/todayinenergy/detail.php?id=25372.

EPA. (2016). *Municipal solid waste Figure 1. MSW generation rates, 1960–2013*, 30 March. Retrieved October 10, 2018, from https://archive.epa.gov/epawaste/nonhaz/municipal/web/html/.

FAO. (2017). *FAOSTAT value of agricultural production*. Retrieved June 13, 2018, from http://www.fao.org/faostat/en/#data/QV.

Farrell, L., Sampat, P., Sarin, R., & Slack, K. (2004). *Dirty metals: Mining, communities and the environment*. Oxfam America.

Forti, V., Baldé, C. P., Kuehr, R., & Bel, G. (2020). *The global e-waste monitor 2020: Quantities, flows and the circular economy potential*. Retrieved August 28, 2020, from (PDF) The Global E-waste Monitor 2020. Quantities, flows, and the circular economy potential (researchgate.net).

Global Footprint Network. (2019). *NFA 2019 national footprint and biocapacity accounts data set (1961–2016)*. Retrieved September 6, 2020, from https://data.world/adejori/ecological-footprint/workspace/file?agentid=footprint&datasetid=nfa-2019-edition&filename=NFA+2019+public_data.csv.

Gourmelon, G. (2015). *Global plastic production rises, recycling lags*. Washington DC, Earthwatch Institute. Retrieved October 10, 2018, from http://www.worldwatch.org/global-plastic-production-rises-recycling-lags-0.

GOV.UK. (2018). *Minister of state for energy and clean growth, the Rt Hon Claire Perry MP*. Retrieved October 15, 2018, from https://www.gov.uk/government/people/claire-perry.

References

Hall, C., Lambert, J., & Balogh, S. (2014). EROI of different fuels and the implications for society. *Energy Policy, 64*, 141–152.

Harrabin, R. (2018). *Is meat's climate impact too hot for politicians?* BBC News 15 Oct. Retrieved October 15, 2018, from https://www.bbc.com/news/science-environment-45838997.

Hawken, P., Lovins, A., & Hunter Lovins, L. (1999). *Natural capitalism.* Little Brown and Co.

Hoornweg, D., & Bhada-Tata, P. (2012). *What a waste: A global review of solid waste management.* Urban development series; knowledge papers no. 15. World Bank.

IPCC. (2018a). *Summary for policymakers of IPCC special report on global warming of 1.5 °C approved by governments.* 2018/24/PR IPCC PRESS RELEASE 8 October 2018 8 October. IPCC Secretariat.

IPCC. (2018b). *IPCC special report on global warming of 1.5 °C frequently asked questions,* 6 October. IPCC Secretariat.

Karp, P. (2018). Australian government backs coal in defiance of IPCC climate warning Deputy PM Michael McCormack says policy will not change based on 'some sort of report'. *The Guardian,* 8 October. Retrieved October 15, 2018, from https://www.theguardian.com/australia-news/2018/oct/09/australian-government-backs-coal-defiance-ipcc-climate-warning.

Kraftwerk Forschung. (undated). *Energy supplies.* Supported by Federal Ministry for Economic Affairs and Energy. Retrieved October 10, 2018, from http://kraftwerkforschung.info/en/quickinfo/energy-supplies/this-is-how-long-the-non-renewable-energy-sources-will-last-information-in-years/.

Meadows, D. H., Meadows, D. L., Randers, J., & Behrens, W. (1972). *The limits to growth.* Universe Books.

Mooney, C., & Dennis, B. (2019). Global greenhouse gas emissions will hit yet another record high this year, experts project. *The Washington Post* December 4. Washington DC, The Washington Post. Retrieved August 12, 2020, from https://www.washingtonpost.com/climate-environment/2019/12/03/global-greenhouse-gas-emissions-will-hit-yet-another-record-high-this-year-experts-project/.

NASA. (2018). *Direct measurements: 2005-present.* NASA Global Climate Change October 11 NASA Jet Propulsion Laboratory, California Institute of Technology. Retrieved October 13, 2018, from https://climate.nasa.gov/vital-signs/carbon-dioxide/.

NOAA (National Oceanic and Atmospheric Administration). (2018). *Trends in atmospheric carbon dioxide: Full Mauna Loa CO_2 record Boulder CO.* Earth System Research Laboratory. Global Monitoring Division. Retrieved October 13, 2018, from https://www.esrl.noaa.gov/gmd/ccgg/trends/full.html.

OECD/FAO. (2015). *OECD-FAO agricultural outlook 2015–2025* "Figure 1.8 Protein intake per capita in least developed, other developing and most developed countries". OECD Publishing.

OECD. (2018). *Industrial production (indicator).* https://doi.org/10.1787/39121c55-en. Retrieved October 10, 2018, from data.oecd.org/industry/industrial-production.htmv.

Pascal, B. (1932). *Pensées* (trans. John Warrington). Dent (Everyman's Library No. 874).

Piketty, T. (2020). *Capital and ideology* [Capital et idéologie.]. Harvard University Press.

Pimentel, D., & Pimentel, M. (2003). Sustainability of meat-based and plant-based diets and the environment. *The American Journal of Clinical Nutrition, 78*(3), 660S-663S.

Rupp, K. (2018, 15 February). 40 years of microprocessor trend data. Retrieved September 6, 2020, from https://github.com/karlrupp/microprocessor-trend-data/commit/47382e2e3c653d71ebae66d8e8aecc088866543d.

Schumacher, E. F. (1973). *Small is beautiful; Economics as if people mattered.* Harper & Row.

Stiglitz, J. (1974). Growth with exhaustible natural resources: Efficient and optimal growth paths. *The Review of Economic Studies, 41*, 123–137.

Tilton, J. (2001). *Depletion and the long-run availability of mineral commodities.* International Institute for Environment and Development and Geneva, World Business Council for Sustainable Development.

Thompson, R. (2017). *Future of the sea: Plastic pollution.* Foresight; Government Office for Science.

Tokarska, K., Gillett, N., Weaver, A., Arora, V., & Eby, M. (2016). The climate response to five trillion tonnes of carbon. *Nature Climate Change, 6*, 851–855.

Turner, G. (2008). *A comparison of the limits to growth with 30 years of reality*. CSIRO Sustainable Ecosystems.

UNFCCC. (2018a). *Kyoto protocol—Targets for the first commitment period*. Retrieved May 28, 2018, from https://unfccc.int/process/the-kyoto-protocol.

UNFCCC. (2018b). *KP introduction*. Retrieved June 8, 2018, from https://unfccc.int/process/the-kyoto-protocol.

UNHCR (The United Nations High Commissioner for Refugees). (2020). *Syria: Stories*. Retrieved August 28, 2020, from https://www.unhcr.org/sy/stories.

United Nations. (2015). *Paris agreement*. Retrieved June 8, 2018, from http://unfccc.int/files/essential_background/convention/application/pdf/english_paris_agreement.pdf.

The World Bank Group. (2018a). *Prevalence of undernourishment (% of population)*. Retrieved October 9, 2018, from https://data.worldbank.org/indicator/SN.ITK.DEFC.ZS.

The World Bank Group. (2018b). *Industrial production, constant 2010 US$, not seasonally adjusted*. Retrieved October 10, 2018, from http://www.worldbank.org/en/research/brief/economic-monitoring.

WID.world. (2019a). Gross domestic product. average income or wealth (dollar, PPP constant, 2019). Retrieved September 6, 2020, from https://wid.world/data/#countriestimeseries/agdpro_p0p100_z/WO/1930/2019/us/k/p/yearly/a.

WID.world. (2019b). Pre-tax national income share. Retrieved September 6, 2020, from https://wid.world/data/#countriestimeseries/sptinc_p90p100_z/WO/1930/2019/eu/k/p/yearly/s.

Worldometers. (2020). *World population by year*. Retrieved August 12, 2020, from https://www.worldometers.info/world-population/world-population-by-year/.

Chapter 2
Current Ideas for Future Built Environments

Such stuff as dreams are made of
Shakespeare: The Tempest

Introduction

Chapter 1 introduced the myth of endless economic growth. Growth, and the technological developments it inspires, forms the main driver of a social and environmental crisis that could lead the world to some form of collapse. Since both economic growth and technological developments occur within built environments, the latter could be linked to the cause and consequences of a future collapse. The design of the built environment requires not only satisfying immediate issues but also ensuring that the resources invested in buildings and infrastructure will be useful in the future.

How many designers deal with future thinking is a moot point and this is where construction traditions come in, revealing not only what is durable but also what is useful. The Georgian terraced houses of England, with their cellular internal planning, a number of storeys designed for stair access and durable materials that can be repaired, such as brick walls and slate roofs, have shown that not only can they be used to house a wealthy family with their servants but also they can be later reconfigured into apartments while serving equally well as office space or even a School of Architecture. In this way, built projects provide concrete evidence of what ideas have worked and what have not worked.

In contrast, unbuilt projects, those that remain on paper, often define trends in thinking about the built environment. In the light of the issues raised in Chap. 1 much of what is discussed critically in this chapter deals with such thinking. This is important because these future thinking ideas are what professionals look for in magazines and internet websites. They also influence the work of students, the aspirations of institutions and the goals of those running cities in forming a response to the perceived social and environmental crises. Basically, these new, innovative and unbuilt ideas are supposed to be the solution to current problems but do they work and could they work in the future?

Inger Andersen, Executive Director of the United Nations Environment Programme (UNEP), suggests three actions are required to save the world: "invest in nature, change our habits and listen to young people" (UNEP, 2019). However, in the face of this simple advice, it seems architects and others interested in the built environment would rather not change their habits, preferring to imagine new castles in the air that would allow modern habits to continue. This chapter sets out to describe some of these visions with a view to seeing whether they offer any solutions to the problems currently facing the world or those problems that may be still to come in the future.

Smart Cities (Even When It Is a Dumb Idea)

One of these designers' visions is the "Smart City". This fashionable idea recognises that cities, through bringing people closer together, are often places where innovation happens and assumes that such innovation can be encouraged because the nature of modern IT links people even more. However, this very general definition is not the only one. Barlow and Levy-Bencheton (2019:1) are much more effusive in their definition.

> The smart city is a modern myth, a dream for our time. It's an archetype and an ideal, formed in the realm of our collective unconscious. It's a magical place we long for, a vision shimmering in the distance and yet embedded deeply in our psyche. For those of us who love cities, the smart city is where we want to live, work, play, raise a family, start a business or simply stroll around on a pleasant day. The smart city inspires genius and originality. It also offers tranquillity and peace.

Put like that and given the noisy and polluted nature of many modern cities in both the developed and developing worlds, the smart city sounds more like a utopia than something achievable. As an idea, this concept seems to offer a solution to both environmental and socio-economic problems in a single package. Its enthusiasts manage to avoid explaining how climate change can be dealt with (Barlow & Levy-Bencheton, 2019:131), or how poverty can be eradicated, beyond the comment that people with a two-year degree are less likely to be poor (Barlow & Levy-Bencheton, 2019:100).

Once an idea like smart cities is portrayed as a good one, definitions and descriptions start to proliferate. The idea of smart cities also covers a wide spectrum, from definitions that focus on the capacity of information and communications technology to make cities economically successful, to definitions that are less technologically oriented and more related to the improvement of city infrastructure and eventually its human capital. Marsal-Llacuna et al. (2015), for example, suggest that the objective of smart cities is to use technology to produce more efficient services and data with which to monitor the performance of the infrastructure of a city. Even though what efficient means in this context remains uncertain, what is clear is that smart cities need to promote new business, in a process that is generally referred to as innovation. Kamel Boulos et al. (2015) suggested that smart cities represent the path to innovation, and innovation "is key to avoiding collapse and promoting sustainability of cities/regions and their infrastructures". How this is supposed to happen is not explained as faith is being put in the notion of innovation. What is evident is that the definition of a smart city is not clear. They can be anything that the market wants, even just a convenient prefix to sell the same thing with a different name.

Bennett et al. (2017) criticised the lack of agreement regarding the definition of smart cities. These authors also suggested the main challenge of smart cities was not technological but political, since market forces need to be shaped by the government to secure a collective benefit. Hollands (2008) suggested that what all smart cities do is to prioritize informational business over people and that they are basically a "high-tech variation of the 'entrepreneurial city'". Rochet (2018:11) highlighted the fact that the marketing of smart cities proposes a return to functionalist ideas "a notion that has already failed over 50 years of urbanism and destroys human communities. It is a voluntary and theorized ignorance of lessons from the past." Raya et al. (2017) raised the issue of the impact of the smart city idea on the value of dwellings, particularly if the location of a dwelling in a smart city affects its selling price, concluding, "living in a smart city increases the value of new dwellings". It seems probable, therefore, that moving towards smart cities will have a negative impact in terms of housing affordability, gentrification and the already large inequality gap affecting cities around the world. This issue is discussed in greater depth in Chap. 7.

The latest trend in smart city thinking is to link them with, and even use them as, a synonym for sustainable cities (Kramers et al., 2014), although one of the criticisms of the smart city concept is that it fails to deal with the same issues as sustainability. This was the subject of a study performed by Ahvenniemi et al. (2017). They tested the importance and role of the social and economic dimensions through a comparative analysis of sustainable and smart city frameworks. They found that even though smart cities claimed to be the medium for creating more energy-efficient environments, most of the time energy-related indicators were under-represented. Moreover, the environmental impact of smart technologies was never fully assessed so as to understand the extent to which they could mitigate their initial energy investment. Ahvenniemi et al. (2017) concluded that smart city frameworks mainly focus on economic aspects with the assumption that these will produce social improvements. This feels like a return to the "trickle-down economics" that has already proven not to work (Chang, 2015). Moreover, how these social improvements will take shape is not that clear.

What is also not clear, given the problems outlined in Chap. 1, is why smart cities seem to show little or no interest in reducing environmental impacts. The passive-aggressive language used to label cities as "smart" implies that if some cities are smart, what are the rest? Are they less smart? Are they stupid? If smart cities are smart enough, they should be tackling climate change better than the "stupid cities". However, when carbon dioxide emissions per capita of the top ten "smartest" cities as ranked by the IESE's Cities in Motion Index (IESE, 2019) are compared with

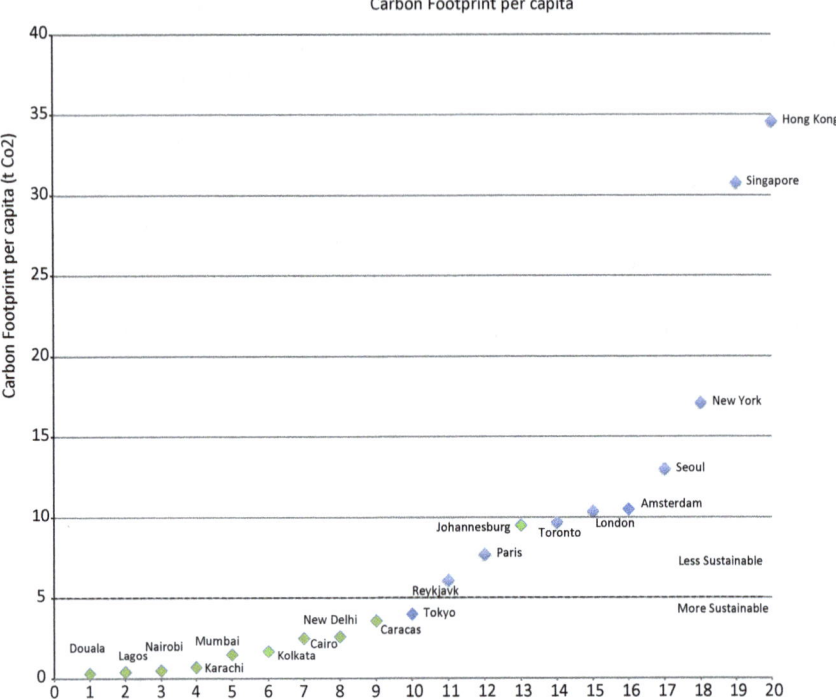

Fig. 2.1 Carbon footprint of cities (per capita). Cities in blue are the 10 "smartest cities" and cities in green are the 10 "stupid cities". The horizontal scale ranks the 20 cities according to their carbon footprints per capita, with 1 being the most sustainable and 20 the least sustainable (adapted from Moran et al., 2018 except for Reykjavik, Douala, Cairo and Nairobi where national averages from 2016 were used from DataBank, 2020)

the carbon dioxide emissions per capita of the ten "most stupid" cities, a different panorama is revealed (Fig. 2.1).

Figure 2.1 shows that almost all the "stupid cities", apart from Johannesburg, are clustered at the bottom left corner of the chart, which is the area that corresponds with the lowest carbon footprints per capita. Tokyo is the only "smartest city" from the top ten that sits in the more sustainable part of the graph. Figure 2.1 also shows three clusters of cities with increasing gaps between them. The first group contains the cities with the lowest impact: Douala, Lagos, Nairobi, Karachi, Mumbai, Kolkata, Cairo, New Delhi, Caracas and Tokyo/Yokohama, all with a carbon footprint below 5 tCO_2 per capita. In the second group are Reykjavik, Paris, Amsterdam, Toronto, Johannesburg, London and Seoul all with carbon footprints between 5 tCO_2 and 15 tCO_2 per capita. In the third group New York, Hong Kong and Singapore have carbon footprints of between 15 tCO_2 and 35 tCO_2 per capita. The gap between the carbon footprints per capita of the first group (the most stupid) and the third group of cities (the smartest) is so big that the carbon footprint per capita of New York is larger than the total carbon footprints per capita of 9 "stupid cities" combined (Moran et al., 2018). This unequal relationship in the use of resources makes it impossible to justify smart cities as sustainable habitats for human beings.

The smartest cities are not sustainable. With the exception of Reykjavik and Tokyo/Yokohama, the smartest cities consistently show higher carbon footprints per capita. At the moment, the smartest cities in the CIMI ranking are just cities that are well dressed for collapse. If all the "stupid cities" become as smart as Singapore, Hong Kong, New York, Seoul and London, carbon footprints per capita in poor cities like Lagos, Karachi or Mumbai will have to increase at least 30 times. The faith in technological and economic development that prevails in the ideology of smart cities is a "stupid" disguise to avoid tackling human-induced climate change issues seriously. Smart cities only look smart in smart city rankings where the only measure that really matters is economic. If smart cities keep on privileging economic development over environmental deficits, they will be choosing the smartest way of committing ecocide.

Buildings All at Sea (a Good Place from Which to Watch Tsunamis)

People have lived on water, whether at sea on-board a ship for many months during expeditions in search for new land to settle, or on rivers and canals on boats used for the transport of resources and goods. Inevitably, however, such ships return to land to replenish stores. One modern idea in the face of climate change and rising sea levels (see Chap. 5 for further discussion) is to create whole self-contained cities that float on the oceans. "*Ocean colonization* is the theory and practice of permanent human settlement of oceans. Such settlements may float on the surface of the water, or be secured to the ocean floor, or exist in an intermediate position" (Bolonkin, 2011:967).

As the human population increases, the demands made on the existing land areas also increase. This has led to the seemingly simple solution of taking to the water since water occupies 71% of the surface of the planet with the oceans making up 96.5% of this (USGS, n.d.). This ignores the fact that water is already the home of many other species, all of which, such as fish and whales, are better adapted for watery living than we are. What is being proposed is the creation of artificial land that floats on coastal waters but that at the same time will be casting a shadow and thus changing the nature of that water for the species that currently see it as "home". Those who espouse floating structures are quick to minimise the idea of any harm to the natural environment, as with this proposal for the Maldives (Gammon, 2012).

> The design disturbs only a small patch of the seafloor while preserving natural currents. And many smaller islands are more ecologically sound than one large one because they cast smaller shadows on the water, minimizing the impact on sea life. Although the company is starting to build an island for 200 luxury residences and another for a floating golf course this year, it is working on plans to construct islands for more affordable housing next.

Quite why the low-lying Maldives, which has a serious problem with sea-level rise (Aslumaimei & Bailey, 2018), should need a floating golf course is unexplained, except that tourism is an important way of making money, and golf does take up a lot of land (Vale & Vale, 2009:255–256).

On the surface, a more seemingly worthy venture is the African Water Cities Project (Adeyemi, 2012). Although the sea-level rise is not seen as the most important effect on Africa from climate change (Hinkel et al., 2012), much wealth and many people are concentrated in the continent's coastal cities. Moreover, "Africa is expected to experience the highest rates of population growth and urbanisation in the coastal zone, particularly in Egypt and sub-Saharan countries in Western and Eastern Africa" (Neumann et al., 2015). In response to these issues NLÉ, an Amsterdam-based architecture and urban design firm with a focus on developing world communities (NLÉ, 2020), developed a floating school for Makoko in Lagos as a model for how houses and other similar buildings could extend this already watery settlement (Adeyemi, 2012) (Fig. 2.2). Makoko is a fishing community located in the Lagos Lagoon, with a third of its 100,000 plus inhabitants living in stilt houses above the water, the latter having also made it a tourist attraction (WorldAtlas, 2020). Makoko is also known as a floating slum and is the cheapest property in the Lagos version of Monopoly (Anon, 2013). Because the water is shallow the timber supports for the houses are simply fixed into the bed of the lagoon (Olafimihan, 2009:22). However, because of impending storm surges and sea-level rise due

Fig. 2.2 Floating school, Lagos (adapted from https://www.dezeen.com/2014/03/25/makoko-floating-school-nigeria-nle/)

to climate change that threaten coastal human developments (see Chap. 5), Makoko was held up as an example of how communities can cope with living in a watery environment (Adeyemi, 2012), despite the fact that there is no proper sanitation or garbage disposal, leading to serious water pollution issues (Olafimihan, 2009:25).

> Nevertheless, NLÉ is learning from models such as these to develop improved prototypes and catalysts for future African coastal cities. Its Floating School proposal, for example, though its potential economic viability in the longer term is recognised, is a social contribution/development not intended for commercial gains. Instead, it will generate a viable and ecologically sustainable alternative construction system for the teeming population of Africa's coastal regions (Adeyemi, 2012).

The floating school was also important in saving the community as at the time of its construction the government had a programme of evicting the inhabitants, not least because the waterfront site was potentially valuable (Collins, 2015). The school was a triangular structure with a base of 10 m × 10 m and a height of 10 m. This makes it much taller than the single storey structures that formed the community. The base sits on a floating pontoon made of wood and recycled plastic barrels, and the structure is also local wood and bamboo. The triangular space contains three floors, the first being an open play area that is also a community space outside school hours, the second floor contains the classrooms and there is a workshop on the third floor (Riise & Adeyemi, 2015).

The school was located in the heart of the community and was meant as an example of how the community could be developed in a healthier and more sustainable way. However, in June 2016 the school collapsed in a heavy rain storm. Fortunately, it was unoccupied at the time as children had been removed from the school some months before because it vibrated in the rain (Okorafor, 2016). The remains of the school were soon carted away by the locals for their own building projects but questions were raised over the fact that had it been single storey it might have been stronger (Okorafor, 2016). Adeyemi claimed it was only a prototype and "The demolition and upgrading of the structure was in planning for a while and it had since been decommissioned...We are glad there were no casualties in what seemed like an abrupt collapse" (Frearson, 2016). An improved version of the triangular floating school was exhibited at the Venice Architecture Biennale 2016, where it won a Silver Lion (Mairs, 2016), and before the prototype collapsed in Makoko.

This well-meaning but sad story raises a number of issues. Climate change will bring more severe storms (see Chap. 5), which means what we build now should be capable of standing up to those storms. The school was not the only building destroyed in this particular storm.

> Classes had already been moved to another location in late March after heavy downpours at the start of the rainy season began to affect classes. "It is not only the floating school that collapsed. It collapsed many houses surrounding the floating school," said David Shemede, a Makoko resident and brother of the school's director (Reuters in Lagos, 2016).

However, unlike the homemade houses of found materials, this was a building designed by the well educated as an example of how the poor need to build to cope with climate change and sea-level rise, and as such it should have been designed not just to cope with the present rainy season but with future more severe ones. The fact it was built at all is in its favour, as it helped to stem the removal of the Makoko community. However, that threat came back with its demolition. While the school was there it was an attraction for tourists, but with perhaps little benefit to the community. "There's a widespread feeling that many of the people who come to Makoko to take photographs do so to make money off it—selling photos or stories to raise funds from which the people of Makoko will never benefit" (Ogunlesi, 2016). When running, the school catered for only 50 children, although it was always claimed as a prototype. Other projects in Makoko, which have not resulted in prizes for architecture, seem to have had greater success. In 2017, Nigerian students at the University of Nottingham launched the "…Liter of Light Nigeria charity drive, an outreach initiative which will make sustainable solar lighting accessible to impoverished regions of Nigeria" (University of Nottingham, n.d.a). This student crowd-funded project aimed at trying to find a way to meet the needs of the 1.5 billion people without access to electric light has been a modest success in empowerment:

> The team of nine students and volunteers have developed solar-powered "light bottles". Fitted with a solar panel and an LED bulb, the bottles can charge during the day and provide light in the hours of darkness.
>
> Last year, they installed around 100 light bottles in Makoko and trained young residents to build the units. Within a month of their departure, those trained on the visit had already trained 20 more residents to assemble the lights (University of Nottingham, n.d.b).

The residents have not yet rebuilt the floating school.

Living in Space (Because We've Made the Earth Uninhabitable)

On 14 April 1950, *Eagle,* a new children's comic, appeared in the newsagents' shops of Britain. The population was still suffering from the rationing of food and other commodities of daily life that had begun in the Second World War and that continued well into the 1950s (Imperial War Museum, 2018). In this profoundly grey era, the full-colour front page of *Eagle* was a revelation. "The new national strip cartoon weekly" featured the adventures of Dan Dare, Pilot of the Future. Colonel Dare was Chief Pilot of the Interplanet Space Fleet, whose headquarters and launch site were in Lancashire. He stood for "good, honest English decency" in contrast to "the muscle-bound, garishly-costumed superheroes from the United States" (Barnett, 2017). Eagle was soon selling a million copies a week (Dan Dare Corporation, 2020), and it became a part of the daily life of many people in 1950s Britain.

One of the attractive features of the Dan Dare stories was that they showed a convincing picture of what life might be like in the future. The first *Eagle* artist, Frank Hampson, and the others who followed him drew pictures of the buildings and vehicles of 1990s Britain that appeared completely plausible, whether it was the red London Transport monorails running along the Thames and through the streamlined blocks of London or the austere twin towers of Space Fleet HQ in Lancashire. The buildings of colonies on nearby planets such as the moon and Mars were also featured. The link between *Eagle* and design was celebrated in an exhibition at the Science Museum in 2008 called "Dan Dare and the birth of High-Tech Britain" which pointed out the influence of Dan Dare's world on a whole generation of British architects (Glancey, 2008).

This desire to colonise beyond the bounds of the planet has also been promoted as something necessary for the survival of the human species (Smith & Davies, 2012:149), and a number of architects, perhaps influenced by the drawings that appeared in the *Eagle,* have responded with designs. In 2012, Foster Associates developed their Lunar Habitation in conjunction with the European Space Agency. This base would house four people and was "first unfolded from a tubular module that can be transported by space rocket. An inflatable dome then extends from one end of this cylinder to provide a support structure for construction" (Foster + Partners, 2020a) (Fig. 2.3). The loose soil and rocks found on the surface, known as regolith, was then to be built up around this in layers using a 3D printing technique that would be operated by a robot. In 2015, this was followed by their Mars Habitat. This was a competition backed by NASA for "…a 3d-printed modular habitat on Mars". This would also be created by robots before the four human astronauts arrived to live in their "…robust 3D-printed dwelling" (Foster + Partners, 2020b).

A similar venture is a later collaboration of Skidmore, Owens and Merrill (SOM) with the European Space Agency and Massachusetts Institute of Technology (MIT)

Fig. 2.3 Lunar Habitation (adapted from https://www.fosterandpartners.com/projects/lunar-habitation/#gallery)

with the aim of designing a permanent village on the moon, on the rim of the Shackleton Crater (Johnston, 2020). The settlement has a focus on "sustainability and resilience", which perhaps goes without saying in such an inhospitable environment. Using solar energy in this part of the moon which is in near-continuous daylight, "Water–ice deposits could be extracted to produce breathable air and rocket propellant for transportation and industrial activities. Clusters of modules would be connected to enable easy mobility between structures…" (Johnston, 2020). The proposed modules would each house four to six people and would be formed of a three-column rigid frame and an inflatable shell.

Neither of these projects is seen as a way of finding new ways to live in an increasingly over occupied planet, at least in terms of human numbers. In the words of Professor Johann-Dietrich Woerner, head of the European Space Agency (ESA), "A Moon village shouldn't just mean some houses, a church and a town hall…This Moon village should mean partners from all over the world contributing to this community with robotic and astronaut missions and support communication satellites" (Hollingham, 2015). No one has yet mentioned how much such a venture would cost—well almost no one. "One of the fundamental obstacles is financial. Even if the moon village concept succeeded, most taxpayers would not understand why it was worth the cost" (Mori, 2016).

Perhaps what is more significant is that some see such exploration as a way of gaining resources over and above those available on earth (Indo-Asian News Service, 2016). The Moon Village Association see the concept as creating "…an environment where both international cooperation and the commercialization of space can thrive in a sustainable manner" (Moon Village Association, 2019).

Grand "Sustainable" Buildings (for the Rich)

The 2019 CopenHill Energy Plant and Urban Recreation Center built in Copenhagen and designed by the architectural firm BIG (Bjarke Ingels Group) has been praised for being state-of-the-art in technology, a bold idea and an icon for the city. From the New York Times to the fashion magazine Vogue, the reviews have been very positive. CopenHill has recently won the "building of the year 2020" in the industrial architecture category (ArchDaily, 2020). This is a recognition awarded by the readers of the website ArchDaily, one of the most popular media outlets for architectural projects and buildings on the internet. Described as a "zero-emission waste-to-energy plant" the attractive feature of this project is that the energy plant does not look like an industrial building. It has the shape of a mountain and the roof is to be used as a ski slope (Fig. 2.4). Considering there are no hills in Copenhagen, this artificial mountain has a strong presence in the city. The building offers the possibility of hiking, climbing and even having a coffee in the rooftop bar.

The idea behind the mountain shape is to accommodate the physical demands of the machinery and processes necessary to turn waste into energy effectively. The interior space contains furnaces, turbines and ventilation shafts but it also makes

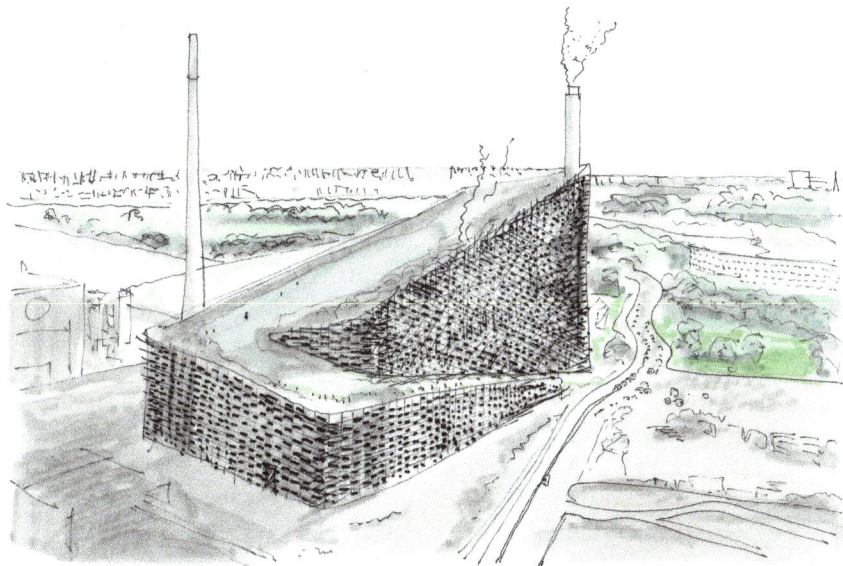

Fig. 2.4 CopenHill Energy Plant (adapted from http://emag.directindustry.com/copenhill-a-waste-to-energy-plant-with-a-ski-slope/)

room for the administrative offices of the Amager Ressourcecenter (ARC), which include an education centre for academic tours and other purposes, like conferences and workshops. These design efforts look towards transforming the place into a family destination and a probable tourist icon for the city. This shows good intentions that could be welcomed by the community are aligned with the goal of the city of becoming a carbon-neutral capital. This is the information promoted on the most popular architectural websites. However, the information omitted in the architectural media is that the planning of CopenHill did not happen without controversy.

According to BIG, CopenHill was designed to incinerate 440,000 tons of waste per year and to supply energy to the electricity grid and the district heating system, benefiting 150,000 homes (BIG, n.d.). This new building replaced a 40-year-old plant that had the same capacity (440,000 tons per year) based on the numbers provided by BIG. The problem is that these numbers are not correct. The CopenHill plant has, in fact, the capacity to process 560,000 tons per year. The old plant only managed to process 400,000 tons per year because the waste for incineration was declining due to the success of recycling projects in Denmark. In 2015, Denmark recycled 68% of its waste and burnt 26%, which accounted for 12% of the heating and 4% of the electricity of the country (Gurzu, 2019). Denmark is one of the most efficient countries in the European Union when it comes to recycling. Therefore, what was the reason for building a bigger plant when the old one was not even working at its full capacity?

This was one of the points raised by the Technical and Environmental Administration of the city of Copenhagen (TEA) when they assessed the CopenHill project in 2011. The TEA stressed that the project could impact negatively on the environment and become an obstacle in the face of the goal of the city becoming the first carbon-neutral capital of the world. The TEA highlighted the fact the waste-fired power plants in Denmark were already suffering from the risk of economic losses due to overcapacity and the fact European Union plans were heading towards reducing waste incineration and promoting more recycling (Kohl, 2019).

The assessments of the TEA were not incorrect. In 2019, CopenHill only processed 451,321 tons of waste, powered 30,000 houses and heated 72,000. This is only 10,000 tons per year above the full capacity of the previous plant without considering that 30,000 tons of garbage had to be imported from the United Kingdom and Ireland to achieve these results. CopenHill is not only performing at 80% of its full capacity but is also dependent on increasing garbage imports. The plan for 2020 is to import 90,000 tons from the United Kingdom, Germany, Italy, Ireland and the Netherlands, all of which comes with a transport carbon footprint. The CEO of the Amager Ressourcecenter, which is the company that runs the plant, is sceptical of the capacity of people to consume less and recycle more and declared that his group's function is to "rectify the consumption of the average citizens" (Gurzu, 2019).

There are so many contradictions in the CopenHill project that is hard to see why it is considered an example of sustainable architecture or even a solution to making cities more sustainable. Perhaps some of the doubts can be clarified when analysing the discourse of Bjarke Ingels, the lead architect of the firm BIG. In a Tedx Talk of 2011 and then in a lecture in the KTH (Royal Institute of Technology) in 2012, Ingels proposed that they are two classes of sustainability, the "sad and depressing", where people make sacrifices to their lifestyles in order to be more sustainable, and his own preferred version the "hedonistic sustainability" which increases life quality and human enjoyment (Ingels, 2012). It is pertinent to observe that the star-architect Ingels does not conceive the remote chance that people can be happy with very little and be proud of making sacrifices for others. As an example, my father was a biochemist who worked for 12 years without any holidays to start his own business and sustain his family. We had a very happy childhood and I do not remember lacking anything even though we were living on a tight budget.

How does the hedonistic sustainability of Bjarke Ingels increase life quality and enjoyment in CopenHill? The intention of the project was to offer a public space to the city and families. However, the public use of the rooftop and its activities is not completely free. The basic ski pass for an adult per hour is 150 DKK (US$23) and the rent of the equipment per adult is another 150 DKK (US$23) per hour (CopenHill A/S, n.d.). Children younger than 10 years old pay 105 DKK (US$16) per hour and 120 DKK (US$18) per hour for the equipment. This means that a family of two adults and two children will need 1050 DKK (US$157) per hour to enjoy skiing down CopenHill. The average wage in Denmark is 110 DKK (almost US$17) per hour, which means that one hour's skiing on CopenHill represents a day and a half of work for families living on the minimum wage. All this without even thinking of patronising the café or taking a guided tour at 80 DKK per person to learn about "the architecture, vision, sustainability...and thoughts behind" (CopenHill A/S, n.d.). The social sustainability proposed by Bjarke Ingels is not that social for everybody. This is a paradox considering that people's taxes were used to build CopenHill and now they need to keep on paying to use it (Kohl, 2019).

CopenHill is the most expensive waste to energy plant ever built with a budget estimated at 500 million euros. It has gone through a couple of malfunctioning issues that have cost another 13 million euros in 2016, during its construction, as well as technical problems since its opening (Nicastro, 2017). The mountain shape of the building, justified by the designers as necessary to accommodate the industrial processes and machinery, has also been questioned. The previous plant was a "high-modernist and functionalistic industrial masterpiece where purpose defined shape" (Kohl, 2019:28). However, in the design of CopenHill, the positioning of the chimney is far from the place where the smoke is generated adding more material to the project. It is the slope of the roof that has defined the placement of the machinery inside the building. The consulting firm Moe stated that there was not enough space for the rafters above the machinery, so the quality of steel had to be upgraded to build the sloping roof. This roof is made of a complex structure of steel and concrete to support 1500 people playing on the roof plus all the safety precautions to secure the people and the plant. None of these structural problems and the consequent demands placed on materials were necessary for the old building. The design of the external chimney was another problem since it weighs 345 tons and hangs from the external wall, 56 m above the ground. To avoid oscillations from the wind they needed to include an attenuator to keep oscillations at a safe level (Andersen, 2017). The transformation of the old incinerator building into CopenHill was done at the expense of investing more resources and capital to satisfy the demands of a complicated design that created big and expensive problems.

Given these comments, is CopenHill a sustainable architecture? Building an oversized, expensive and unnecessary project should be enough reasons to answer the question. A sustainable architecture should promote and facilitate the accomplishment of a sustainable lifestyle. "A sustainable building design is no more than the interface between people and their choice to live in a sustainable way" (Garcia & Vale, 2017:26). Another point that challenges the perception of CopenHill as sustainable architecture is the difference between incinerating and recycling waste. The economist and environmental advisor Morris (1996) compared recycling and incineration in 25 materials usually found in municipal solid waste. He found that for "24 out of 25 solid waste materials, recycling saves more energy than is generated by incinerating mixed solid waste". CopenHill favours the idea of incinerating waste, which saves less energy and is more expensive than promoting policies to recycle waste, something that people in Norway have been doing for decades, including recycling 97% of all plastic bottles via a deposit scheme (Cassella, 2019).

The blind faith in economic growth and technological development helps to advertise CopenHill as a positive solution. However, when looking at the facts, it has only generated more complex and expensive problems. It is a monument to a hedonistic and decadent lifestyle that people did not request. It defines a new trend—green consumerism. People will be encouraged to consume more, recycle less and still feel good about it because they would have the feeling of turning their waste into electricity and heat for the city.

The Sustainability of "Sustainable" Houses (also for the Rich)

Because the built environment tends to be long-lived, designers need to consider how a building behaves in the long term. However, all too often it is the photographs and condition of a building immediately after completion that receive the attention, not how well that same building looks some 30 years later, nor how it performs over the same period.

Investigating the Technical Aspects of a Building That Is Claimed to Be "Sustainable"

The Royal Institute of British Architects' (RIBA) 2018 house of the year was a single-storey Lochside House located in the West Highlands of Scotland (RIBA, 2020). "The off-grid, low energy home was designed by HaysomWardMiller Architects and incorporates a complex floorplan with three different floor levels and eight roof areas across a single-story building" (Anon, 2018a). The house is off the grid and has been described as "…a modest, sustainable home which sits in a magical location on the edge of a Scottish lake" (RIBA, 2018). The low energy aspect was achieved using structural insulated panels (SIPs) formed of a core of expanded polystyrene between two layers of oriented strand board (OSB) (JML SIPs, 2020a). The remote location meant that the SIPs were specially prefabricated so they could be brought to the very remote site using a Land Rover and trailer (JML SIPs, 2020b). SIPs panels have thermal advantages in terms of making an airtight timber structure with very low thermal bridging. However, the core is a polymer and polymers like foamed polyurethane (102 MJ/kg) and expanded polystyrene (117 MJ/kg) have more energy put into them in their manufacture than glass fibre (26 MJ/kg) (Moradbistouni et al., 2020). Without a life-cycle energy study, it is impossible to say whether over the life of the house using SIPs panels with their polymer core to make a very low energy house is worthwhile.

In a study of a small SIPs house in New Zealand, designed to be zero energy in operation, Moradbistouni et al. (2020) found that polymers in the house, including paint and wire insulation as well as the insulating materials, made a significant contribution to the overall life-cycle energy. "Looking at the total EE [embodied energy] of polymers in the house over 50 years, foam polyurethane inside the SIPs, paint, extruded polystyrene beneath the slab, and others accounted for 51%, 29% 12%, and 8% respectively." This same group of commonly used polymers in the building accounted for 59% of the life-cycle embodied energy and 89% of the embodied life-cycle CO_2 emissions of the house. Although a technically efficient solution, SIPs panels may be reducing the energy to operate the building at the cost of increasing the embodied energy and emissions over the life of the building.

Building a house is often seen as the be-all and end-all, but the impact of a house goes much wider than its materials. The makers of the panels used to build the house stress how remote it is and the difficulty of getting to it (JML SIPs, 2020b). The occupants of the house will not be able to pop round the corner to the shops, they will have to bring in their food and other supplies from their nearest town. They will not be going to town and back on the bus, nor in an electric four-wheel drive, and the overall carbon emissions from their choice to live in a "sustainable" house in a remote location will be much greater than those of someone living in a perfectly ordinary house closer to the shops. But ordinary people in ordinary houses never win architectural awards.

In all the published descriptions of the Lochside House, there is very little said about how it works. We are told that it is "off-grid" and "low energy" but no details are given. The photographs show a fireplace in the living room and a few PV panels on the roof. However, the plan shows that at one end of the building there is a 10 m^2 "generator house" (Frearson, 2018). You would not spend money on 10 m^2 of a building that you did not plan to use, so it is reasonable to assume this off-grid house is only able to operate because it has a fossil-fuelled generator. Given the small area of solar panels, the generator will need to run quite a lot of the time. It is quite depressing how architects assume that if something is described as "off-grid" it is more sustainable than something which uses the grid.

The sustainability of the award-winning Lochside House comes into further question with the realisation that all this effort has gone into making what the SIPs panel manufacturers describe as a holiday house rather than a permanent dwelling (JML SIPs, 2020b) (for further discussion of holiday homes, see Chap. 10).

Urban Design

Sometimes projects promise to be technically more efficient but their designs are not sustainable at an urban scale. The "Meridian First Light House" started as a project created by four students from the School of Architecture and Design at Victoria University of Wellington (Fig. 2.5). The project gained recognition when it achieved third place in the 2011 Solar Decathlon competition, an event organised by the U.S. Department of Energy. It also won the New Zealand Architecture Awards in the international architecture category. The jury commented that the building "reconciles a typological narrative and mandated sustainable features with architecture" (First Light Studio Ltd., n.d.), in simpler words, that it looks good. The prefabricated prototype was first built on the Wellington waterfront and then transported to Washington DC for the competition. The building was a single bedroom house of 70 m^2, which for New Zealand is a very luxurious unit for just one person or even a couple.

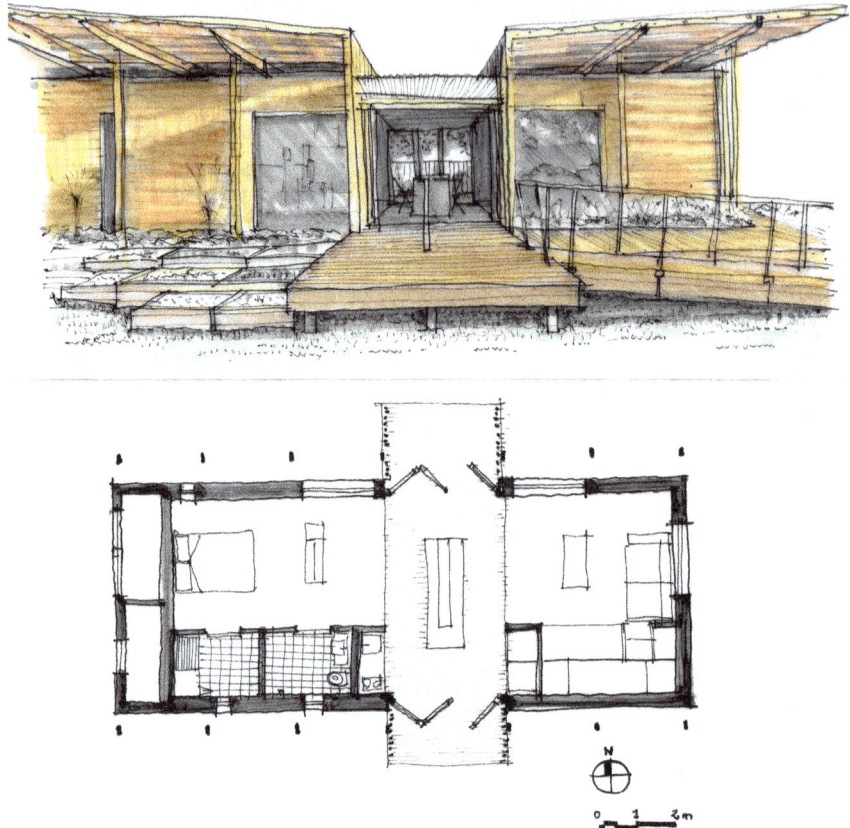

Fig. 2.5 Meridian First Light House (adapted from https://www.firstlightstudio.co.nz/the-meridian-first-light-house)

Even though the building used design strategies incorporating photovoltaic panels, wool as insulation (in New Zealand there are 26,821,846 sheep (Livestock, 2019) but as an insulation material wool has a greenhouse gas problem because sheep are ruminants that emit methane), and timber from the region, it is a house that can only work in an isolated beach situation or in a plot without neighbours. The design of the house opens in all directions; therefore it needs a lot of space around it to be functional.

The Meridian First Light House represents a great effort and very good intentions that could have been guided in a different way but were restricted by the nature of the Solar Decathlon competition and the need to be able to guide visitors around the house. However, it was the effort of a group of students and understandably they could lack an urban vision.

The case of the ZEB Multi-Comfort House in Norway is a little bit different. The project, built in 2014, was a collaboration between the Research Center on Zero

Emission Buildings and the architectural firm Snøhetta, all experienced professionals in Oslo. The very large house of 220 m^2 has a C-shape with an interior courtyard and also features a set of external amenities, like a garage, pool, and services that can only happen in a very wide plot (facing the street) which is the type of plot that tends to be very expensive (Fig. 2.6).

In 2011, 72% of all dwellings in Oslo were apartments, making this house a rarity that could only be built on the periphery of the city and only for very wealthy families. Even though the architects proclaimed this to be a zero-emission house, it could not operate efficiently in an urban environment surrounded by or attached to houses of its same kind. Such a house would only work if the council allowed suburban sprawl around the whole city. The reality is that 60% of all dwellings in Oslo are 80 m^2 or less, almost a third the size of this house (Statistics Norway, 2020). The house can use geothermal energy in a sustainable manner but it does not use land, which is the most valuable natural resource in cities, in a sustainable manner. This is the case of hundreds of projects that are promoted in websites and prestigious architectural magazines as "sustainable houses".

Another factor that confuses the perception of what sustainable means in architecture is the gap between the theoretical performance of a project on paper and its performance after it is built and people use it. There are many factors that affect this gap. One of them is the "belief" that a project will perform well even though common sense tells the contrary, such as the "off-grid" example above. The focus on technological aspects of a "sustainable building" cannot be separated from the behaviour of its inhabitants. If you have to drive a long way to your remote sustainable home, you will certainly be using more energy than the home will ever save.

Politically Correct Solutions (Even Though They Are Not Correct)

The idea of giving to the poor in a way that is visible so that people can see how generous you have been, rather than just handing over the money, has a long history. In past centuries alms-houses, where the poor could live, were as likely to be named after their founders as much as saints, such as the two houses of John Isbury (Lambourn, n.d.) and Jacob Hardrett (National Archives, n.d.) in Lambourn in Berkshire, UK. Such practices are still around.

Refugee Camps

By the end of 2019, globally the mass migrations produced by social and political conflicts had forced 79.5 million people to leave their homes (UNHCR, 2020a), which is like having more than the entire population of the United Kingdom displaced

Fig. 2.6 ZEB Multi-Comfort House (adapted from https://www.archdaily.com/773383/zeb-pilot-house-pilot-project-snohetta)

and homeless, as the 2019 population of the UK was 67 million (Population UK, 2020). The UN Refugee Agency (UNHCR) states there are 25,000,000 refugees with a further 3.5 million still seeking asylum. Syria is the country with the largest population of displaced persons (6.6 million) while Turkey (3.6 million), Pakistan, Germany, Uganda, Lebanon, Bangladesh and Jordan are the countries with the biggest refugee population components. The way hosting countries have dealt with refugees in the built environment offers notable contrasts. Germany decided to host displaced people in cities while Australia uses the island of Nauru in Papua New Guinea as a detention centre for asylum seekers arriving by sea (Harrison, 2018).

One of the formal responses to this situation has been to provide temporary shelter for displaced populations in the so-called refugee camps. As a result, some countries like Jordan or Bangladesh have developed city-size refugee camps (UNHCR, 2020b). Kutupalong in Bangladesh hosts around 600,000 refugees from Myanmar. This is a population larger than that of Canberra (Dingle, 2019). Due to the impact of climate change, and according to reports prepared by the International Organization for Migration (IOM) and the UN, the number of migrants could escalate to figures that range from 1 to 1.5 billion by 2050 (United Nations University, 2017). Therefore, future climate change migrants could turn refugee camps into the new megalopolises of the twenty-first century. However, designers have been quick to see an opportunity emerging from such suffering, with the design of prefabricated emergency shelters. One such venture is that of IKEA, the Swedish furniture and household goods chain.

Design for Refugees; the IKEA Better Shelter

The IKEA Better Shelter was perhaps not unsurprisingly, given what the firm sells, a flat pack. This made a tent-like structure of plastic that could be recycled once its three-year design life was over. It came with a solar light, had a lockable door and was fitted with interior curtains for some privacy for its design target of five occupants. It could also be assembled in four hours without the need for tools (Tubertini, 2018) (Fig. 2.7).

In 2017, after IKEA's refugee shelter was awarded the Beazley Design of the Year, the United Nations High Commission for Refugees raised concerns about the vulnerability of the Better Shelters to fire. The UNHCR bought 15,000 shelters in 2015 at a price of US$12.5 million (each tent was around US$1250). In 2017, a third of them had been deployed, those remaining waiting for fire safety and other issues to be solved. The design has been subject to a series of criticisms. The ventilation is insufficient, it does not provide a groundsheet and the accessibility for people in wheelchairs is complicated (Fairs, 2017a).

A scholar from Oxford, Tom Scott-Smith (2019), sees the IKEA project as an example of the stubbornness of humanitarian agencies that insist on developing universal solutions through design and technologies that do not account for the sociopolitical and environmental differences of migrants and settlements. These shelters denied the possibility of working with the communities involved from the bottom-up,

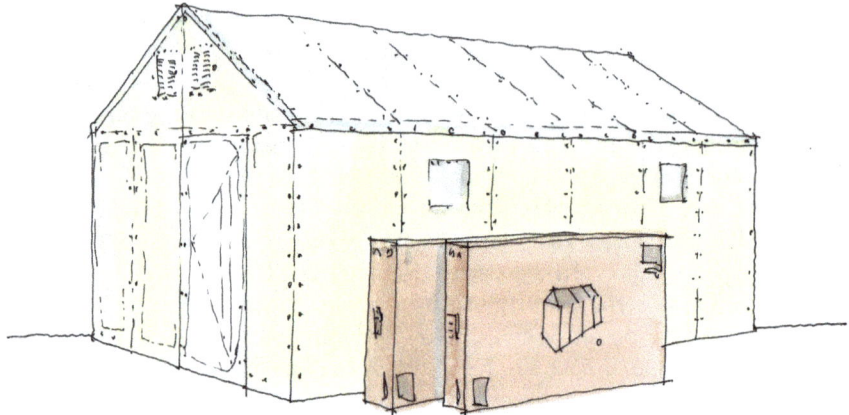

Fig. 2.7 The IKEA Better Shelter (adapted from https://bettershelter.org/)

producing perhaps more adaptable and suitable solutions. In comparison with other initiatives, for him, the IKEA shelters were too expensive and complex because they "provided a solution to a problem that does not exist" (Scott-Smith, 2019). Ordinary tents are much cheaper and deliver almost the same services. A standard tent can be built in 30 min in comparison to the four hours of the IKEA shelter, a process that also needs training to avoid the risk of injuries to those constructing them.

Housing Refugees

When Ricky Gervais, the English comedian, hosted the 2020 Oscars Awards ceremony he provided a recommendation to all the award winners. "So if you do win an award tonight, don't use it as a platform to make a political speech. You're in no position to lecture the public about anything. You know nothing about the real world. Most of you spent less time in school than Greta Thunberg" (White, 2020). Similar advice could be helpful in assessing the design of firms and companies that are trying to tackle humanitarian causes without necessarily having the success suggested by their websites and design awards. One of the common problems derived from the naivety of these designers is the belief that everything is about design and that the world can be saved through design. This is, of course, not an ideal situation when the clock is ticking and we have less than 30 years to produce the radical changes suggested by the discussion in Chap. 1. These really need to be changes in behaviour but we prefer not to think about that, believing instead that we can design them. Humanitarian causes open the door to the production of designs whose intentions looked politically correct in the media but end up being less than useful in real life.

When Kilian Kleinschmidt, the Director of the Zaatari refugee camp, who worked for the UNHCR was interviewed at the Dutch Design Week, he made a call to stop designers from designing shelters for refugees. "They're not a species. So, there is no need for tech for refugees. Or design for refugees, or architecture for refugees" (Fairs, 2017a, b). The alternative could be to provide people with housing and jobs instead of disconnecting them from such opportunities in refugee camps. The Chilean architect Aravena proposed that designers could play an important role only if long-term solutions were to become the focus instead of investing in tents as emergency shelters that are quick responses but are the equivalent of throwing money away, or in his words, they represent "money that melts" (Fairs, 2015).

The future impact that climate change could have in communities will make the aspirations to eradicate refugee camps a lot harder to achieve but unless a permanent solution is found to housing refugees, refugee shelter will remain a "band-aid" rather than a solution.

The Solution Must Be Digital and Employ Robots (Even When the Analogical Works Fine and We Have a Lot of People)

Much has been made of the idea of 3D printing buildings, and this has already been mentioned in connection with creating shelters on the moon and Mars. One approach to 3D printed buildings is that machines in the guise of such printers could reproduce components that have typically been used to make architecture, such as bricks. As Rael and Fratello (2018:12) stated in describing their enterprises, "By 3D printing small, fundamental architectural components, we aim to make 3D-printed architecture accessible, interactive, and related to the craft traditions of the past but with all the yet-to-be-explored potential that this emerging technology has to offer." Bricks are already machine-made, so 3D printing using clay hardly seems a step forward and such bricks can only be used in dry climates unless they are fired. It is the firing that uses the energy in brick making, not running the machine to make the brick (Sunil Kumar et al., 2016). Moreover, unfired brick or adobe construction is a way of empowering those without homes to learn to build for themselves using inexpensive materials, with the "cost" being in the self-built labour, and this has been proposed as a way of moving people out of slum conditions into better housing (Venter et al., 2019). It seems more sensible to make use of people than machines in such situations since training introduces skills into a community rather than transferring the skills to a machine.

One 3D printing building project aims to print a three-storey house in Amsterdam using printed polymers in the form of blocks measuring $1.8 \times 1.8 \times 3.0$ m. An investigation into the materials that might be used for the project produced the following recommendation (Van de Veen et al., 2015).

> In case one actually wants to build a house with the currently applied printing material, the material is only applicable as a permanent mould for concrete. In that case, the inner holes of a printed geometry can be filled with reinforced concrete and insulating materials. On

the outside of the permanent formwork, at least a heat-resistant façade-covering needs to be added.

Effectively in this project high embodied energy plastics are being used as permanent formwork for concrete, while traditional wooden formwork would probably be cheaper and would certainly embody less energy. Concrete is the material most often used for 3D printed buildings, such as the 65 m^2 3D printed house in Austin, Texas where the crane-like printer poured concrete in layers as it revolved around the predetermined plan. At a cost of under US$10,000, this is claimed to be the answer to affordable housing (Ramirez, 2018), though whether that cost is for the complete house including site works and fixtures and fittings or just the bare concrete shell and timber roof is not clear. It is true that the house offers "…a continuous thermal envelope, high thermal mass, and near zero-waste" (Ramirez, 2018) but no mention is made of insulation, so like the 3D printed clay brick house, this would have to be an abode for a warm, dry climate (Fig. 2.8).

3D printing is not the only attempt to mechanise the building industry, which historically has been focused on manual trades. Some advances are already accepted practice, such as the replacement of the human treadmill powered crane for lifting, as used in mediaeval times, with the modern motorised crane. More recent advances include a machine that can lay bricks at the rate of 2000–3000 a day, compared with the 400–500 of the human bricklayer, robotic glazing machines, drones for surveying the site and works under construction, and autonomous earth moving vehicles (Bogue, 2018). Such ideas raise issues about the energy it takes to make and run the machines and the cost invested in them, which in turn will affect the

Fig. 2.8 3D printed house, Austin, Texas (adapted from https://singularityhub.com/2018/03/18/this-3d-printed-house-goes-up-in-a-day-for-under-10000/)

types of project for which they are suitable. It also seems perverse in a world with an increasing population to seek solutions that reduce the need to employ people.

Prefabrication, which has also sought to transfer building from the site to the factory, has not had a good history of being cost-effective (for more discussion, see Chap. 5). The use of robots also raises architectural issues. The brick developed as a fit to the human hand, and one delight of a brick wall is to see that it has been hand-built. The next development will probably be the robot bricklayer that produces a "wobbly" character in imitation of what a person does naturally. In a world where people are abundant and sources of energy that do not contribute to climate change much less so, it would seem more sensible to be moving away from machines towards more use of human labour, but then that would add to cost and so would be totally unacceptable.

Climate Change Solutions Must Sound "Scientifically Plausible" (Even Though They Are not Feasible or Even Necessary)

If we do not want to change our lifestyles to reduce our carbon emissions perhaps we can use technology to remove carbon from the atmosphere or mitigate its effects on the climate? Currently, the two main lines of research in geoengineering and climate change mitigation are carbon dioxide removal (CDR) and solar radiation management (SRM). CDR focuses on carbon capture and storage and SRM tries to find ways to reflect sunlight away from the earth to reduce the greenhouse effect (Aouf, 2018). The first carbon-capture plant, funded by Climeworks, opened in 2017 near Zurich in Switzerland. The plant was designed to extract 900 tons of carbon dioxide annually "…or the approximate level released from 200 cars—and pipe the gas to help grow vegetables" (Marshall, 2017). In another attempt to extract CO_2, the London-based firm ecoLogicStudio in collaboration with Climate-KIC has designed a shading system called *Photo.Synth.Etica* that filters air by capturing CO_2 and other pollutants (Walsh, 2018).

> The filtration process involved urban air introduced to the bottom of the façade, causing air bubbles to rise through the watery medium within the bioplastic. CO_2 and other pollutants are captured and stored in the algae, and grow into biomass. The biomass can be harvested and used in the production of bioplastic, which is in turn used as the main building material of the photobioreactors themselves. The process culminates with freshly filtered oxygen released from the top of each façade unit.

The shading system, composed of 16 clear flexible polymer modules each 2 m × 7 m, was displayed at the Printworks Building in Ireland's Dublin Castle and it is supposed to capture one kilogram of CO_2 per day, which is what might be captured by 20 mature trees. Mature trees are not as good as young trees at carbon sequestration, although this view has been questioned (Jacob et al., 2013). Poletto, one of the co-founders of the ecoLogicStudio, criticises the mentality of architects that want to control and understand nature as a machine. "Now, we have a new notion of a machine that is much more biological, much more inspired by the notion of complexity" (ecoLogicStudio, 2014). In describing the project, the media consistently highlights the claim that one kilo of CO_2 per day represents the amount absorbed by 20 large trees in order to provide a sense of the scale of the project's contribution. However, the price of the plastic prototype and the CO_2 emissions invested in its creation are never mentioned. The cost of young trees would be low and they could be planted by unskilled labour but such a simple idea could not be patented nor would it make millions of dollars in profit. The biomass from the trees could also be harvested and used to make buildings, furniture, paper or fibre, just as proposed for the *Photo.Synth.Etica*. What is also clearly omitted is that one kilo of CO_2 emissions could easily have been saved if one user of the building took the bus to work instead of driving or just used half a litre less of petrol. What the media fails to mention is that it is easier to avoid one kilo of CO_2 emissions than capture it after it has been released, although the firm ecoLogicStudio has been named by the magazine Metropolis among the top ten global talents in architecture in 2014.

The history of urban and architectural projects that look forward to purifying their environments and how they failed to be built or deliver what they promised deserves a chapter in another book. Is it possible to have a sustainable building in an unsustainable city? Is it possible to have a sustainable city on an unsustainable planet? In 2015, Dutch designer Daan Roosegaarde launched a campaign to build the largest air purifier in the world (Studio Roosegaarde, n.d.). The idea was to create pockets of clean air in cities that are heavily polluted, presumably so that rich people would not be inconvenienced, but such efforts seem to be missing the point. The goal should be to work out how to behave so that cities do not become heavily polluted.

The faith in geoengineering as the solution to climate change issues implies that humanity must remain passive and expect machines to do their job. Even if these technologies are feasible, what are their implications? In other fields, the dependence of cities on machines as a way of persisting has already been attempted. The Sultanate of Oman relies on desalination plants to provide potable water to its population. In the Governorate of Muscat, the capital of Oman, 90% of the residential water consumed depends on desalination plants (Kotagama et al., 2017:907). In 2018, the Sultanate of Oman invested US$1.1 million in water subsides per day but one gallon of water in Oman is still far more expensive than a gallon of water in the United States of America. Since the population of Oman keeps on growing and the modern lifestyles have increased consumption, the government is obliged to keep on building new plants that need to be more sophisticated to produce more potable water using fewer resources. The latest acquisition, Barka 4, costs more than US$300 million (James, 2019). This is a very expensive and energy-dependent way of providing water that may well be unsustainable in the short term, particularly if the impacts of climate change are considered. The dependence on machines to deal with climate change issues could be to the earth the equivalent of a ventilator for a patient in a coma. If the machine is turned off, the patient may well die, as we have been seeing with the COVID-19 pandemic.

It seems we are supposed to believe that changing human behaviour is not a variable that can lead to results in the short term because through design, machines and other artefacts can deliver better and more predictable outcomes than people. After all, designers design things not people. People's behaviour cannot be trusted. The corollary of these types of assumption is that even a vacuum cleaner is trustier than your neighbour when it comes to cleaning the carpet. For professionals involved in the development of the built environment, this means that buildings are more important than people. More than the triumph of technology this looks like the defeat of the belief in humankind, as shown by the global rush to make self-driving cars rather than getting people onto public transport and bicycles.

The Role of Design

Of course, designers will not accept being on the bench in the race for the idea that saves the world. Marcus Fairs, the editor-in-chief of the popular design website Deezen, proposed that designers "can shape the anthropocene era to prevent global catastrophe" (Hobson, 2018). Stories about new technologies are for designers what methamphetamine is for a junkie. The architecture studio PARQ (Anon, 2018b) believes that a series of megastructures of volcanic shape can be used as cloud makers to soften the effect of the sun over icy regions and to precipitate rain over desert areas. Two of the most important challenges of humanity seem to be solved with a single genius design! The idea is to place these artificial volcanos in the oceans so they can extract hydrogen and oxygen from water via electrolysis and then mix them in a combustion chamber to finally inject the gas into the atmosphere. Pedro Ramalho, one of the partners at PARQ, states that "this is just an idea, as I do not have a scientific background or calculations to back this up" (Anon, 2018b). To an older generation, admitting this level of incompetence by someone who presumably wishes to be taken seriously seems breathtaking.

However, not all architects think in the same way. Spencer De Grey, the head of design at Foster + Partners, declared that their winning Stirling Prize building for ecological contribution, the European headquarters for stock-market data giant Bloomberg, is still a "three-degree" building (Harper, 2018). De Grey stated that even the world's "most sustainable" office building produces emissions that are double the amount that we can afford if we are to have no more than a one-and-a-half degree rise in average temperature. The head of design at Foster + Partners made it clear that making buildings more efficient is not enough because it only helps to prolong the present unsustainable situation. Without a change in the paradigm that redesigns the economic system in which architecture is embedded, speculations about sustainable architecture are just futile. De Grey suggests that there are minimal (and extremely humble) actions that can help designers to have a more important role, like upgrading the present built environment instead of replacing it with a more sustainable one. The humbleness of this idea is doubtless the reason why designers, who are taught to be arrogant, will never adopt it. Having said all this, it is important to acknowledge that Foster + Partners is the same studio that proposed extra-terrestrial habitats, as discussed earlier in this chapter. In the meantime, for those that will not be able to get to Mars by bus, having a plan B to collapse gracefully on planet Earth might not be a bad idea.

Conclusion

This chapter has looked at a series of projects that represent some of the current trends in the design of the built environment to deal with the social and economic crisis described in Chap. 1. In all the cases discussed there is a common theme: the myth that climate change issues can be dealt with through economic growth and technological development. There is also a common outcome in that all the projects contribute to worsening the social crisis, the environmental crisis or both. It seems it does not matter what designs really do but how they sound and what they promise to do. As in any regular political campaign, in the design of buildings what matters most is the size of the promise. All that projects and designers need is to sound scientifically plausible even though their designs might not be feasible or good for everybody. This creates the need for innovation, which benefits economic growth and this, in turn, justifies all actions. This vicious circle demands increasing quantities of material resources and energy. Therefore, it is not coincidental that carbon dioxide emissions have increased from 1990 to the present, regardless of this being the period in which the development of sustainable architecture has reached a global scale. The negative impact of the symbiosis between technology and economic growth is the real measure of an unachieved ideal of progress and this is expressed in the built environment and ideas about its future.

What has been exposed in the projects presented is that economic greed and the use of inappropriate technologies tend to produce two types of designs. The first type consists of designs that provide solutions to problems that do not exist, and the second are designs that tend to make problems bigger. In both cases, the problem-solving mechanism of design ends up creating new and/or bigger problems instead of solving them. Creating bigger problems is one aspect and consequence of increasing complexity, which is discussed further in Chap. 3, where it is linked to theories of collapse.

References

Adeyemi, K. (2012). African water cities. *Architectural Design, 82*(5), 98–101.
Ahvenniemi, H., Huovila, A., Pinto-Seppä, I., & Airaksinen, M. (2017). What are the differences between sustainable and smart cities? *Cities, 60*, 234–245.
Andersen, U. (2017). *'Crazy idea' at Amager Bakke wins steel award*. Retrieved July 18, 2020, from https://ing.dk/artikel/skoer-ide-paa-amager-bakke-vinder-staalpris-205390.
Anon. (2013). Lagos is new site of Monopoly game. *African Business, 394*(February), 6.

References

Anon. (2018a). *JML Contracts hails role on RIBA House of the Year project.* Scottish Construction Now. Retrieved July 19, 2020, from https://www.sips.org/downloads/JML-Contracts-hails-role-on-RIBA-House-of-the-Year-project---Scottish-Construction-Now-2018.pdf.

Anon. (2018b). *PARQ's volcanic machines release steam clouds to reverse the effects of global warming.* Retrieved July 22, 2020, from https://www.designboom.com/design/parq-volcano-structures-steam-cloud-maker-global-warming-10-24-2018/.

Aouf, R. S. (2018). *Five geoengineering solutions proposed to fight climate change.* Retrieved July 21, 2020, from https://www.dezeen.com/2018/10/18/five-geoengineering-solutions-climate-change-un-ipcc-technology/.

ArchDaily. (2020). *Building of the Year 2020.* Retrieved July 18, 2020, from https://boty.archdaily.com/us/2020.

Aslumaimei, A. A., & Bailey, R. T. (2018). Quantifying threats to groundwater resources in the Republic of Maldives: Part 1 future rainfall patterns and sea-level rise. *Hydrological Processes, 32*(9), 1137–1153.

Barlow, M. A., & Levy-Bencheton, C. (2019). *Smart cities, smart future: Showcasing tomorrow.* Wiley.

Barnett, D. (2017). *Dan Dare: how the British superhero survived to make the digital age.* Independent. 1 September.

Bennett, D., Pérez-Bustamante, D., & Medrano, M. (2017). Challenges for smart cities in the UK. In M. Peris-Ortiz, D. R. Bennett, & D. Pérez-Bustamante Yábar (Eds.), *Sustainable smart cities: Creating spaces for technological, social and business development* (pp. 1–14). Springer International Publishing.

BIG. (n.d.). *ARC: CopenHill/Amager Bakke.* Retrieved July 18, 2020, from https://big.dk/#projects-arc.

Bogue, R. (2018). What are the prospects for robots in the construction industry? *Industrial Robot, 45*(1), 1–6.

Bolonkin, A. A. (2011). Floating cities. In S. D. Brunn (Ed.), *Engineering earth: The impacts of megaengineeing projects* (pp. 967–983). Springer.

Cassella, C. (2019). *Norway's insanely efficient scheme recycles 97% of all plastic bottles they use.* Retrieved August 11, 2020, from https://www.sciencealert.com/norway-s-recycling-scheme-is-so-effective-92-percent-of-plastic-bottles-can-be-reused.

Chang, H. (2015). Defying gravity. *New Scientist, 226*(3018), 1.

Collins, J. (2015). Makoko Floating School, beacon of hope for the Lagos 'waterworld'—A history of cities in 50 buildings, day 48. *The Guardian.* Retrieved March 8, 2020, from https://www.theguardian.com/cities/2015/jun/02/makoko-floating-school-lagos-waterworld-history-cities-50-buildings.

CopenHill A/S. (n.d.). *Prices.* Retrieved July 18, 2020, from https://www.copenhill.dk/en/ski365#/pricing.

Dan Dare Corporation Ltd. (2020). *The Eagle Comic: The official home of Dan Dare.* Retrieved March 12, 2020, from http://www.dandare.com/eagle-comic.htm.

DataBank. (2020). *World development indicators. CO_2 emissions.* Retrieved September 9, 2020, from https://databank.worldbank.org/reports.aspx?source=2&series=EN.ATM.CO2E.PC&country=#.

Dingle, S. (2019). *Kutupalong refugee camp, home to more than 600,000 Rohingya, faces daily challenges.* Retrieved July 20, 2020, from https://www.abc.net.au/news/2020-07-20/intensive-care-units-icu-coronavirus-covid19-victoria/12467050.

ecoLogicStudio. (2014). *BioCities/MetaFollies/algae/magazines/responsive systems/urban design.* Retrieved July 21, 2020, from http://www.ecologicstudio.com/v2/project.php?idcat=10&idsubcat=71&idproj=142.

Fairs, J. (2015). *Refugee tents are a waste of money, says Alejandro Aravena.* Retrieved July 20, 2020, from https://www.dezeen.com/2015/11/30/alejandro-aravena-humanitarian-architecture-refugee-tents-waste-money-emergency-shelter-disaster-relief/.

Fairs, M. (2017a). *Ten thousand IKEA refugee shelters left unused over fire fears, UN admits*. Retrieved July 20, 2020, from https://www.dezeen.com/2017/04/29/united-nations-admits-10000-ikea-better-shelter-refugees-mothballed-fire-fears/.

Fairs, M. (2017b). *"Don't design yet anther shelter" for refugees, say experts*. Retrieved July 20, 2020, from https://www.dezeen.com/2017/12/18/dont-design-shelter-refugees-kilian-kleinschmidt-rene-boer-good-design-bad-world/.

First Light Studio Ltd. (n.d.). *The Meridian First Light House*. Retrieved July 19, 2020, from http://www.firstlightstudio.co.nz/the-meridian-first-light-house.

Foster + Partners. (2020a). *Lunar Habitation*. Retrieved March 11, 2020, from https://www.fosterandpartners.com/projects/lunar-habitation/.

Foster + Partners. (2020b). *Mars Habitat*. Retrieved March 11, 2020, from https://www.fosterandpartners.com/projects/mars-habitat/.

Frearson, A. (2016). Kunlé Adeyemi's floating school suffers "abrupt collapse". *Dezeen*. Retrieved March 9, 2020, from https://www.dezeen.com/2016/06/08/kunle-adeyemi-nle-makoko-floating-school-nigeria-destroyed-abrupt-collapse/.

Frearson, A. (2018, November 28). "Breathtaking" off-grid Lochside House wins RIBA House of the Year 2018". *Dezeen*. London, Dezeen. Retrieved August 14, 2020, from https://www.dezeen.com/2018/11/28/lochside-house-scotland-haysom-ward-millar-architects-riba-house-year-2018/.

Gammon, K. (2012). Hope floats: Artificial islands that rise with the seas. *Popular Science, 281*(2), 32.

Garcia, E. J., & Vale, B. (2017). *Unravelling sustainability and resilience in the built environment*. Routledge.

Glancey, J. (2008, April 28). "Sufferin' satellites! We've built the future!" *The Guardian*. Retrieved March 12, 2020, from https://www.theguardian.com/artanddesign/2008/apr/28/architecture.

Gurzu, A. (2019). *Not enough's rotten in the state of Denmark*. Retrieved July 18, 2020, from https://www.politico.eu/article/denmark-garbage-gamble-amager-bakke-plant-waste/.

Harper, P. (2018). *The world's most sustainable office building isn't enough to save the planet*. Retrieved July 21, 2020, from https://www.dezeen.com/2018/10/12/opinion-foster-partners-bloomberg-sustainability-climate-change-phineas-harper/.

Harrison, V. (2018). *Nauru refugees: The island where children have given up on life*. Retrieved July 20, 2020, from https://www.bbc.com/news/world-asia-45327058.

Hinkel, J., Brown, S., Exner, L., Nicholls, R., Vafeidis, A., & Kebede, A. (2012). Sea-level rise impacts on Africa and the effects of mitigation and adaptation: An application of DIVA. *Regional Environmental Change, 12*(1), 207–224.

Hobson, B. (2018). *Watch our talk about design and the Anthropocene at Dutch Design Week 2018*. Retrieved July 21, 2020, from https://www.dezeen.com/2018/10/20/livestream-talk-good-design-bad-world-anthropocene-dutch-design-week-2018/.

Hollands, R. G. (2008). Will the real smart city please stand up?: Intelligent, progressive or entrepreneurial? *City, 12*(3), 303–320.

Hollingham, R. (2015). *Should we build a village on the moon?* Retrieved March 11, 2020, from https://www.bbc.com/future/article/20150712-should-we-build-a-village-on-the-moon.

IESE. (2019). *IESE Cities in Motion Index 2019*. Retrieved July 17, 2020, from https://media.iese.edu/research/pdfs/ST-0509-E.pdf.

Imperial War Museum. (2018). *What you need to know about rationing in the Second World War*. Retrieved March 11, 2020, from https://www.iwm.org.uk/history/what-you-need-to-know-about-rationing-in-the-second-world-war.

Indo-Asian News Service. (2016). *'Villages on the Moon' can be reality by 2030: Study*. Retrieved March 11, 2020, from https://gadgets.ndtv.com/science/news/villages-on-the-moon-can-be-reality-by-2030-study-788110.

Ingels, B. (2012). *Hedonistic sustainability*. Retrieved July 18, 2020, from https://www.youtube.com/watch?v=M34fqaMPrv4.

References

Jacob, M., Bade, C., Calvete, H., Dittrich, S., Leuschner, C., & Hauck, M. (2013). Significance of over-mature and decaying trees for carbon stocks in a Central European natural Spruce forest. *Ecosystems, 16*(2), 336–346.

James, I. (2019). *In the middle east, countries spend heavily to transform seawater into drinking water.* Retrieved July 22, 2020, from https://www.azcentral.com/story/news/local/arizona-environment/2019/11/29/middle-east-oman-water-desalination-reliance-costs/2123698001/.

JML SIPs. (2020a). *Supply only.* Retrieved July 29, 2020, from https://www.jmlsips.co.uk/services/supply-only.

JML SIPs. (2020b). *Lochside holiday house, Scottish Highlands.* Retrieved July 19, 2020, from https://www.jmlsips.co.uk/projects/lochside-holiday-house-scottish-highlands.

Johnston, G. (2020). *Visionary designs for living on the moon from SOM and ESA.* Retrieved March 11, 2020, from https://www.worldarchitecturenews.com/article/1673406/visionary-designs-living-moon-som-esa?bulletin=daily-review-bulletin2&utm_medium=EMAIL&utm_campaign=eNews%20Bulletin&utm_source=20200310&utm_content=News%20Review%20-%20Tuesday%20(8)::&email_hash=.

Kamel Boulos, M. N. K., Tsouros, A. D., & Holopainen, A. (2015). 'Social, innovative and smart cities are happy and resilient': Insights from the WHO EURO 2014 International Healthy Cities Conference. *International Journal of Health Geographics, 14*(3), 1–9.

Kramers, A., Höjer, M., Lövehagen, N., & Wangel, J. (2014). Smart sustainable cities—Exploring ICT solutions for reduced energy use in cities. *Environmental Modelling and Software, 56*, 52–62.

Kotagama, H., Zekri, S., Al Harthi, R., & Boughanmi, H. (2017). Demand function estimate for residential water in Oman. *International Journal of Water Resources Development, 33*(6), 907–916.

Kohl, U. (2019). *The Copenhill crisis. The dark side of planning the greenest waste-fired power plant ever seen.* Malmö Universitet.

Lambourn. (n.d.). *Isbury or Esbury Almshouses.* Retrieved August 11, 2020, from https://lambourn.org/isbury-or-esbury-almshouses/.

Livestock Numbers by Regional Council. (2019). Retrieved September 8, 2020, from http://nzdotstat.stats.govt.nz/wbos/Index.aspx?_ga=2.164575031.987461716.1599534146-1682620751.1598411173&_gac=1.39985942.1598411774.EAIaIQobChMIh5iyyPG36wIVBmoqCh23CQQREAAYASABEgLVPvD_BwE.

Mairs, J. (2016). Kunlé Adeyemi docks Makoko Floating School at the Venice Biennale. *Dezeen.* Retrieved March 9, 2020, from https://www.dezeen.com/2016/05/31/kunle-adeyemi-docks-makoko-floating-school-venice-architecture-biennale-2016/.

Marsal-Llacuna, M., Colomer-Llinàs, J., & Meléndez-Frigola, J. (2015). Lessons in urban monitoring taken from sustainable and livable cities to better address the smart cities initiative. *Technological Forecasting & Social Change, 90*, 611–622.

Marshall, C. (2017). *In Switzerland, a giant new machine is sucking carbon directly from the air.* Retrieved July 21, 2020, from https://www.sciencemag.org/news/2017/06/switzerland-giant-new-machine-sucking-carbon-directly-air.

Moon Village Association. (2019). *Moon Village implementation.* Retrieved March 11, 2020, from https://moonvillageassociation.org/about/moon-village-implementation/.

Moradbistouni, M., Vale, B., & Isaacs, N. (2020). Evaluating the use of polymers in residential buildings: Case study of a single storey detached house in New Zealand. *Journal of Building Engineering, 32*, 1–9.

Moran, D., Kanemoto, K., Jiborn, M., Wood, R., Többen, J., & Seto, K. C. (2018). *Global gridded model of carbon footprints (GGMCF).* Retrieved September 9, 2020, from http://citycarbonfootprints.info/.

Mori, T. (2016). *To make a Moon village, think beyond science and engineering (Op-Ed).* Retrieved March 11, 2020, from https://www.space.com/31985-space-settlements-require-input-from-everyone.html.

Morris, J. (1996). Recycling versus incineration: An energy conservation analysis. *Journal of Hazardous Materials, 47*(1–3), 277–293.

National Archives. (n.d.). *Hardrett's Almshouses, Lambourn.* Retrieved August 11, 2020, from https://discovery.nationalarchives.gov.uk/details/c/F92093.

Neumann, B., Vafeidis, A. T., Zimmerman, J., & Nicholls, R. J. (2015). Future coastal population growth and exposure to sea-level rise and coastal flooding—A global assessment. *PLoS One, 10*(3). Retrieved March 7, 2020, from https://journals.plos.org/plosone/article/file?id=10.1371/journal.pone.0118571&type=printable.

Nicastro, C. (2017). *Copenhagen goes all in on incineration, and it's a costly mistake.* Retrieved July 18, 2020, from https://zerowasteeurope.eu/2017/10/copenhagen-goes-all-in-on-incineration-and-its-a-costly-mistake/.

NLÉ. (2020). *NLÉ.* Retrieved March 7, 2020, from http://www.nleworks.com/.

Ogunlesi, T. (2016). Inside Makoko: Danger and ingenuity in the world's biggest floating slum. *The Guardian.* Retrieved March 9, 2020, from https://www.theguardian.com/cities/2016/feb/23/makoko-lagos-danger-ingenuity-floating-slum.

Okorafor, C. (2016). Does Makoko Floating School's collapse threaten the whole slum's future? *The Guardian.* Retrieved March 8, 2020, from https://www.theguardian.com/cities/2016/jun/10/makoko-floating-school-collapse-lagos-nigeria-slum-water.

Olafimihan, K.-O. (2009). *Improvisation in architecture: A critical study of the Makoko fishing community of Lagos, Nigeria.* Professional Master of Architecture Thesis, Carlton University, Ottawa

Population UK. (2020). *UK population 2020.* Retrieved July 20, 2020, from http://www.ukpopulation.org/.

Rael, R., & Fratello, V. (2018). *Printing architecture: Innovative recipes for 3D printing.* Princeton Architectural Press.

Ramirez, B. V. (2018). *This 3D printed house goes up in a day for under $10,000.* Retrieved July 20, 2020, from https://singularityhub.com/2018/03/18/this-3d-printed-house-goes-up-in-a-day-for-under-10000/.

Raya, J. M., Estévez, P. G., Prado-Román, C., & Pruñonosa, J. T. (2017). Living in a smart city affects the value of a dwelling? In M. Peris-Ortiz, D. R. Bennett, & D. Pérez-Bustamante Yábar (Eds.), *Sustainable smart cities: Creating spaces for technological, social and business development* (pp. 193–198). Springer International Publishing.

Reuters in Lagos. (2016). Nigerian floating school collapses due to heavy rains seven months after opening. *The Guardian.* Retrieved March 9, 2020, from https://www.theguardian.com/world/2016/jun/08/nigerian-floating-school-collapses-seven-months-after-opening-due-to-heavy-rains.

Riise, J., & Adeyemi, K. (2015). Case study: Makoko floating school. *Current Opinion in Environmental Sustainability, 13*, 58–60.

Rochet, C. (2018). *Smart cities: Reality or fiction.* Wiley.

RIBA. (2018). *Off-grid Scottish hideaway named UK's best new house.* Retrieved July 19, 2020, from https://www.architecture.com/knowledge-and-resources/knowledge-landing-page/off-grid-scottish-hideaway-named-uks-best-new-house#:~:text=Lochside%20House%20is%20a%20modest,by%20a%20traditional%20drystone%20wall.

RIBA (Royal Institute of British Architecture). (2020). *Scottish Lochside House named House of the Year 2018.* Retrieved July 19, 2020, from https://www.architecture.com/knowledge-and-resources/knowledge-landing-page/house-of-the-year-winner-2018.

Scott-Smith, T. (2019). Beyond the boxes. *American Ethnologist, 46*(4), 509–521.

Smith, C. M., & Davies, E. T. (2012). *Emigrating beyond Earth: Human adaptation and space colonisation.* Springer.

Statistics Norway. (2020). *Population and Housing Census, dwellings (discontinued), 10 November 2011.* Retrieved July 19, 2020, from https://www.ssb.no/en/befolkning/statistikker/fobbolig/hvert-10-aar/2013-02-26.

Studio Roosegaarde. (n.d.). *Smog Free Tower.* Retrieved July 21, 2020, from https://www.studioroosegaarde.net/project/smog-free-tower.

Sunil Kumar, C. P., Parvathi, S., & Rudramoorthy, R. (2016). Impact categories through life-cycle assessment of coal-fired brick. *Procedia Technology, 24*, 531–537.

Tubertini, C. (2018). *Good design that's doing good*. Retrieved July 20, 2020, from https://www.ikea.com/ms/en_JO/this-is-ikea/ikea-highlights/2017/better-shelter/index.html.

United Nations Environment Programme. (2019). *UN75: UN Environment Programme's leader shares three actions to save the world*. Retrieved July 16, 2020, from https://www.unenvironment.org/news-and-stories/story/un75-un-environment-programmes-leader-shares-three-actions-save-world.

UNHCR (United Nations High Commission for Refugees). (2020a). *Figures at a glance*. Retrieved July 20, 2020, from https://www.unhcr.org/figures-at-a-glance.html.

UNHCR. (2020b). *Saving lives at the world's largest refugee camp*. Retrieved July 20, 2020, from https://www.unhcr.org/news/latest/2019/7/5d2eefd74/saving-lives-worlds-largest-refugee-camp.html.

United Nations University. (2017). *Climate migrants might reach one billion by 2050*. Retrieved July 20, 2020, from https://unu.edu/media-relations/media-coverage/climate-migrants-might-reach-one-billion-by-2050.html.

University of Nottingham. (n.d.a). *Students' 'Liter of Light' initiative to help light Nigerian communities with solar lamps*. Retrieved March 9, 2020, from https://www.nottingham.ac.uk/news/pressreleases/2017/march/students'-'liter-of-light'-initiative-to-help-light-nigerian-communities-with-solar-lamps.aspx.

University of Nottingham. (n.d.b). Let there be light. Retrieved March 9, 2020, from https://www.nottingham.ac.uk/impactcampaign/news-and-views/items/news/alt/2018/let-there-be-light.aspx.

USGS (U. S. Geological Survey). (n.d.). *How much water is there on Earth?* Retrieved March 6, 2020, from https://www.usgs.gov/special-topic/water-science-school/science/how-much-water-there-earth?qt-science_center_objects=0#qt-science_center_objects.

Vale, R., & Vale, B. (2009). *Time to eat the dog? The real guide to sustainable living*. Thames and Hudson.

Van der Veen, A. C., Coenders, J. L., Veer, F. A., Nijsse, R., Houtman, R. & Schonwalder, J. (2015, August 17–20). The structural feasibility of 3D-printing houses using printable polymers. In *Proceedings of the International Association for Shell and Spatial Structures (IASS) Symposium "Future Visions"*, Amsterdam, The Netherlands (12pp.).

Venter, A., Lochner, M., & Morgan, H. (2019). Informal settlement upgrading in South Africa: A preliminary regenerative perspective. *Sustainability, 11*(9), 15.

Walsh, N. P. (2018). *ecoLogicStudio's bio-digital curtain fights climate change by filtering air and creating bioplastic*. Retrieved July 21, 2020, from https://www.archdaily.com/905595/ecologicstudios-bio-digital-curtain-fights-climate-change-by-filtering-air-and-creating-bioplastic?ad_source=myarchdaily&ad_medium=bookmark-show&ad_content=current-user.

White, A. (2020). *Golden Globes: Read Ricky Gervais' Scathing Opening Monologue*. Retrieved July 20, 2020, from https://www.hollywoodreporter.com/news/transcript-ricky-gervais-golden-globes-2020-opening-monologue-1266516.

WorldAtlas. (2020) *What and where is the Makoko Slum?* Retrieved March 7, 2020, from https://www.worldatlas.com/articles/what-and-where-is-the-makoko-slum.html.

Chapter 3
What Can We Learn from the Collapse of Societies in the Past?

*Is an empire
the light that goes out
or a firefly?*
Jorge Luis Borges

Introduction

Built environments remain while the societies that created them are no longer recognisable. From the c. 2580–2560 BC Pyramids of Giza in Egypt to the fifteenth-century Inca citadel of Machu Picchu in Peru, these extraordinary structures have outlasted the societies that created them. This, in turn, has led to speculation about why societies that would now seem primitive in terms of their technology, but that were powerful enough to create such megastructures, no longer exist. So important is the built environment that the seven wonders of the ancient world—the Great Pyramid of Giza, the Hanging Gardens of Babylon, the Statue of Zeus at Olympia, the Temple of Artemis at Ephesus, the Mausoleum at Halicarnassus, the Colossus of Rhodes and the Lighthouse of Alexandria—are all built artefacts, not the civilisations that created them. This fascination with artefacts has led many to speculate why powerful societies disappear. Looking at the modern built environment will it become a ruined built object that signals some kind of collapse of modern developed society, like (spoiler alert if you have yet to see the 1968 film *Planet of the Apes*) the abandoned Statue of Liberty?

This chapter sets out to look at the literature on the collapse or loss of societies in order to define the type of collapse that might happen, given the issues raised in Chap. 1 about current ways of living and their impact on the only planet we have. The literature is drawn from the fields of archaeology and anthropology in an attempt to see what can be learned from the past as well as what others have predicted for the future.

Collapse and Survival

The first thing to note is that collapse is not the same as extinction. Dinosaurs are extinct but exactly why has been the cause of much scientific argument (Prothero, 2009:121–143). The reasons for the collapse of certain human societies can be equally contentious, especially as people survive even if their way of life has to change because of a drastic change to their circumstances. However, understanding why things happen is the first step in making plans to avoid the same thing happening again, and hence the fascination with why societies collapse. However, this raises another issue.

The problem with all theories of collapse is that they attempt to find patterns in what appears to be chaos. This is no more than working with the human condition. Parents of new-born babies are faced with the chaos of a very small human being with wants that he or she cannot communicate easily. This has led to a raft of advice in books and on the internet on how to settle your baby into a routine—or pattern of living—that is supposed to make life easier for both baby and parents. People like routines or patterns in their existence. These make life more tenable and theories are no different. They are there in an attempt to explain why things happen in the way they do, but they are possibilities and not certainties. That said, there are plenty of theories as to why societies collapse, some of which are examined below.

Theories Behind the Collapse of Civilisations

The American anthropologist and historian Joseph Tainter (1988) proposed that collapse is a process that happens over time, not just a catastrophic event like the destruction caused by the 1666 Great Fire of London. For those people who lost their homes and businesses in the fire but not their lives, their way of life had collapsed and it had to be rebuilt, just as the destroyed part of the city had to be rebuilt. So the collapse of a civilisation or society is unlikely to happen because of some type of catastrophic event, whether natural like the 1556 Shaanxi Earthquake in China that killed over 800,000 people, making it the world's deadliest earthquake, or man-made such as the far greater number of people killed by the Holocaust, a figure estimated to be 5.7 million (Gilbert, 2002:245). Appalling though this latter figure is, Gilbert (2002:11) stated that "…at least an equal number of non-Jews was also killed, not in the heat of the battle, not by military siege, aerial bombardment or the harsh conditions of modern war, but by deliberate, planned murder". Millet (2017:1) in writing about the Holocaust makes the point that such devastation does not lead to total collapse. The stated purpose of her book on the victims of slavery, colonisation and the Holocaust was to "…compare the concepts and principles they used to survive persecution".

A catastrophic event leading to significant loss of human life was only one of many theories of the collapse of civilisations reviewed and criticised by Tainter (1988:44–90). Theories ranged from resource depletion to social dysfunction. Having discussed the inadequacies of these various theories Tainter developed his own, which perhaps simplistically, can be described as the point at which the marginal returns in terms of solving a societal problem are less than the investment being made by society to solve that problem (Tainter, 1988:118). Marginal return is a concept used by economists to explain why increasing one of the factors of production—classically they are labour, capital and land—will not necessarily result in an equal increase in profits from the outputs of production. Any increase in either capital or labour, such as adding more workers, can lead to a reduction in anticipated returns. Imagine you had a radio factory and had the bright idea of increasing profits by hiring extra workers so as to make more radios. You will not simply double your profits if you double the number of workers in your radio factory. Putting more workers into the same factory could well mean they get in each other's way and fail to make the radios as quickly as when the factory had fewer workers. Making more radios could also flood the market so that prices drop and the returns also drop, which is a problem since more workers cost more money. This suggests there might be an optimum situation in which the investment in capital (factory, machinery) and labour (workers) in the case of the radio factory yields the maximum profit from selling radios. This leads to the classic graph (Fig. 3.1).

Figure 3.1 shows that if only one variable changes, to begin with, production increases. This would be the situation where the radio factory had few workers to begin with but as more were hired, production increased. However, more workers could disrupt the production process, eventually leading to the point where there are negative returns. This would be a classic example of diminishing marginal returns. Another familiar example is adding fertiliser to land. Initially, this may raise output which will cover the cost of the extra fertiliser, but adding more and more fertiliser only leads to a marginal increase in production while the price of the extra fertiliser

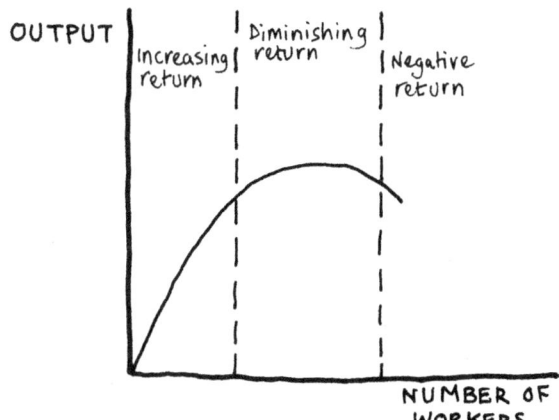

Fig. 3.1 Law of diminishing returns (adapted from https://bohemianeconomics.wordpress.com/2018/11/01/a-criticism-of-diminishing-returns/)

stays the same. The discussion of Energy Return on Investment in Chap. 1 provides another example of a diminishing marginal return.

Tainter's view of collapse is thus centred on the moment the human ability to solve problems stumbles because the solution to the problem only offers at best a diminishing return and at worst a negative return.

> Our unique combination of historical ignorance and historical arrogance joins with the pervasive influence of neoclassical economic theory to perpetuate the notion that all problems are solvable if only there is resolution and incentive to do so. Yet the experiences of the Romans or any of the other societies that have collapsed suggests that there are conditions under which problem-solving breaks down (Allen et al., 2003:60).

As an example from Chap. 2, the effort and impact on the environment involved in making a machine to capture and store carbon, including extracting and processing all the materials required for its construction and maintenance, and the energy to run it need to be assessed against the effort and impact on the environment involved in not emitting that carbon in the first place.

The concept of diminishing returns also has a parallel in architecture that may help to illustrate it further. In late nineteenth-century America, there was the desire to build high, especially in Chicago. Describing Chicago in the 1880s, Giedion (1967:370) states:

> At the time its great buildings went up, Chicago was the real point of concentration for the products of the West and Middle West, and not merely an enormous stock exchange, like New York. The rapid growth of this centre led to a sudden enlargement of its needs … Large office buildings … were the first to appear … immense hotels for the travellers who passed through the city … the problem of the modern apartment house … was attacked with great consistency and daring.

These demands coupled with the desire to be in the city centre and at the heart of business meant taller buildings. However, the then current technology of load-bearing walls of masonry meant there was a limit to how high it was possible to build and still have useable space at ground level because of the thickness of the supporting walls. A building of brick or stone eventually reached a point where the floor area of every storey added to the top level was offset by the floor area lost due to the extra thickness of the walls needed at ground level to support the extra floor. The built limit was the simple 16-storey 1891 Monadnock Building, which also had small windows because of the need to not make large holes in the structural masonry. The Monadnock Building is a perfect example of a diminishing marginal return—anything above 16 storeys meant the rentable floor area added at the top was simultaneously lost at the bottom.

At this point in human history there was a way round this diminishing marginal return by inventing a whole new way of building—the steel frame structure, which first appeared in William Le Baron Jenny's 1885 10-storey Home Insurance Building with a height of just over 42 m, somewhat less than the 67 m of the very much earlier 1056 wooden Pagoda of Fugong Temple in China (Fig. 3.2). This demonstrates that problem-solving ideas also need time to develop and mature. The 110-storey Willis

Fig. 3.2 Pagoda of Fugong Temple and Home Insurance Building (adapted from https://en.wikipedia.org/wiki/Pagoda_of_Fogong_Temple and http://architectuul.com/architecture/home-insurance-building)

Tower, completed decades later in 1973 and also in Chicago, is claimed to be the world's tallest steel-framed building (ITP Media, n.d.).

Hence, the real question is why can a society sometimes cope with a problem or threat better than at other times? Is there a limit to problem-solving abilities and can this limit be recognised? Through the ability of a society to solve problems, in the example above it was the problem in Chicago of how to put more people and their activities into a limited land area, societies have changed from roaming foragers to urban settlements, and in doing this, they developed agriculture (Jacobs, 1969:13–54). In the simplest of these early agricultural societies, people worked as farmers and grew food. If they produced more food than they needed to feed themselves and their families, they could sell this surplus. The ability to buy food allowed other people to give up farming and become producers of tools or pots that they could sell to the farmers in exchange for some of that surplus food. At this stage all the people were producing goods to sell, whether these were food, pots, clothes or jewellery. As the system grew in scale other types of people were needed. Originally farmers would take their produce to a market to sell but this meant there was a limit on how far they could walk to do this, and hence on the size of any settlement growing up around this market. To overcome this problem and allow settlements to grow, the idea evolved of selling produce to a middle man who then sold the products. This meant farming had to become more productive to feed both the producers of goods and the middlemen. As long as the effort to do this did not exceed the marginal returns gained in terms of surplus products to feed the non-farmers, the system was stable, although a bit more complex. Further complexity might then be introduced as there was a need to have people oversee the system so that no one was cheated and there was fair play. In turn, agriculture had to become more productive to feed this extra tier of people, and the society became more complex. An example of how agricultural productivity was enhanced in the past came with the invention of the horse collar that permeated Europe in the twelfth century (Encyclopaedia Britannica, 2020). This allowed for much more effective pulling power as the weight rested on the shoulders of the horse, so the windpipe was not restricted, as had happened with earlier harnesses. Provided the inventions happened to make agriculture more productive, then society could become increasingly complex in terms of the way it was organised.

The big jump in productivity, including that of agriculture, came with mechanisation and the increasing use of fossil fuels. As with the example of tall buildings in Chicago, this mechanisation not only allowed for the production of such buildings through the development of steel but was also the thing that necessitated them to deal with the influx of goods from West and Midwest America. What should not be forgotten is that underlying all these changes in society is the need to deal with an increasing human population. If adjoining societies are annexed to provide the resources needed to sustain a level of complexity, such as going to war to obtain a fuel source, then these additional people also need to be subsumed into the original society and their needs have to be supplied. Thus at first having the extra fuel source may seem a benefit but the marginal returns are soon cancelled by the need to deal with the extra population and the society is back where it started (Tainter, 1988:120).

The decision to become more complex is not necessarily taken by choice but by necessity because problems keep on emerging and their solutions imply making an investment in more complexity. If problems keep on arising and constant benefits are sought in solving them, then it will be difficult for a society to reduce its consumption of resources over the long term (Tainter, 2011). It can be argued that much of the complexity of modern society has arisen not from the solving of real problems but from the creation of perceived problems in order to promote the selling of products. In the home kitchen, an electric whisk does not do a better job of beating eggs, but it costs more than a hand whisk and so it makes more profit although it uses more resources and energy. Societies cannot grow in complexity indefinitely, and examples of this can be found in the collapse of ancient societies like the Mayan or Roman civilisations (Tainter, 1988:127–178). Both these empires collapsed after reaching levels of complexity where the return on further investment in social and economic complexity started to diminish. Tainter maintains that a collapse emerges when the difference between the cost of becoming more complex and the cost of maintaining the complexity already acquired arrives at a breaking point where negative marginal returns can no longer be absorbed. The result is the loss of complexity of the system. This implies a return to a less complex state, which might involve less centralised control, and having diminishing specialisation and smaller settlements. Tainter (1988:198) concludes:

> Collapse then is not a fall to some primordial chaos, but a return to the normal human condition of lower complexity ... To a population that is receiving little return on the cost of supporting complexity, the loss of that complexity brings economic, and perhaps administrative, gains.

This view that a sudden loss in complexity can bring gains also appears in a study of what happens after natural catastrophes, such as earthquakes. Soden and Lord (2018) state such an event "…enables us to consider crisis as a relational or ontological, rather than epistemic, phenomenon and offer approaches to recovery that don't take the wholeness or permanence of pre-crisis socio-material relations for granted". In other words, the point of collapse is a chance to rethink the way a society was organised prior to the devastating event. The city of Christchurch in New Zealand offers a good example of this. Following the 2010 and 2011 earthquakes, it was suggested that rather than rebuilding there was the opportunity "…to design urban landscapes which provide environmental, economic, social and cultural sustainability" (Tavares et al., 2013). In the event this failed to happen in the way envisaged post-disaster, not least because those at the bottom fighting for sustainability initiatives tended to be quashed by the top-down directives of the Canterbury Earthquake Recovery Minister, leaving Brundiers (2018) to note that "Overall, the sustainability change initiatives did little to bring about the necessary far-reaching sustainability change." The result is that Christchurch is being re-built as a twentieth-century city and the opportunity for change has been lost.

Complexity and Societal Hierarchy

Before leaving Tainter to look at more recent theories of collapse, it could be useful to discuss the link between complexity in a society and the hierarchy of its citizens. Tainter (1988:91) explains:

> As societies increase in complexity, more networks are created among individuals, more hierarchical controls are created to regulate these networks, more information is processed, there is more centralisation of information flow, there is increasing need to support specialists not directly involved in resource production, and the like.

This hierarchy could become reflected in the built environment to the extent that the size of a person's dwelling and the materials it was made of reflected their place in society, very similar to the "sustainable" houses discussed in Chap. 2. A typical example is the country gentleman's house, such as described by Jane Austen in her novel *Emma,* which would be the largest in an English village, followed in size by that of the vicar, and then came the houses of the professionals like the doctor and lawyer, followed by those of the tradesmen and the smaller cottages and hovels of the labourers. Going back further in history, the *insulae* of Rome that date back to at least the first century AD (Harvey, 2016:157) were six to seven-storey blocks of flats where the smallest and poorest apartments were in the upper storeys, which were built of wood and flammable, while the wealthier Romans lived in the lower and street-level storeys (Aldrete, 2004:79). This gave the wealthy easy egress in case of fire (Strabo, c18AD) and also meant they had fewer stairs to climb. This pattern has been reversed in modern apartment blocks served by lifts where the largest and most expensive apartments are generally found in the penthouses of the top storeys, where the inhabitants can enjoy the view away from the noise and bustle of the life of the city, but the introduction of the lift makes the services of the apartment building more complex and if the lift fails there is a long climb ahead.

The effect of specialisation can be seen in architecture and this has also affected modern architectural education. Whereas before the last quarter of the twentieth century, the architect was not only the designer of the scheme but also supervised the construction, this task has now been allotted to the project manager. Such division of labour was necessary for large construction projects, but it has now filtered down to the small ones as well. Whereas previously the architect would have designed the services (heating, plumbing and drainage) for at least a small building, this may no longer be the case. In some instances, the architect is now only the overall designer and may not even prepare the working drawings. This fragmentation of a task that was previously done by one person means that architectural education is no longer focused on producing the architect but on preparing students for a variety of possible roles in the construction industry. This has led to a proliferation of courses, making the whole process more complex and more expensive, in turn leading to stress for both academics and students (Bachman & Bachman, 2006).

The big issue with Tainter's theory that increasing complexity in a civilisation or society can lead to its collapse is why has the modern developed world, with its complexity, not collapsed (at least not yet)? Tainter's answer would be that we are still problem-solving without having reached the stage of diminishing marginal returns on the investments made. However, as outlined in Chap. 1, this same modern developed world is facing some big problems to which we know the answers. To avoid problematic global warming, societies have to decarbonise. Although small efforts are being made in this direction, they are nowhere near sufficient to solve the problem. This at first sight seems a position of knowing what to do and not doing it, rather than being faced with the need to solve a problem. However, this may not be the case. Manzzanti and Rizzo (2017) in outlining the innovations needed to decarbonise include "behavioural innovations", thereby suggesting that there is a problem that has not yet been solved in modern societies, the simple one of not knowing how to get people to change their behaviour. Such change would not be just at the level of individual behaviour but would have to embrace the behaviour of human systems, for example implying changes, and probably significant changes, to the economic basis of human societies and other societal issues such as population growth. The important word here is change, not collapse. Society may not look the same but it is still there. It is perhaps significant that Middleton (2017:12) noted the following: "Roman, Maya, and Ancient Puebloan civilisations did not collapse at all—all three survive, transformed, to this day. It is specific political regimes that collapsed, social systems that changed, and religious and ideological systems that were transformed and/or rejected."

More Recent Views of Collapse

Jared Diamond, physiologist and Professor of Geography (2005:3) defines collapse as: "a drastic decrease in human population size and/or political/economic/social complexity, over a considerable area, for an extended time". The idea of including the social, political and economic complexity in the definition probably comes from Diamond's reading of Tainter. Diamond's main point is that past societies have collapsed due to environmental damage, whether this was produced by human actions or climate change, despite Tainter (1988:89) dismissing this as a sole reason for the collapse. However, Diamond also acknowledges that environmental problems, such as a drastic, anthropogenic change to the climate, cannot be the single cause of the collapse, as societies could still trade for resources. So if crops fail in one country because of drought caused by climate change, exchanging other goods for food with countries whose climates were favoured by climate change would still be possible. This would, of course, depend on how friendly the neighbours were and on the scale of the environmental problem and such trading would not be possible in the face of unfavourable changes to the climate for food growing on a global scale.

The English academic Guy Middleton included Diamond's proposal in a group of views that he calls "ecocide", namely, societies that almost intentionally run into collapse by depleting their key resources. He goes on to state (Middleton, 2017:285):

> The view that collapse is a phenomenon simply determined by unexpected environmental or climatic shifts, resource degradation, and maladaptation fails to recognize that reactions to problems and challenges of all kinds are not simple cause and effect; rather they are bound up with and refracted by social realities, priorities, and motivations that may be far from unified or singular.

For Middleton, collapse is essentially a "human story", since there is no such thing as collapse in nature. It is simpler to regard collapse as an adaptation phase that creates opportunities, as suggested above for Christchurch, and as happened for some in the example of the 1665 London plague and the great fire the year afterwards. The plague saw people leave the city to flee from the disease, which meant that there were no merchants and no work for the poor, who either starved or succumbed to the disease, or both. Nevertheless, the empty city offered entrepreneurial opportunities for those brave enough to stay (Moote & Moote, 2008:161–176). However, although the crisis produced disaster for some (100,000 Londoners died of the plague) it brought others opportunities. The built environment, however, remained unaffected in London but was in turn responsible for the plague spreading to other areas of the country as people fled the city and took the disease along the highways with them (Moote & Moote, 2008:200). This was to change with the great fire in September of the following year, where much of the city of London was destroyed in an area stretching along the Thames from the Tower of London almost to the Temple stairs (Wiles, 2016). Such destruction and disaster still produced opportunities, especially for people with transport to hire.

> The streets were crowded with people and carts, to carry what goods they could get out; they who were most active and had most money to pay carriage at exorbitant prices, saved much,

the rest lost almost all. Carts, drays, coaches, and horses, as many as could have entrance into the city, were laden, and any money is given for help ... (Gideon, 1769:9).

The destruction also produced opportunities for those looking on as immediately after the fire plans for rebuilding were submitted, including one by Christopher Wren (Wiles, 2016). The city was rebuilt in brick in place of the timber frame which, with the wind, had allowed the fire to spread so quickly. Some 9000 brick houses of four types were constructed instead of the original 13,000 of many shapes and sizes, and with this fall in population, the nature of the city of London changed. No longer did the rich live close by the poor but the city became a commercial centre with the rich rehousing themselves to the west, up-wind of the smoke and smells and the poor to the east, down-wind, setting up a London pattern that was to last for centuries (Davis, 1923).

The point of these two examples is to show two things. The first that a crisis (for many) in both cases produced opportunities (for some). The second is to show that the form of the built environment can influence what happens when disaster strikes. In the case of the plague, it was the transportation network that aided the spread of the disease, and in the fire it was the wooden houses that burned so well, and that produced a complete change in building materials. However, neither disaster produced collapse at the scale of London, which as a city persisted. For both the plague and the fire, the collapse was at the scale of the personal and for these people it meant a change to the path of life. However, the fire and the consequent devastation of the city did effect a change in its nature with a loss of diversity in terms of who lived there and their occupations and a focus on it as a commercial centre after the crisis.

Archaeologists have understood collapse in ways other than the economic understanding of the collapse of Tainter and the ecological viewpoint of Diamond. Schwartz (2006:5–6) notes that in archaeology definitions of collapse include some or all of "…the fragmentation of states into smaller political entities; the partial abandonment or complete desertion of urban centres, along with the loss or depletion of their centralising functions; the breakdown of regional economic systems; and the failure of civilisational ideologies". This definition describes the effects of collapse, namely what happens after societies collapse, but it does not help in understanding the logic of collapse or why it happens. Butzer and Endfield (2012) state that "Societal collapse represents transformation at a large social or spatial scale", which has an impact in the long term and is a combination of environmental, demographic, cultural, socioeconomic and political variables. This definition is very inclusive and highlights the importance of understanding collapse as a large-scale phenomenon but it is perhaps not specific enough for designers wanting to know what to build in the face of current human problems (or even whether to build at all).

Middleton (2017:18) suggests that the whole area of collapse is difficult and should be approached with caution and with a critical mind. However, he also goes on to say:

> Many archaeologists agree that collapse often affected the elite members of societies most—or at least most visibly—and that those least affected (at least in some ways) would be the

peasant farmers that made up the bulk of the population in pre-industrial societies. Thus collapse could be 'socially intolerable' to the elite but welcomed by 'middle classes' or others.

This suggests that collapse for one group is not collapse for all, and Middleton (2017:30) goes on to suggest that "…the simpler societies that arise out of collapse are better adapted, and they may themselves start along the path of increasing complexity, suggesting a cyclical process of collapse and regeneration (on some level)". Following this line, the American anthropologist Norman Yoffee (2006:222) suggests, "collapse seldom denotes the death of a civilisation as opposed to the end of a particular form of government". Even after apparently apocalyptic events, cultures continue. The trams were running in Hiroshima only three days after the atomic bombing, and two of the trams damaged in the blast are still running in the city. Interestingly, in view of Tainter's ideas on complexity, the man in charge of maintenance says of one of the old trams, "Since it uses analog technology, it can be kept operational almost eternally if proper maintenance is carried out" (Hashimoto, 2018). This again comes back to the question of the scale of a collapse and the fact that collapse for some may lead to opportunities for others. From this point of view, collapse might mean different things in different circumstances. Sometimes collapse could be a change or transformation that involves an abrupt change of scale, from living in big cities to small villages; a change in the organisation of the built environment, from centralised urban areas to dispersed rural settlements; a change in the way of governing, from hierarchical organisations to more equalitarian ones; a change in the "sophistication" of the built environment, from the development of big institutional buildings and monuments to the predominance of smaller, domestic scale constructions. At least this was the lesson learnt from the archaeological studies of Bronze Age urban centres in the Euphrates Valley of Syria. After the fall of the Akkadian Empire and the Third Dynasty of Ur, many urban centres were abandoned in the Euphrates Valley. However, Cooper (2006:18–37) hypothesised that the revival of urban centres in the Euphrates Valley during the Middle Bronze Age would have been difficult without the permanence of smaller, loosely organised communities that preserved their structures and provided cultural continuity during the regeneration years.

Yoffee (2006:222) also posited that while collapse usually refers to a change from complex societies to less complex, smaller and more fragmented ones, such as moving away from urban centres to more decentralised settlements, is there any reason that it could not work the other way? He states, "Thus, if stability connotes village life, then the appearance of urban sites in the region—which were based, in part, on connections with outsiders and were unstable—could be called a collapse!" This idea of regrouping and rearrangement is discussed next in relation to resilience theory.

Resilience and Collapse

While theories of collapse have grown from the study of what happened in past human societies, the theory of resilience grew out of the study of what happens in natural ecosystems. The innovation in thinking that emerged from these studies is that change and even collapse in terms of moving to a different state are both sometimes necessary for the persistence of an ecosystem. A forest fire can destroy a mature ecosystem but there will eventually be regeneration and the establishment of a new system. In other circumstances external pressures, such as drought, might force changes in an ecosystem. This ability of an ecosystem to adapt to change has been termed ecological resilience. This idea was first developed by Holling (1973) who stated resilience is "the persistence of relationships within a system and is a measure of the ability of these systems to absorb changes of state variables, driving variables, and parameters, and still persist". Within a lake, which is a confined system, fish populations will fluctuate, as in the natural world there is no stable state. Ecological resilience is thus different from engineering resilience, which is normally defined as the ability of a material under some pressure to change to return to its original state. A rubber band that quickly returns to its original shape after being stretched thus has a high resilience in engineering terms. Pliable materials like bamboo or certain plastics can be bent considerably or deformed without losing their shape, making bamboo a material that has found many uses from furniture to bridges and even scaffolding for high rise buildings. However, even engineering resilience has limits. The limit to the resilience of a material is first met when the material is permanently deformed, and ultimately after it reaches a breaking point when it collapses. In this way, the mechanical properties of materials have clear and measurable phases. Engineering resilience thus implies a stable state which gets disturbed. Some writers have suggested that the natural world is more like the world of engineering. Pimm (1991:18) proposed that resilience was a measure of the time a population density in an ecosystem takes to come back to its previous state. The quicker a population came back to its starting point the more resilient it would be. This implies the ecosystem has a stable state like in engineering, whereas Holling's ecological resilience theory implies a system that is always in a state of flux as it responds to external pressures.

Ecologists have also questioned why our physical environment has not yet collapsed in spite of all humanity's attempts to destroy it. This will be discussed in more detail at the end of Chap. 4. This echoes the descriptions of the collapse of civilisations and societies by those interested in collapse theories. Governments may fall and societies may fragment but life goes on. What the ecological resilience approach offers to collapse theory is the idea that the process might be managed to achieve a stable outcome—not a stable state—such as the ability annually to take similar levels of wood from a coppice. In a coppice some trees will be cut whilst others are growing, so the coppice changes in appearance but the quantity of wood can be managed as a near-constant output that would only vary with whether the season had been good for growth or not. Holling (1973) introduced the concept of resilience in ecology as an alternative way of managing resources. This was not based on the idea of trying to avoid change so as to keep ecosystems stable, but rather to deliver the same outcomes consistently. Holling further proposed that the management of ecosystems should be focused on enhancing their resilience capacity instead of trying to resist change to keep stability. If after a disturbance, like a flood, fire or drought, the ecosystem continues to perform the same functions, the key processes keep on going and some at least of its communities of flora and fauna survive in some form, then the system has shown some resilience to the disturbance.

Since the initial work of Holling, the definition of resilience has included the idea that the relationships, driving variables and parameters of a system are linked to its identity (Walker et al., 2004). Therefore, if an ecosystem keeps its identity after a disturbance then it has shown some resilience. From this point of view, an ecosystem collapses when it is no longer recognisable and has changed its identity. However, it is hard to find total changes of identity when events that cause change happen in the built environment, just because buildings are normally built to last.

The city of Manchester in the UK was once known as Cottonopolis (Schofield, 2018). When it became cheaper to make cotton overseas nearer to the site of the production of the raw material the city went into economic decline. The cotton mills that were a feature of the city became redundant and some of the mill buildings were demolished. In the revival of the city's economy as a centre for service industries, some mills were converted. However, what makes Manchester recognisably Manchester is the persistence of the street layout, now adapted once again to trams, and its red-brick housing, along with other features such as the Manchester Ship Canal, which now has a leisure rather than commercial function. The built environment has changed in response to the changing economy on which the city is based but sufficient heritage is left that the image of the city of Manchester persists in the mind. This is different from the Luftwaffe bombing of Coventry that destroyed its former medieval centre. Although fragments remain a short walk from the centre, the city has been rebuilt in a way that looks very different from its medieval heritage before the "crisis" event. Nevertheless, there is still a city called Coventry on the site of medieval Coventry, so Coventry has persisted as a city.

The point raised by Holling and colleagues is that without resilience there is little chance of survival or, in other words, a greater chance of collapse. In the same way that Tainter links a failure in the problem-solving capacity of institutions to their collapse, Holling links the resilience capacity of a system to its chances of surviving and therefore avoiding collapse. The intention was this would lead ecosystem managers to focus more on "creating the conditions for survival" (Holling, 1973) in an ecosystem, instead of trying to avoid disturbances to its stability. Of course, this idea of "ecosystem managers" is to a large extent a concern of modern times, native Americans, for example, tended to live with their environment as it was before the Europeans arrived and began to manage it (Oswald et al., 2020). Ecosystem management appeared in 1970 and aims "to restore and sustain the health, productivity, and biodiversity of ecosystems and the overall quality of life through a natural resource management approach that is fully integrated with social and economic needs" (Szaro et al., 1998).

Holling's theory implies not only thinking of the ecosystem being managed (in accordance with social and economic needs, of course) but also of the environment in which it operates since it is the latter that creates the conditions for failure or success. The change in viewpoint proposed by ecological resilience implies that managers should accept disturbances as a part of the dynamics of change in a system if they want to provide continuity in the system's stability. In this way, natural systems appear to cheat collapse by using "threats", like scarcity, volatility of resources, competition, and invasions, as sources for their own development and persistence.

Tainter (1988:38) also discusses stability and its relationship to collapse, calling collapse "…a process of major, rapid change from one structurally stable level to another." This accords with Holling's definition above of conditions being created for survival. Societal collapse implies change, and change such that the people who made up that society will survive in one form or another. This is both good and bad news. The good news is that whatever happens as a result of the significant problems facing societies in the modern developed world, some people will survive and continue. However, the bad news is that understanding *why* things collapse does not really explain *how* the collapse of these same modern developed world societies could be avoided. The big clue comes from complexity, and the warning that eventually investing in further complexity to solve problems will not provide a sufficient return on investment and this leads to a situation of potential collapse.

In discussing the relationship between population, energy, innovation and complexity, Taylor and Tainter (2016) state:

> High populations generate scalar stress—social, economic and resource challenges that are met by increasing complexity in problem solving. Population, consumption, and complexity require still further energy, generating the energy-complexity spiral. To date we have relied on innovation to offset the resource depletion brought on by population and consumption. As the best resources are consumed and the easiest science developed, investment in both areas produces complexity, costliness, and diminishing returns.

This is a warning of what is to come in a world with an increasing population and an increasing thirst for energy. It can also be viewed as echoing the warning given by Taylor's re-examination of the Limits to Growth model, and finding modern human society is on track for collapse around the middle of the twenty-first century (see Chap. 1). It is true that for those in the developed world human society has continued to develop since the Industrial Revolution, but there have been some near collapses on the way—the Great Depression from 1929 to 1939, which affected most countries in the world; the 2008 financial crisis (see Chap. 10), which in turn led to the collapse of the Greek economy—so there have been warnings that life could be turned upside down in a short time.

In the past when societies collapsed there was always somewhere else to go. In the modern developed world finding somewhere to start afresh in a less complex society is going to be difficult for many people, not least because of the increasing number of the human population living in cities. This brings the argument back to the kinds of built environments that are needed to make sure the "energy-complexity" spiral does not lead to collapse.

Cities are part of ecosystems and depend on the availability of natural resources, particularly in their hinterlands, to keep on going. For this reason, part of the persistence of cities relies on the continuous availability of natural resources, like water and food—things you expect to find in the supermarket when you go shopping or something that you expect to get reliably out of the kitchen tap. Any discontinuity in this flow has historically had dramatic implications for the persistence of a population in a city. Predictability over resource availability is essential for cities because it allows them to persist and keep on working. Cities are dependent on a continuous flow of natural resources to serve their often increasing numbers of citizens. However, thinking of cities as part of ecosystems suggests that they might also have resilience in terms of the ability to adapt to internal and external pressures and thereby persist. Buildings and infrastructure tend to be long-lived items that do not change much and that do not all come to the end of their lives simultaneously. A site will be cleared and a new building put up or a new sewer line will be installed in an existing suburb, but the surrounding buildings and streets remain the same. The built environment is constantly changing in these small ways.

Perhaps resilience in the built environment can be more usefully thought of as a property that helps with the development of an identity. The Western Roman Empire may have collapsed in 476 AD but the city of Rome, which was the heart of that empire, still persists. The buildings that were once in use every day by those living in Rome before the collapse have now found a new use as tourist attractions, such that "…Rome's modern identity is intimately bound up with mass tourism and the millions of tourists that visit it each year" (Hom, 2010).

Collapse Theories and the Built Environment

Since the focus of this book is what we should be doing to the built environment now in the face of current problems to avoid dramatic changes to—and hence the collapse of—the modern way of life in the developed world, it might be useful to look at what has happened in the past when a civilisation declines and disappears and the effect of this on the built environment. Tainter (1988:20) notes that those in urban areas reuse the built environment in what he describes as "…a characteristic manner. There is little new construction and that which is attempted concentrates on adapting existing buildings. Great rooms will be subdivided, flimsy facades built, and public space will be converted to private". Tainter also postulates that monuments will be seen as a useful source of building materials. Any monument no longer in use might suffer this fate. The Romans quarried stone from the Colosseum for other purposes once the games had ended in the sixth century AD (Mueller, 2011), and its stones were also used in the cathedrals of St. Peter (1626) and St. John Lateran (dedicated AD 324 but rebuilt several times) as well as the Palazzo Venezia (1469) (A&E Television Networks, 2019). In a more recent and smaller-scale way, the remains of buildings in Liverpool bombed by the Luftwaffe were used to form coastal protection along the estuary of the River Mersey.

These observations suggest that new additions to the built environment should be flexible in that they can be adapted to many uses, and also use construction methods that allow for easy deconstruction and materials that can be reused or recycled. The fact that concrete is globally the most prevalent material in modern construction is not a good starting point (Crow, 2008). Concrete accounts for more than twice the amount of all other building materials (CCANZ, 2011). Concrete is also a contributor to climate change through the carbon dioxide emitted during the manufacture of cement although it has useful properties such as its durability and ability to withstand severe winds and floods (Georgopolous & Minson, 2014:11–17). Mention has already been made of stones from the buildings of the Roman Empire finding their way into other buildings but the Pantheon in Rome with its concrete dome "…stands virtually intact" (Cartwright, 2018). Concrete can be crushed and used as aggregate in new concrete, and in some European countries taxes on waste and imported aggregates "…have made it viable to recycle concrete generally into a 'low-grade' road-base material" (CCANZ, 2011:6). This is fine as long as there is a need for roads and as long as there is fuel for the vehicles using them, but the needs following a significant change or collapse might be different.

To explore this further it might be useful to return to some of the proposed built environment solutions to current problems outlined in Chap. 2 and see to what extent they might fit into a post-collapse situation of reuse and subdivision.

Proposed Solutions Post-collapse

The problem with most of the design ideas put forward in Chap. 2 is that they are big and will tend to add to complexity. Smart cities require all the same servicing as current cities in terms of bringing in food and other supplies and dealing with a water supply and sewage. There is also the assumption that somewhere food can be grown and waste dealt with. Floating cities face similar problems, to which is added the complexity of trying to create a buoyant built environment, given that most building materials are heavier than water unless people live in ships rather than houses. The buoyancy of a ship is a problem solved. Serious inundation following climate change (see Chap. 5) may see people using the principles of boatbuilding to develop their own ways of living on the water but this will be in small settlements where resources are to hand, not in floating mega-cities.

There is also the issue of the resources involved in some of the future thinking proposals. Living in space, if feasible, will be for the privileged few rather than the many. Geoengineering projects also require a large investment in resources, often for little gain. As discussed in Chap. 2, it would be simple (less complex) and easier to avoid putting the CO_2 into the atmosphere rather than designing large pieces of equipment to extract it afterwards. It is a bit like discovering the bath is overflowing when the first move should be to turn off the taps and not just pull out the plug.

The only idea from Chap. 2 that might be useful in a post-collapse situation is teaching people to do things for themselves, whether this is installing solar lighting in Lagos, or building shelters in refugee camps using local materials, such as adobe, rather than being given a flat-pack tent to assemble. Introducing skills into local society is a way of reducing the complexity, since there is less reliance on bringing in skills, often in the form of pre-made resources such as refugee tents, from outside. This changes the local hierarchy as the local community is more dependent on itself and less on a series of "middle men" including designers, manufacturers, packing suppliers, transportation companies, and the people who write the instructions for assembling the flat-pack. All of these middlemen will also want to be paid for their services.

What Can Be Learned from Collapse Theories

Before moving on to look at collapse and the built environment in more detail in the next chapter, it may be useful to summarise what can be learned from collapse theories. A similarity between the theories of resilience and collapse is that both say that complex systems cannot grow in complexity endlessly. As proposed in Chap. 1 and exemplified in Chap. 2, adding complexity to the built environment in an effort to lessen the environmental and social crises can create more and bigger problems. An increase in complexity as a means of solving a problem needs to be viewed in terms of whether it provides a marginal return for the effort involved in making the change. When marginal returns are negative then collapse is a real possibility. A further issue is that resilience theory shows that change is inevitable and what we should be looking for is a near-constant output from a system that will fluctuate according to changes in its context. This brings into question the whole issue of growth and what should grow, which is explored in more detail in later chapters.

References

A&E Television Networks. (2019). *Colosseum*. Retrieved February 9, 2020, from https://www.history.com/topics/ancient-history/colosseum.
Aldrete, G. S. (2004). *Daily life in the Roman city*. Greenwood Press.
Allen, T. F. H., Hoekstra, T. W., & Tainter, J. A. (2003). *Supply-side sustainability*. Columbia University Press.
Bachman, L., & Bachman, C. (2006). Student perceptions of academic workload in architectural education. *Journal of Architectural and Planning Research, 23*(4), 271–304.
Brundiers, K. (2018). Disasters as opportunities for sustainability: The case of Christchurch, Aotearoa New Zealand. *Sustainability Science, 13*, 1075–1091.
Butzer, K. W., & Endfield, J. H. (2012). Critical perspectives on historical collapse. *Proceedings of the National Academy of Sciences of the United States of America, 109*(10), 3628–3631.
Cartwright, M. (2018). *Pantheon*. Retrieved February 15, 2020, from https://www.ancient.eu/Pantheon/.

CCANZ (Cement and Concrete Association of New Zealand). (2011). *TR 14 best practice guide for the use of recycled aggregates and materials in new concrete.* Cement and Concrete Association of New Zealand.

Cooper, L. (2006). The demise and regeneration of bronze age urban centres in the Euphrates valley of Syria. In G. M. Schwartz & J. J. Nichols (Eds.), *After collapse: The regeneration of complex societies* (pp. 18–37). The University of Arizona Press.

Crow, J. M. (2008, March). The concrete conundrum. *Chemistry World* (pp. 62–66). Retrieved February 15, 2020, from http://www.rsc.org/images/Construction_tcm18-114530.pdf.

Davis, E. J. (1923). Historical revisions: XXV—The great fire of London. *History, 8*(29), 40–44.

Encyclopaedia Britannica. (2020). *Horse Collar*. Retrieved February 9, 2020, from https://www.britannica.com/technology/horse-collar.

Gideon, H. (1769). *An historical narrative of the great and terrible fire of London, Sept 2nd 1666: with some parallel cases, and occasional notes.* Printed for W. Nicoll, in St Paul's Church-Yard.

Giedion, S. (1967). *Space, time and architecture.* Harvard University Press.

Gilbert, M. (2002). *The Routledge atlas of the holocaust* (3rd ed.). Routledge.

Georgopolous, C., & Minson, A. (2014). *Sustainable concrete solutions.* Wiley.

Harvey, B. K. (2016). *Daily life in ancient rome: A sourcebook.* Hackett Publishing Company Inc.

Hashimoto, H. (2018, January 1). 2 streetcars that survived A-bomb are still running in Hiroshima. *The Asahi Shimbum.* Retrieved August 10, 2018, from http://www.asahi.com/ajw/articles/AJ201801010023.html.

Holling, C. S. (1973). Resilience and stability of ecological systems. *Annual Review of Ecology and Semantics, 4*, 1–23.

Hom, S. M. (2010). Consuming the view: Tourism, Rome, and the Topos of the eternal city. *Annali d'Italianistica, 28*, 91–117.

ITP Media. (n.d.). *Top 10 world's tallest steel buildings.* Retrieved February 9, 2020, from https://www.constructionweekonline.com/article-9180-top-10-worlds-tallest-steel-buildings.

Jacobs, J. (1969). *The economy of cities* (1972 ed.). Penguin Books.

Manzzanti, M., & Rizzo, U. (2017). Diversely moving towards a green economy: Techno-organisational decarbonisation trajectories and environmental policy in EU sectors. *Technological Forecasting & Social Change, 115*, 111–116.

Middleton, G. D. (2017). *Understanding collapse: Ancient history and modern myths.* Cambridge University Press.

Millet, K. (2017). *The victims of slavery, colonisation and the holocaust: A comparative history of persecution.* Bloomsbury Publishing plc.

Moote, A. L., & Moote, D. C. (2008). *The great plague: The story of London's most deadly year.* The John Hopkins University Press.

Mueller, T. (2011, January). Secrets of the Colosseum, *Smithsonian Magazine.* Retrieved February 9, 2020, from https://www.smithsonianmag.com/history/secrets-of-the-colosseum-75827047/.

Oswald, W. W., Foster, D. R., Shuman, B. N., Chilton, E. S., Doucette, D. L., & Duranleau, D. L. (2020). Conservation implications of limited native American impacts in pre-contact New England. *Nature Sustainability, 3*, 241–246.

Pimm, S. L. (1991). *The balance of nature? Ecological issues in the conservation of species and communities.* University of Chicago Press.

Prothero, D. R. (2009). *Greenhouse of the dinosaurs: Evolution, extinction and the future of our planet.* Columbia University Press.

Schwartz, G. M. (2006). From collapse to regeneration. In G. M. Schwartz & J. J. Nichols (Eds.), *After collapse: The regeneration of complex societies* (pp. 3–17). The University of Arizona Press.

Schofield, J. (2018). *A short history of Manchester: The rise and fall of Cottonopolis.* Retrieved August 8, 2018, from https://confidentials.com/manchester/a-short-history-of-manchester-the-rise-and-fall-of-cottonopolis.

Soden, R., & Lord, A. (2018, November 1). Mapping silences, reconfiguring loss: Practices of damage assessment & repair in post-earthquake Nepal. In *Proceedings of the ACM on Human-Computer Interaction* (Vol. 2, pp.1–21).

References

Strabo (c18AD). *The geography.* Book 5, Chapter 3. Retrieved July 31, 2018, from http://penelope.uchicago.edu/Thayer/E/Roman/Texts/Strabo/5C*.html.

Szaro, R., Sexton, W. T., & Malone, C. R. (1998). The emergence of ecosystem management as a tool for meeting people's needs and sustaining ecosystems. *Landscape and Urban Planning, 40*(1–3), 1–7.

Tainter, J. A. (1988). *The collapse of complex societies.* Cambridge University Press.

Tainter, J. A. (2011). Energy, complexity, and sustainability: A historical perspective. *Environmental Innovation and Societal Transitions, 1*, 89–95.

Tavares, S. G., Swaffield, S. R., & Stewart, E. (2013). Sustainability, microclimate and culture in post-earthquake Christchurch. *LEaP research paper no. 19.* Retrieved July 22, 2020, from https://researcharchive.lincoln.ac.nz/bitstream/handle/10182/5422/LEaP_rp_19.pdf;sequence=3.

Taylor, T. G., & Tainter, J. A. (2016). The nexus of population, energy, innovation, and complexity. *American Journal of Economics and Sociology, 75*(4), 1005–1043.

Walker, B., Holling, C. S., Carpenter, S., & Kinzing, A. P. (2004). Resilience, adaptability and transformability in social-ecological systems. *Ecology and Society, 9*(2), 5.

Wiles, K. (2016). The map: The great fire of London. *History Today, 66*(9), 20–21.

Yoffee, N. (2006). Notes on regeneration. In G. M. Schwartz & J. J. Nichols (Eds.), *After collapse: The regeneration of complex societies* (pp. 222–228). The University of Arizona Press.

Chapter 4
The Modern Built Environment and Its Relationship to Collapse

The destiny of every wall is to be torn down
Alejandro Dolina

Introduction

In the discussion in Chap. 2, it became apparent that some of the current ideas about the future built environment might not solve the problems facing the modern developed world, and by inference, the whole of humanity. This chapter looks at the built environment and how it might be linked to the idea of the collapse as discussed in Chap. 3.

Because the word "built" implies something that has been physically created, our mental images of the built environment tend to involve buildings and other structures like roads and bridges. In the context of this discussion of the collapse, it might be better to think of a "human altered" rather than "built" environment, to include the many changes that people have made to rural areas to enable them to produce the food that those living in dense urban areas can no longer grow for themselves. These rural areas have not been built in the sense of being covered with buildings, but they have certainly been changed by human hands. These altered rural areas have also been defined as the "cultural landscape", a term which the human geographer Carl Sauer used to describe a landscape that has been changed from a natural landscape by a particular cultural group (James & Martin, 1981:321). Changes might be the introduction of fields and fences as much as streets and houses, thus the cultural landscape is a term covering both cities and their rural hinterlands. Whatever the term we use to describe them, both urban and rural are subsets of the same "built" environment and need to be considered in a discussion of collapse. The purpose of this is to consider what needs to be done now in all aspects of the "built" environment to increase the ability of those living there to "collapse gracefully" in the light of the impending shortages and other problems outlined in Chap. 1. There are naturally those who believe such a crisis will not happen, including many politicians who ought to know better. As proposed in Pascal's wager, this book is based on the precautionary principle that it is better to act now, just in case.

To explore a link between collapse and the built environment, this chapter looks at this relationship under three main headings. These are the perception of collapse, the reciprocity between culture and habitat, and the limits imposed by ecosystems.

Perceptions of Collapse in the Built Environment

The built environment plays a big role in the perception of collapse in everyday life. When buildings fall during an earthquake or a bridge is destroyed by a flooded river, we usually read in the newspaper that they have collapsed. In the built environment, the perception of collapse is deeply linked with the failure of physical structures. If they fall, break or fail, they have collapsed. The collapse of physical structures could be associated with negligence, latent defects, lack of maintenance, terrorism, wars, natural hazards and also unusual events like aircraft, trains, road vehicles or even ships crashing into a building (Craven-Todd, 2018).

Collapse can be also linked to the malfunctioning of networks and systems that provide key services to the built environment, such as breakdowns in the supply of electricity, water and natural gas or breakdowns in the removal of sewage and waste, and even problems with the connection, speed or availability of the internet. As an example, there was an infrastructure failure in Havelock North, New Zealand.

> Safe drinking water is crucial to public health. The outbreak of gastroenteritis in Havelock North in August 2016 shook public confidence in this fundamental service. Some 5,500 of the town's 14,000 residents were estimated to have become ill with campylobacteriosis. Some 45 were subsequently hospitalised. It is possible that the outbreak contributed to three deaths, and an unknown number of residents continue to suffer health complications (Department of Internal Affairs, 2019).

In Havelock North, the authorities had relied on the aquifer not being contaminated. Had they exercised the precautionary principle and treated the raw water taken via boreholes from the aquifer in case it became contaminated, which it did following heavy rains, then the infrastructure would not have failed.

Structural and network failures are not the only ways of perceiving collapse in the built environment. When cities are abandoned after a natural hazard or due to an economic or social change in their context, it is usually stated that the city, town or village has collapsed. In this case, the collapse of the built environment is not linked with buildings falling apart, since the structure of the town may remain intact in the same place, but the collapse is related to the abandonment of a place through a dramatic change to the society that had formally lived there. The built environment without people is just ruins. Hashima Island in Japan was once occupied by a population of more than 5000 in only 6 hectares (Hansen, 2018) (see also Chap. 10). The island was economically dependent on the exploitation of undersea coal. When the resource was exhausted the mine was closed and the people departed but the built environment was left almost intact. The abandonment of places is not always driven by scarcity of natural resources. The entire city of Pripyat in northern Ukraine was abandoned in 1986 after the explosion at the nearby Chernobyl nuclear power station. In a classic example of entrepreneurs taking advantage of the collapse, now international tourists are offered an "eye-opening experience of post-Apocalyptic world" (Chernobyl Tour, 2018).

Whether the perception of collapse is linked to the fall of buildings, the breakdown of infrastructure or the dispersion of a community, it seems to be associated with the abrupt, unexpected, undesired, failure or disappearance of something in the built environment. This could mean the end of an established community or the end of the physical habitat that supports it, which are two different concepts. The disappearance of a building from the landscape is not necessarily linked with its collapse. In Japan, traditionally, houses are completely rebuilt after a period, around 20 years for timber buildings and 30 years for concrete. In this case, the end of the particular building in a particular place happened, but it did not mean that the house collapsed since the change was expected and desired, as in Japan land holds its value but the house on it does not (Rosalsky, 2014). Something similar occurs with the ice hotels rebuilt every year in Sweden, Norway and Finland (Icehotel, 2020).

The temporary presence of buildings in the landscape, which can be more or less ephemeral, depending on whether we are considering a pop-up coffee stall or a cardboard cathedral (Barrie, 2014), makes the definition of collapse in the built environment more complicated, particularly if it is quantified in relationship to a fall in population or change in the occupation density of built areas. It gets even harder when it comes to assessing the collapse of a built environment due to its failure in delivering quintessential qualities and services to people, like safety, liveability, transportation and business opportunities. All this suggests that collapse is a difficult term to define although we all know, or think we know, what it means.

Reciprocity Between Habitat and Culture

The built environment materialises the habitat of a culture; it is an outcome of the habits, rituals, social structures, economy and political organisations of a civilisation. Anthropologists, archaeologists, historians, geographers and architects have re-created the ways of living of ancient societies by looking at their physical cultural legacy, from cloth, tools and pottery, to buildings. The geographer Michael Conzen (1960) showed in his analysis of the development of the northern English town of Alnwick how the history of a place is imprinted in its urban landscape. Amos Rapoport (1969) in his book *House, form and culture* discussed the close links between a built environment, its form and layout, and the culture that produced and lived in it, an idea which is aligned with the idea of a cultural landscape.

The cultural landscape is also a sustainability issue since much vernacular architecture relies on the resources available in the immediate vicinity and therefore is limited by these resources. Resource availability has an effect on what can be built, although Rapoport observes that what is built is probably much more affected by the culture of the society creating the buildings than by the availability of resources. However, the availability of resources and environmental performance can affect culture and this influence can then become codified, as in Chinese *feng shui*. The ideal form for a Chinese village is in a cluster with a hill to the north and facing the water to the south. Clustering the houses in a terraced form is good for energy conservation and facing the buildings to the south ensures solar heat gain. The hill behind protects from the cold winds and the water in front reflects the sun, enhancing the solar gain. In this instance, the environmental performance of buildings has become part of the culture.

Teotihuacan, established around 200 BC, was an ancient Mesoamerican settlement north of Mexico City. It forms a good example of a cultural legacy where the built environment tells the history of the changing values of a community. The site was originally a religious place that started after a peculiar cave was found. On top of the cave, a pyramid was built whose profile is very similar to the slopes of the surrounding mountains and it was dedicated to the Sun, which was understood to be a god. The nearby development of Tenochtitlan, the capital of the Aztec empire, on the site of the modern Mexico City, led to the extension of Teotihuacan with the creation of a Moon Pyramid and a complex of buildings connected to the Sun Pyramid by an axis in the form of a road along which religious processions took place (Brunius et al., 2003:23). Through the years the site kept on developing and people started to live there. This led to the further extension of the main axis and the creation of a new plaza. The market that emerged on the main road as a result was very famous. Eventually, Teotihuacan was to become "…a city-state that dominated Mesoamerica until about A.D. 700" (Kurtz et al., 1987). What is interesting to observe in this example is that the site shifted from being solely a religious place to become the commercial centre of a town complete with multi-level apartment buildings (Brunius et al., 2003:40–45). The changing levels of sophistication and complexity of the society were recorded in the built environment through more intricate ornamentation (Figs. 4.1 and 4.2).

Moreover, the types of buildings also differed. The religious identity of the place was linked to the building of pyramids, but later the society preferred plazas and gathering spaces over pyramids because they wanted more spaces for trading and amusement. Teotihuacan's success led to it becoming a multi-ethnic society, but this most complex and sophisticated state of existence was the prelude to internal unrest between the groups of citizens leading to fires and riots and its eventual collapse (Manzanilla, 2015). The changes and development of the society were mirrored and given physical form in the design and development of the built environment of Teotihuacan.

The relationship between the built environment and the culture that produced it is not a one-way street, where we can change our built environment without being affected by the changes made. Even though the famous words of Winston Churchill during a debate on 28 October 1943 over replacing the bombed House of Commons "We shape our buildings; thereafter they shape us" have been quoted by designers thousands of times, it still seems common sense, while at the same time opening

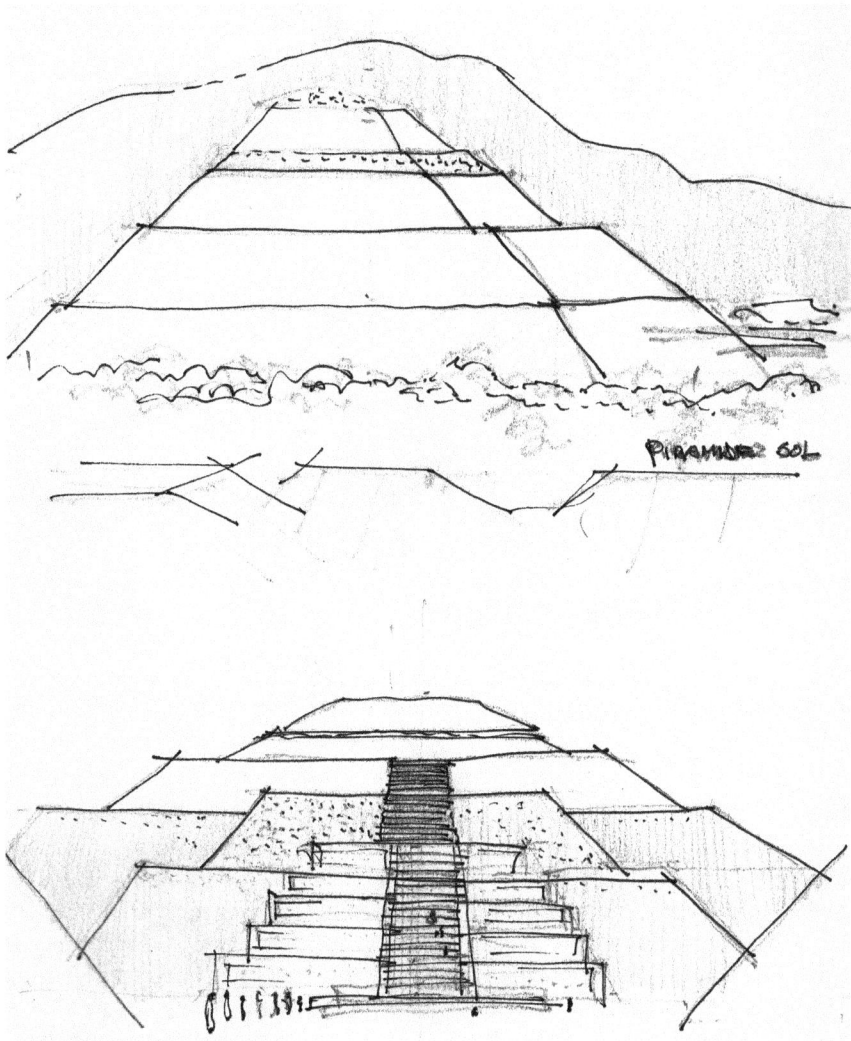

Fig. 4.1 The Pyramid of the Sun, Teotihuacan, Mexico City. At the top is the Pyramid of the Sun. It is the oldest and the largest building in the complex (200 AD). The profile of the Pyramid of the Sun seems to copy the profile of the mountains. Architectural ornaments are humble (adapted from author's photograph)

philosophical questions. We know the impact that the built environment has on the consumption of natural resources and energy and its contribution to carbon emissions and pollution of the environment. Buildings account "…for approximately 31% of global final energy demand, approximately one-third of energy-related CO_2 emissions" (Űrge-Vorstaz et al., 2012:649). At a rather more personal level, there is a connection between the level of "walkability" offered by a built environment and the

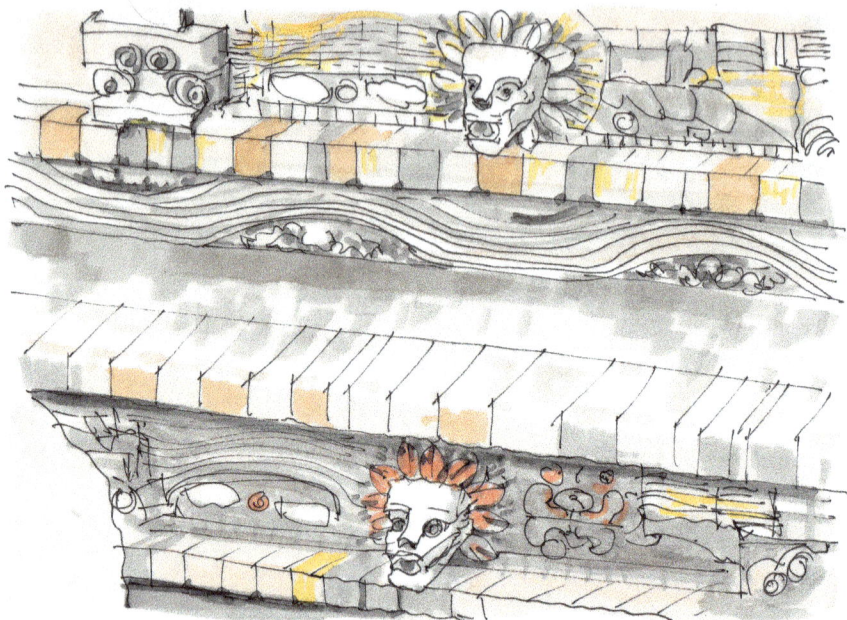

Fig. 4.2 Details of complex carvings in the Temple of Quetzalcoatl (Feathered Serpent) that were added to the building during the high period of development of Teotihuacan (AD 350–650) (adapted from author's photograph)

obesity present in its population (Shahid & Bertazzon, 2015). The built environment has also been linked with mental health issues by environmental psychologists (Evans & Ferguson, 2011). This evidence simply helps us to state that changes made to the built environment to solve one problem may lead to problems in another area. The classic example of this is moving families in post-war UK into high-rise housing, in a well-intentioned move to clear the slums and improve housing conditions. In some places, this was subsequently found to lead to problems, especially for mothers at home with young children who could not be let out to play unless accompanied by an adult (Gillis, 1977). In other parts of the world, such as Asia, where many more activities happen outside the home, high-rise housing has been more successful, again reinforcing the cultural aspects of the built environment.

The reciprocity between culture and habitat implies that built environments collapse when their cultures collapse, and vice versa. This reciprocity allows us to speculate about the information a built environment can provide to describe the state of the culture that produced it. This same reciprocity has helped archaeologists understand the level of complexity of a society by analysing the density of buildings in an area and their sizes. In ancient cities, buildings were clustered in ways, in terms of proximity and type, that were not found in previous villages, such as the apartments of Teotihuacan. Cities had many buildings close together. Monumental public buildings, like temples, represented the wealth of a settlement and were key for the identity of a city (Childe, 1950). If a society collapsed many of these monumental buildings were abandoned or used for different purposes, as happened with the recycling of buildings like the Colosseum in Rome (see Chap. 3). When cities were not abandoned, after a collapse the built environment would be dominated by small domestic buildings (Schwartz & Nichols, 2012). Therefore, for archaeologists changes in size, development and investment related to the built environment can help identify the collapse of a civilisation.

If ruins have helped archaeologists and historians identify and understand the collapse of past civilisations, could the built environment be used to identify symptoms of a future collapse? One possible example is the "skyscraper index" of Lawrence who mapped the building of skyscrapers to the boom and bust cycle of the local economy. This led him to propose that building the tallest skyscrapers heralded a downturn in the economic cycle (for further discussion of the skyscraper index, see Chap. 10). There is also the issue of the extent to which the configuration and design of a built environment might contribute to its potential success or collapse. Has the collapse of the built environment ever produced the collapse of a society? To answer these questions, we need to step back and first consider what the collapse of the built environment might mean.

Built Environments, Ecosystems and Collapse

Cities are part of ecosystems and depend on the availability of natural resources, particularly in their hinterlands (which may be the whole of the rest of the world), to keep on going. For this reason, part of the stability of cities relies on the continuous availability of natural resources to keep the citizens fed, watered and sheltered. Any discontinuity in this flow has historically had dramatic implications for the persistence of those living in a city, as noted in the example of Havelock North discussed above, where contaminated water from the taps caused extensive illness.

The unpredictability and changes produced by internal problems in cities, like socio-political malfunction or negligence, and external threats, like earthquakes, tsunamis, and climate change might challenge the demand of an urban population for constant and stable inputs, like food, water, air quality and safety. Therefore, predictability about what resources will be available is essential for cities because this allows them to persist and keep on working. Cities are dependent on a continuous flow of natural resources so the people who live in them can continue their social, economic, political and belief systems. The ancient city of Angkor in Cambodia depended on water supply from a system of canals and dams to divert water from rivers such that "…generations of Khmer engineers coped with a water system that grew ever more complex and unruly" (Stone, 2009). When in the early 1300s the weather pattern was disrupted leading to the Little Ice Age, it seems the city was abandoned because the water supply was no longer consistent and secure. This is a city collapse through a loss of a vital resource.

The important point is that the persistence of cities depends on continuous deliveries of energy and matter. As mentioned in Chap. 1, the environmental crisis will not provide more predictability or certainty about the availability of resources. For this reason, it is important to incorporate aspects of sustainability and resilience into cities. These can help to deal with the current fragility and future uncertainty of city viability by learning how to manage and profit from changes in built environments in order to avoid collapse. The following sections provide a deeper look into sustainability and resilience in the built environment, with a particular focus on enriching the understanding of collapse in the built environment.

Dealing with Collapse Through a Better Understanding of Sustainability and Resilience

The goal of sustainability is to avoid collapse as the desire is to sustain a particular system or state of a system. However, as discussed in Chap. 1, the reason for investigating the issues introduced in this chapter is because human society as a whole is faced with some serious environmental problems that could, at the very least, lead to significant changes in the way things are organised, and that in turn might affect the built environment. From a sustainability perspective, the interest is in how these changes might happen and whether it is possible to avoid a sudden collapse of the way things currently work. In turn, if it is possible to control the predicted major changes so as to avoid sudden collapse, what should we be doing now to affect this and, in particular, what sort of built environment will help to support, rather than thwart, change?

The idea of ecological resilience was introduced in Chap. 3, and there has been much talk about making modern cities and communities more resilient, leading to the Rockefeller Foundation in 2013 creating an index of "100 Resilient Cities" (Rodin, 2013), although this initiative was wound up in 2019. In ecological resilience terms, a resilient city would be one that is able to persist despite the change that occurs through either inside or outside pressures, something that Teotihuacan was unable to absorb. However, the more resilient cities listed in the modern index were those in the developed world that had the financial resources to deal with any crisis, while poorer cities in the developing world were placed low in the ranking. As Vale and Garcia (2016:556–557) noted,

> …the aim of 100 Resilient Cities is to make Bangkok more like a developed world city, and thus more like Vancouver, with its own problems when it comes to sustainability. Suggesting that urban poverty is potentially more sustainable and resilient than wealth is not going to be a popular idea but in any investigation of urban resilience it has to be faced. In terms of human settlement the very persistence of urban slums suggests they have something to teach about built environment resilience.

The very idea that a far from ideal life in a slum may be more sustainable or resilient than life in a developed world city comes about because one can survive on very little, whereas the other needs far more resources for its maintenance. This suggests a clear link between sustainability, resilience and collapse in the built environment.

From an environmental sustainability perspective, the focus is on what needs to be done to avoid collapse more than the idea of collapse itself. Collapse is assumed as the end of something while the aim of sustainability is the system should continue in some form. The contribution from ecological resilience theory (see Chap. 3) is to explain that change and even collapse are sometimes necessary for the persistence of a system. Instead of resisting change, to persist a system needs to adapt to pressures that induce change. As in chess, sometimes you sacrifice a piece to open lines to your opponent's king. The other understanding stemming from this is that systems can become victims of their own success. There has been an example of this on the remote West Coast of New Zealand where a rise in tourism and the money it brings has led to the need to spend a lot more money on upgrading infrastructure in terms of carparks and waste water treatment as well as providing more accommodation (Wright, 2016). Following the COVID-19 pandemic (see Chap. 9) and the collapse of international tourism in New Zealand, these upgrading efforts may well be wasted, at least in the short term. The problem is that problems facing a system are often unpredictable.

The approach of sustainability focuses on controlling the use of resources and the flow of material and energy to ensure life can continue into the future, and in this way avoid the collapse of current ways of organising society, especially in the developed world. In this context, resilience has emerged as a concept that complements sustainability but offers an alternative understanding of collapse. A better understanding of what resilience means in the built environment can be useful as it suggests collapse is a phase in a process of change. This is pursued in the next sections.

Engineering Resilience and Collapse in the Built Environment

As noted in Chap. 3, the resilience approach comes in two different versions. The engineering approach emphasises the capacity of a system to recover quickly, while the ecological approach emphasises the adaptive capacity, through using change as an opportunity that leads to persistence through adaptation. The former is the approach taken by the 100 Resilient Cities index. However, in both cases, resilience can explain the process that leads to facing or avoiding collapse in relation to the capacity of a system to buffer disturbances and the role of change in this. In the built environment the concept of resilience has traditionally been associated with what Holling (1973) described as engineering resilience (see Chap. 3). What is relevant to this discussion is that the diagrams produced by Tainter of complexity and the collapse of old societies and the graphical plots of engineering resilience of materials have some similarities (Fig. 4.3). Both have a tipping point, after which the collapse happens.

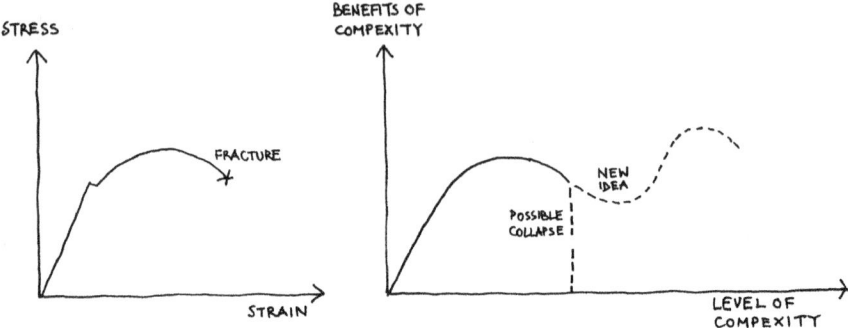

Fig. 4.3 Collapse curves (adapted from Tainter, 1988:125)

From the choice of materials to the design of seismic resistant building structures or the construction of urban systems to control floods, the understanding of resilience has been intrinsically linked with avoiding some form of a physical collapse in the built environment. Without mastering engineering resilience, it would be impossible to imagine the initial existence and the continuing stability of a skyscraper in the city or the possibility of living safely next to the sea protected by sea walls or having efficient storm water systems that help to avoid floods. All these structures were created to contain and to buffer disturbances, and in return, they provide stability unless overwhelmed by forces beyond their design limits. From this point of view, the more stability a built environment achieves, the less the chances it has of collapsing.

Engineering resilience is based on predictable behaviours and responses to predictable threats. Resisting the changes and disturbances produced by a threat is essential for engineering resilience as well as the ability to fix a problem quickly if something goes wrong. This makes sense in the built environment, particularly when we are discussing what is built, which has an intrinsic relationship to the mechanical properties of materials. Nobody in a suburb next to the beach expects the sea wall to behave in a temperamental way during high tides or storms. It must be reliable and stable, it must resist change. Nonetheless, what makes the seawall stable is also what makes it very vulnerable. The sea wall depends on a few key factors: the calculations of engineers relating to specific high tide levels, the response of the materials of the sea wall (that have limited toughness) and a well-mannered nature that behaves according to what was predicted. When all these factors are not met, seawalls may not resist an inrush from the sea, as happened with the 2011 tsunami in Japan, leading to "…Japan's worst nuclear accident" (Onishi, 2011).

There is a strong link between technology and engineering resilience since the latter is a property that was discovered to improve technology. In the built environment, engineering resilience depends heavily on technology, particularly high technology and large-scale infrastructures. It would be very challenging to try to build a skyscraper with vernacular techniques and materials (though the high-rise scaffolding of bamboo challenges this idea to an extent), and to offer the same stability and predictability that can be delivered by building with materials of known properties, such as steel and concrete. Knowledge of such materials and the technologies of using them help to support the stability in the built environment resulting from the increasing pressures of growing cities. However, since technology is also limited, the stability and security provided are both built on what is predictable and also in the belief that, no matter what happens, some engineers might have thought about it, and therefore we are safe. The 2012 Hurricane Sandy superstorm in the United States provided a good example of both such dependence on technology and of the weakness of cities. From a developed world perspective, it seems that the only places where floods are expected to happen are in developing countries. It seemed impossible that floods could also affect developed cities like New York, leading one resident to comment "…the city itself suddenly seemed vulnerable: the sense of certainty that underlies quotidian urban life had been dramatically interrupted" (Ashley, 2017:44).

The dependence of the built environment on technology (see Chap. 5) is not only linked to the materiality and physicality of the built environment but also to the increasing need to solve issues stemming from the development of cities. However, technology is also driven by both market forces and the imagination (or should that be nightmares?) of designers. Designers do not wait to have the appropriate technology before imagining a city in a single building, like Paolo Soleri (Soleri, 1969), or drawing a mile-high skyscraper, like the tower that Frank Lloyd Wright proposed in his book *A Testament* (Wright, 1957:239–240). Historically, advances in technology have not been an obstacle to the imagination of designers. Designers have a long history of imagining things long before they were feasible and often of building them without concern for whether they would work (see Chap. 2), leaving others to pick up the pieces.

With the increasing frequency of social and ecological hazards, some of which were discussed in Chap. 1, the concept of resilience has exceeded the boundaries of engineering and it is perhaps more useful to apply the idea of ecological resilience to the built environment so that when crises happen normal life can go on; in other words, the human-built environment system can adjust to pressures from both within and without. This adds another layer to the imperative of stability discussed before, by suggesting there is the need to fix a problem speedily so as to get back to normal as quickly as possible, like the trams in Hiroshima. Considering that cities are extremely dependent on many services and resources that need to be delivered without disruption, to keep them functioning resilience has become the property that makes such recovery happen. However, whether this is engineering resilience or ecological resilience or something else entirely is still to be determined.

The Ecological Resilience Approach and Collapse

As noted in Chap. 3, Holling (1973) introduced the concept of resilience in ecology as an alternative way of managing resources, along with the idea that an ecosystem has shown resilience if after a threat that might cause change, it is still recognisably the same ecosystem. In other words, it has retained its identity (Walker et al., 2004). This link between identity and resilience has been used to understand changes in the built environment, and this creates a link with the notion of heritage (Garcia & Vale, 2017:203–204). From this point of view, a disturbance would collapse the resilience of a system when it changes its identity.

The point raised by Holling and colleagues is that without resilience there is little chance of survival, or in other words, the greater chance of collapse. In the same way that Tainter links a failure in the problem-solving capacity of institutions to their collapse, Holling links the resilience capacity of a system to its chances of surviving and therefore avoiding collapse. In this way, natural systems appear to cheat collapse by using "threats", like scarcity, volatility of resources, competition and invasions, as sources for their own development and persistence.

Tainter questioned why the same problem in one society could be the cause of a crisis and in another the cause of the collapse of a civilisation. He attributed this to the state of the political development of a culture, and particularly the costs and benefits that this produces. In the case of resilience, the state of the system matters a lot but also the timing of the disturbances. The same problem might affect an ecosystem differently at discrete phases of its development. Holling hypothesised that changes in ecosystems are cyclical, namely they are recurrent and follow the familiar developmental pattern of growth, maturity and decay with the addition of one more phase, that of reorganisation. This hypothesis of cyclical change was termed the "adaptive cycle" and explains the development of systems according to changes in their stability and resilience (Holling & Gunderson, 2002:47). There are four phases in the adaptive cycle: exploitation, conservation, release and reorganisation (Fig. 4.4).

In the exploitation or development phase, a system is undergoing rapid change but it is not very stable and its identity is not yet consolidated. The conservation phase can be considered as establishing the identity of the system. In built environment terms the exploitation phase might be the establishment of a few houses at a ford where

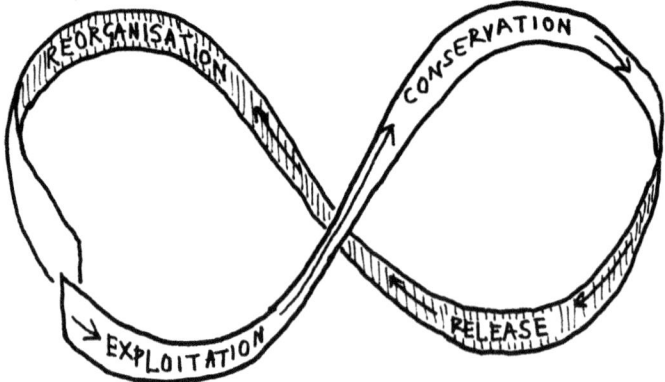

Fig. 4.4 The adaptive cycle (adapted from https://www.resalliance.org/adaptive-cycle)

a river could be crossed. Because of the usefulness of this point, these few houses gradually grow into a complete city but at the cost of increasing the complexity of the systems required to keep it going, or to persist. If there is a storm that creates a huge flood as the river bursts its banks, destroying much of the physical structure of the city to the extent that the complexity of its systems can no longer be maintained, a change is inevitable—this would correspond to the release phase of the adaptive cycle. After the release phase, many survivors might leave but a few might stay on and once again establish life on boats and buildings built on rafts at a useful river crossing but one that might well flood again. This would correspond to the reorganisation phase. The former city has gone, its identity is lost, and something new is developing in its place. This idea is linked with both the proposals of the Austrian economist Joseph Schumpeter about the possibility of a crisis being either constructive or destructive (McCraw, 2007) and the concept of open systems being capable of self-organising criticality proposed by theoretical physicist Per Bak. Bak (1996:2–3) states that change does not happen smoothly but rather through "catastrophic events", using the example of a child building a pile of sand on the beach. As the pile becomes higher and the sides steeper, small avalanches of sand occur until some equilibrium is reached providing the child stops piling on the sand. Bak (1996:3) goes on to say that "Self-organized criticality is so far the only known general mechanism to generate complexity". This is useful in explaining that complexity is part of how natural systems work. For the pile of sand, however much you know about the properties of a single grain, these will not explain the behaviour of the pile, which has become a complex system with its own properties.

In the adaptive cycle, the reorganisation phase is key to the resilience approach since it provides meaning to the release phase because it makes the crisis something important for the future of the system. In other words, the crisis can be seen as an opportunity to keep on developing or as a new beginning. What is important for collapse is the similarity between the curve associated with just the conservation phase of the adaptive cycle and Tainter's (1988:125) curves of the growth and collapse of empires (Fig. 4.5). The conservation phase involves a steep building up of capital in the system which then slows up to the point where some form of disturbance moves the cycle on to the next—release—phase. The second curve depicts the area

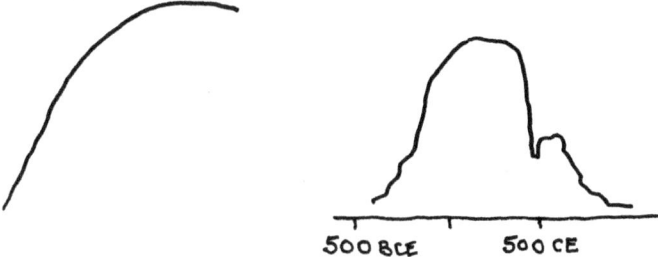

Fig. 4.5 Shape of the conservation curve in the adaptive cycle (see Fig. 4.4) and area covered by the Roman empire (adapted from Tainter, 1988:125)

of land occupied by the Roman Empire, again showing a steep increase in the area which slows and then starts to decline, followed by a steep decline. An increase in complexity in both cases reaches a tipping point that cannot be sustained and therefore the system collapses.

Regardless of the collapse of the Western Roman Empire, the city of Rome is still there. This fact says something about the relationship between collapse and resilience. The city of Rome did not disappear nor did its culture. What collapsed with the Western Roman Empire was the imperial system. This is coherent with Tainter's definition of collapse as a socio-political change. However, after the collapse, the Roman cultural legacy continued to be inherited through generations. In Europe, eleventh and twelfth century Norman architecture with its round arches is still referred to as the Romanesque style (Clifton-Taylor, 1967:32). This cultural legacy nurtured the development of the Roman identity. From the resilience point of view, the imperial collapse could be viewed as the end of one and the beginning of a new adaptive cycle that started with a reorganisation process after Rome was sacked (Fig. 4.6). The much-reduced city population remained relatively stable during the Dark Ages but began to rise in an apparent "exploitation phase" during the Renaissance and to begin to build steadily in the "conservation phase" once Italy was unified. This qualitative assessment, using the population growth of the city of Rome as a reference, has a parallel in the growth of the urban landscape. This cycle highlights the tight bond between collapse and resilience within the built environment.

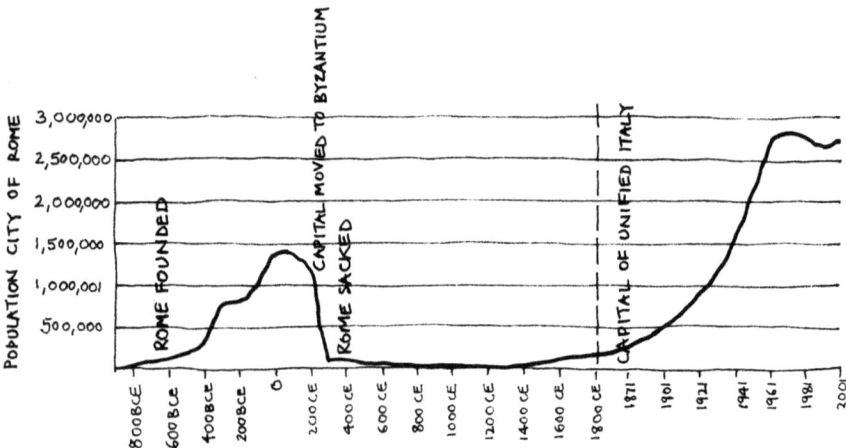

Fig. 4.6 Growth in the population of Rome (the dashed line indicates the change in scale) (adapted from https://romabyrachel.weebly.com/the-timeline.html)

The Importance of Scales in Resilience: Panarchy

Comparing the collapse of a building in a city with the collapse of civilisations seems like comparing a street fight with the Second World War. The previous discussion has linked collapse to the development of complex societies, which are usually understood as societies that have been able to develop political states. Where the scale of analysis is an entire civilisation, this is something that is completely out of the scope of designers who very often find themselves designing at the domestic scale of a house on a plot, only sometimes at the scale of an institutional building, rarely at that of a neighbourhood, and almost never that of a whole city.

Scale matters because it helps to define the boundaries when looking at a problem or an object. Limits and boundaries are important for managing the persistence of systems. Archaeologists have also dealt with this issue. They infer a great deal of information from relatively small-scale elements (in comparison with the implications of their analysis). Tainter (1988:133–137) used data about the change of weight in the denarius (the basic unit of Roman currency) to explain one reason behind the collapse of the Roman Empire. Archaeologists are obliged to recreate the puzzle of history piece by piece and to make a lot from very little because the excavation of an entire archaeological site could take years.

In ecological resilience, Gunderson and Holling (2002) posit that changes in ecosystems happen not only in adaptive cycles but also within and across scales. They called this multiple scale approach Panarchy, a made-up word which is a mix between the suffix -archy, as in monarchy, anarchy and hierarchy, referring to a type of rule or organisation and Pan, both the Greek god of the wild and shepherding and a prefix meaning "many", as in pantheism. In this way, a Panarchy is presented as a more dynamic version of a regular hierarchical organisation. In ecosystems the concept of scale is linked with the rate of change in space and time. A leaf, a tree and a forest belong to different scales in the analysis of the vegetation of a landscape since they experience different speeds, quantity and quality of change. A leaf dies annually, a tree may live a thousand years, and a forest may have been in the same place for millennia. A firefly, a mouse and a cheetah living in the same location have different sizes, weights, foraging areas and consume different amounts of resources; therefore they can be considered within different scales of an analysis. In a Panarchy every scale goes through an adaptive cycle, with different speeds of change, which means that in a landscape, smaller things might experience collapse more frequently than bigger things. The loss of foliage that some species of tree experience in winter does not mean that the tree has collapsed. Nor does the collapse of one tree mean the collapse of the forest. Indeed, deciduous trees with ephemeral foliage depend on the seasonal loss of their leaves for persisting. In a Panarchy, changes happen not only within but also across scales due to two processes: "revolt" and "remember" (Gunderson & Holling, 2002:69–72).

In a "revolt" changes are the product of bottom-up processes, namely, changes that start at a small scale and cascade up to affect the large scales of the system. In a city, revolts can be linked to the behaviour at the domestic scale of a single plot and its larger implications for the urban fabric. As an example, houses have grown with the addition of extensions in many developed cities, meaning the plot has become more built over with less open space. This can affect the permeability of the ground so that during a heavy storm the resultant water draining off all the impermeable built surfaces can overload the storm water system, leading to city-wide flooding. "Newcastle city centre, for instance, is around 92% impermeable, and has suffered major flooding in the past" (Cambridge University, 2020).

The process of "remember" is a top-down process, which means that it starts at the large scale of a system and cascades down to affect the small scales. City planning policies form an example. For instance, Wellington City Council introduced a policy of urban intensification along an identified transport spine (Gray, 2007). This could have the effect of reducing permeable areas by covering land with more houses and places for car parking with the increased risk of flooding after heavy rain events in an era of climate change.

The Panarchy approach helps in understanding that collapse at some scales can even be necessary for the persistence of the entire system. For this reason, Gunderson and Holling (2002) proposed that scales are essential to the resilience of ecosystems because they help to create "discontinuities" in the landscape which are largely responsible for the heterogeneity and richness of an ecosystem.

Complex adaptive systems, like ecosystems or cities, are organised at multiple scales, and so have more chances to buffer disturbances than systems with fewer scales. The more scales and diversity within a system, the more complex the system becomes. This complex organisation enables ecosystems to develop a resilience capacity that allows them to persist. However, increases in complexity can also lead to the collapse of human societies as defined at the end of Chap. 3, so this issue needs further investigation.

Panarchy and the Scales of Collapse in the Built Environment

When one house is affected by a flood in a small town, this event might not mean too much for the rest of the population in a country. However, it means everything for the family that has suddenly lost all its belongings. On a household scale, it is a critical situation, which in turn, might not at all affect the functioning and everyday life of the rest of the population, even in its closest neighbourhood. This makes it vital in any investigation of why things collapse, and how this might be avoided, to know the scale of the investigation. In the built environment this can range from a single building to a whole highly built-up country, like the Vatican City or Singapore. Landscapes, buildings and even whole neighbourhoods have collapsed. Collapse can thus be contextualised at different scales in the built environment so that the seriousness of a collapse greatly depends on the scale of the building or habitat affected.

It is important to remember that the human-altered—or built—environment is not just urban and that rural examples have also been through what might be termed a collapse, which has changed both their appearance and the way they operate. As one example, the clearances in the Highlands of Scotland in the late eighteenth to mid-nineteenth centuries (Richards, 2007:45–65) changed the appearance of the landscape. Celtic Life (2019) states that prior to the clearances and the introduction of large-scale sheep farming, Highland families lived in townships of around 100 people that acted as a tenanted collective farm. The buildings of these townships "were substantial" and made from local materials such as stone, clay and turf for walls and "heather, broom, bracken, straw or rushes", for the thatched roofs. With the coming of the sheep, nature took over and these buildings soon became ruins. The clearances were based on the belief that sheep farming would be much more profitable for the landowners than letting their land to the many mixed farming, subsistence Highlanders (Whyte, 2013). Those displaced left to seek work elsewhere, many to go overseas to the colonies, causing a local collapse of the previous clan-based social system, which in turn altered the landscape (Dodgshon, 2017:192–226). A second example can be seen in the history of the part of North America lying between the Rocky Mountains and the Mississippi River, which was once primarily grassland where the buffalo roamed, with few trees except for those alongside rivers. Unlike the Highland clearances it was the influx of settlers from Europe, some of whom may well have been displaced Highlanders, and their desire for agriculture, that changed this rural landscape. Cook et al. (2009) state that "During the 1920s, agriculture in the United States expanded into the central Great Plains. Much of the original, drought-resistant prairie grass was replaced with drought-sensitive wheat." Following a series of droughts in the 1930s crops failed, vegetation died and the soil was exposed to the wind creating the Dust Bowl, described as "…one of the greatest ecological disasters in the history of the United States" (Engle et al., 2008:255). This is an example of a changed or collapsed landscape caused by human intervention.

As a civilisation, our idea of collapse is in great part related to the perception of collapse in our habitats, which for more than half of the world's population means how collapse is manifested in cities. The concept of civilisation is historically (and etymologically) linked to the concept of the city and the idea of urban life, and the urban development that occurs when societies develop complex forms of government, like nation-states. The Highland township described above was run by the clan chieftain so the governance was simple, you did what the chieftain told you. For the township to survive food had to be grown and the directions of the chieftain were needed to ensure that this happened. Once settlements grew beyond the family group, then more complex systems of governance were necessary, not least to find a way to decide who was going to make the necessary decisions for survival.

A collapse in the built environment can affect an entire neighbourhood. There is the example of the failed housing complex of Pruitt-Igoe in St Louis, Missouri that was opened in 1954–1955 and demolished in 1976, some 21 years after its opening. For students of architecture this is often held up as a failure of modernist design, with the 11-storey blocks being seen as inhuman in their scale and repetitiveness, and as such its ruin should be preserved (Jencks, 1977:9). Others, however, have argued that this "collapse" was not a design failure but a failure of the economic system of the city of St Louis that did not manage to produce the projected growth and need for housing on which Pruitt-Igoe was based (Crawford, 2015:462–463). This in turn led to under-occupation and insufficient rental income to cover the costs of the planned maintenance of the buildings. As a city, however, St Louis survived these economic problems and is now "…one of the world's great river cities" (GLOBOsapiens, 2018). The city has persisted even though a neighbourhood has collapsed. This question of scale and at what scale collapse happens is vital to any discussion of collapse in the built environment.

Collapse at the small scale could even be seen as a vital ingredient of persistence at the larger scale as considered in ecological resilience (see Chap. 3). Conzen (1960) studied the medieval burgage, an ancient form of tenure in England and Scotland by which land or property in a town (the burgh) was held in return for some form of service or payment of annual rent. His study of maps of the town of Alnwick in northern England revealed that plots held under the burgage system acquired more buildings until a point was reached when they could acquire no more and still function. This in turn eventually led to clearance and starting again, creating what he called the burgage cycle. In this case, the collapse was marked by the clearance of all the buildings in one plot, itself a radical change, although this phase of the burgage cycle makes room for new buildings in the landscape. The town persisted throughout these changes (and still does) but "collapse" was observed at the level of the individual plot. This suggests that many systems make up a built environment so that, even when physical attributes collapse, the urban area persists. The other thing to note is that a crisis in terms of having to clear a plot produced an opportunity for rebuilding.

The Links Between Sustainability, Resilience, Collapse and the Built Environment

Links have been made between sustainability and resilience (Marchese et al., 2018) but in any effort to tease out a relationship between these two terms the first thing to note is that neither is a priori a target. The question to ask of sustainability is "what is to be sustained?" When it comes to ecological resilience the question is "the resilience of what to what?" For engineering resilience, the question might be somewhat different, "how fast can you recover?".

It is perhaps simplest to link engineering resilience to sustainability in the light of the question posed above. An elasticated waistband can be stretched many times but eventually the elastic will come to the point where it can no longer recover its original shape and starts to stretch. At this point, two things can happen. Either the garment with the elasticated waist, say a polyester skirt, is thrown away or a new elastic is inserted. There is perhaps a third option whereby the waistband is gathered and remade, so it now closes with a zip or button. It is worth looking at these different results. Waiting for failure and then throwing the skirt away is a waste of resources and hence might not be considered part of sustainability, where the goal is to try to reduce the use of exhaustible resources like oil (the basis of the polyester garment) and to ensure that the benefits from using these resources are invested in human and institutional capital (Acar, 2017:77). If the skirt is repaired with a new elastic, then it has returned to its original form although one of its internal components has changed (new elastic for old). In the third option, the skirt is still recognisably the same skirt but it now looks a bit different. In the last two examples, the resources have been "saved" through an initial design that allows for both straight repair (new elastic) and adaptation (new waistband).

This ability to persist through the renewal of part of the system (in the example above the skirt is the system) is something intrinsic to the built environment. Buildings and infrastructure have to be designed for regular maintenance and easy repair because they are made up of many components that have different "lives". Timber windows need repainting at least every 10 years, so they will then last 50 years. At this point, the building may need replacement windows, which is similar to fashioning a new waistband for the skirt. It is still recognisably the same building although it may look slightly different because of the new windows. This example suggests that engineering resilience is something that can be anticipated and planned for so that the resources that go into a building or a road can be made to last as long as possible. However, there will always come a point where it will be cheaper in resource terms to start again. The link here with sustainability is planning to use the available resources to create something like a built environment in the best possible way so that it lasts as long as possible. The change will happen at a small scale but the identity of the built environment will persist. This is perhaps the opposite of the unsustainable, throw away consumerist society where users are encouraged to want the new and the different rather than repairing the old (Assadourian, 2010). Even 20 years ago only 1% of the total materials flow in the United States was still in use in products six months after their sale (Hawken et al., 1999:81). This situation seems unlikely to have improved. The built environment is different because its products still are expected to, and indeed do, have a very long life. Eighty-year-old cars tend to be in museums but eighty-year-old houses are still lived in. The Royal Albert Bridge in England, designed by the famous engineer Isambard Kingdom Brunel and opened by Prince Albert after walking across it on 2 May 1859, still carries the main railway line from London to Cornwall (Royal Albert Bridge, undated). Even when houses are built of non-durable materials, they still last a long time. New Zealand's wooden houses last over 100 years before it becomes cheaper to demolish them and rebuild (Johnstone, 2001).

If the engineering definition of resilience tends to suggest a built environment that persists, the ecological approach, at first sight, seems to suggest the opposite. As stated above, perhaps we can understand resilience better by looking at how built environments use change, both to adapt to disturbances and even to produce innovation. In the built environment, resilience could be a property that encourages persistence through change. At first sight, this seems like a paradox as buildings and infrastructure tend to be long-lived items but, of course, they do not all come to the end of their lives simultaneously, unless there is a serious disruption such as an earthquake or a bombing raid. The built environment is constantly changing in small ways as a street is widened or a site cleared to create a pocket park.

Perhaps resilience can be better understood by looking at how built environments use change to adapt to disturbances and even produce innovation (Garcia & Vale, 2017:53). Resilience in the built environment is also linked to maintaining identity so that small changes do not affect the identity of an area, be it a street, a neighbourhood or the whole city. Resilience contributes to keeping key functions viable while allowing change. In this way, built environments could avoid collapsing due to internal or external pressures and could even use these threats as an engine for persistence. The disturbance of the 1666 Great Fire of London (see also Chap. 3) produced change not only through rebuilding in brick in place of the earlier timber but also in the organisation of London, with the city becoming the commercial centre, with wealth migrating to the west and poverty to the east. This migration can still be seen in the current built environment of London (Bloomfield, 2018; Peace, 2015). In this case, there was a change in the way London functions at the economic level and a change in building materials but not a great change in the street layout. Reorganisation at one scale did not, in this example, produce reorganisation at every scale. In the history of change in an ecosystem, this is common as is proposed in the concept of Panarchy.

The fauna of New Zealand, apart from one bat, is avian rather than mammalian. The introduction of people and mammals, first by the Maori and a thousand years later by European settlers led to the extinction of some birds, such as the Moa, eaten to extinction by the Maori 500 years ago (Worthy, 2015) and the Huia collected to death for its skins and feathers by the Europeans early in the twentieth century (Szabo, 2013), but other birds survived this collapse of individual species, at least initially, because there was a lot of country and few people. As the human population of New Zealand has increased so has the pressure on bird habitats. This has led to the call for the creation of a predator-free environment, but as someone observed "if you get rid of the rats the mice just get bigger…" (Associated Press, 2017), and of course the greatest predators are people, as both the Moa and the Huia discovered.

An ecosystem is an ever-changing system within which there are, in human terms, winners and losers and where it is not possible for a win–win situation. Making "sustainable" buildings has been claimed as win–win as such buildings become evidence of a business commitment to the sustainability agenda (Anon, 2013) and there is evidence that such buildings improve productivity and hence profits (Anon, 2017). However, the components for making such buildings, even if the buildings really are in some way sustainable and not merely attempts at "greenwashing", might not all be sourced from sustainable producers, so in terms of exhaustible resources such as copper there may be losers (future generations). This is where the resilience approach has much to offer sustainability thinking. The resilience approach suggests that more sustainable societies will still be ever-changing, and within such changes, there will be losers. This is something we shall return to in Chap. 6.

The environmental problems described in Chap. 1 have encouraged a move towards sustainability, or at least towards thinking about sustainability, or perhaps at least towards no longer denying sustainability, by confronting humanity with several possibilities for a potential collapse. Sustainability is about avoiding collapse, which certainly depends on not passing the limits to growth or consuming what the environment cannot sustain. On this basis, our global civilisation has been collapsing since the day humanity's Ecological Footprint overshot the earth's biocapacity in 1970 (Global Footprint Network, 2018). However, this "collapse" does not seem to have caused a sudden loss of complexity or an abrupt change in life in the developed world. Is this perhaps a graceful version of collapse? Or do we need a different definition of collapse? Or are we still travelling along the top of the curve with the big drop to come?

As shown in Chap. 3, archaeologists, historians and ecologists have been considering and challenging the idea of collapse in order to enrich our understanding of it. The concepts of sustainability and resilience have been used by archaeologists (Redman, 2005) to get a "deep time perspective" of how we arrived at the present point. Sustainability is closely linked to history, since "factors that make a society sustainable, or vulnerable to collapse, can rarely be discerned within a human lifetime. Sustainability challenges develop over periods of decades, generations, or centuries" (Tainter, 2014). However, divergent opinions arise at the moment of explaining collapse as the result of unsustainable choices or practices. The hypothesis that overshooting has inevitably been the prelude to the collapse of ancient civilisations is still contested in archaeology. "There does not presently appear to be a confirmed archaeological case of overshoot, resource degradation, and collapse brought on by overpopulation and/or mass consumption" (Tainter, 2006). Butzer and Endfield (2012), however, see this contested area as leading to public confusion.

> Unfortunately there are insufficient empirical, rather than simulated, data on the nature of societal response to cross-disciplinary inputs, triggers, or tipping points. The public is confronted by metanarratives of global change or by semipopular works that suggest oversimplified causal correlations. Such hypotheses can readily be misunderstood as fact.

This is, of course, not a reason to believe that collapse due to overshoot is impossible in our civilisation, not least because the population is now so much larger but the planet is still the same size, and previous human civilisations had not really set about changing the global climate, but it certainly shows that historically the explanations behind the collapse of ancient societies go beyond attempts to correlate environmental events to collapse. This comment is in line with the suggestion of Middleton (2017:3) that collapse is a story, subject to influence from multiple biases, and one telescoped in time. These factors, paraphrasing Borges' opening quote in Chap. 3, make it difficult to distinguish the collapse of an empire from the brightness of a firefly. However, what we may now be facing involves, in Clive Hamilton's words, a "requiem for a species" (Hamilton, 2010) rather than the end of an empire. The species is humankind.

The built environment is also a historical product and a legacy of a culture, and for this reason, it represents a source of evidence for archaeologists. Collapse in the built environment of contemporary cities could be very tangible at a small scale, especially when this is related to the physical infrastructure, but looser and more difficult to understand when it is linked to a larger scale. The role of the design of currently built environments in the collapse of cities is uncertain since part of the evidence lies in the future. Moreover, the idea of collapse when applied to contemporary megacities, megalopolises, regions and nations would be difficult to compare with much smaller, ancient urban centres.

Similarities and Differences Between the Understanding of Collapse in History, the Built Environment, Sustainability and Resilience

The definition of collapse in history and archaeology has been approached in multiple ways. However, a common ground in the definitions of collapse is that the subject of analysis is people coming together to form a society. Even though the idea that collapse implies the end of civilisation has been challenged, its meaning remains closely linked to the end of a phase or era and a major change in the society formed by the same people.

Theories can help to explain the collapse. From the historical point of view, Tainter relies on an approach that has a parallel in thermodynamics, where collapse happens due to the accumulation of complexity (entropy) produced by our need, as a society, to resolve problems so as to keep on developing. Sustainability is based on a similar thermodynamic approach. Collapse happens when the limits imposed by the availability of natural resources are exceeded by the demands of growth and patterns of consumption. The marginal returns described by Tainter and the overshooting described by the Ecological Footprint measure are conceptually very similar.

When it comes to collapse the subject of analysis is important because it describes what is being observed and therefore what is measured. In history and archaeology, the literature dedicated to theorising collapse is focused on the analysis of ancient civilisations, and particularly how complex societies suddenly shift to less complex forms of organisation. For designers, the subject of analysis is the built environment. However, this subject of analysis varies across scales; it could be a wall, a building, or a city. In sustainability and resilience, the subject of analysis is the relationship between societies and nature. Therefore, in sustainability, the flows of energy and matter along with the patterns of consumption that drive these flows are quintessential for the persistence of 'societies. The version of resilience that we are interested in comes from the consideration of ecosystems. The application of the ideas from ecological resilience to other complex systems, like cities, is still developing, and this book aims to be a small part of that development. Table 4.1 sets out the similarities and differences between the ideas discussed in this chapter.

Looking at the similarities in the table, collapse emerges as a type of change that could be perceived as abrupt, dramatic and undesirable or as a necessary stage in the development of a system, and that can even create new opportunities. These contrasting perceptions of collapse are possible because it is a type of change that implies a transition or transformation from one state where things are known, familiar, and apparently ordinary to another state, where reality is unfamiliar and extraordinary.

Table 4.1 Views of collapse

	History/archaeology	Sustainability	Resilience	Built environment
Subject of analysis	Complex societies, particularly ancient civilisations, states and empires	Management of natural resources to sustain the development of humanity	Behaviour of ecosystems and social-ecological systems	Human habitat, its physical infrastructure and community that can be either rural or urban
Causes of collapse	Internal: collapse is a socio-political phenomenon caused by problem-solving dynamics. External: collapse is mainly driven by ecological factors/natural hazards. Multiple: collapse is produced by cultural and environmental changes	Collapse is produced by internal causes (behavioural) that are cultural (consumption/growth) and environmental (human-induced climate change)	Collapse is linked to internal causes (increase in complexity and specialisation) and external causes that are social-ecological (natural hazards)	Collapse could be caused by internal factors (failure in structure or cultural changes) external factors (social-ecological hazards) or both
Direction	Unidirectional: abrupt change from more complex societies to less complex societies. Multiple directional: collapse might happen when societies either gain or lose a significant level of complexity	Sustainability is not a static state but involves change as natural resources are also always changing	Collapse has multiple directions. It can be produced by top-down changes in the Panarchy (remember) or by bottom-up changes (revolt)	Change tends to happen slowly in the built environment and can be both top-down and bottom-up
Scale	Collapse is linked to the large scale, like civilisations or an entire society	Collapse could be quite rapid and large scale. Sustainability involves global and local scales	Collapse happens at multiple scales, within the adaptive cycle and across scales in the Panarchy	Collapse happens at multiple scales, from houses to cities
Change: Process or event? Fast or slow?	Looking back makes change appear abrupt	Looking forward requires planning for change through understanding energy and resource flows	It is ambiguous. Collapse is linked to the release phase, part of the adaptive cycle. However, the release phase should happen in a short time	Change can happen abruptly but rebuilding can be slow

(continued)

Table 4.1 (continued)

	History/archaeology	Sustainability	Resilience	Built environment
Predictability	N/A	There are reasons like overshoot that suggest the need for change	Change will always happen	How should the built environment change?

Another point of controversy is whether collapse is an event or a process. This is an important point because it defines when the collapse starts and where it ends. Currently, there is no agreement about this between fields of research. In the built environment collapse can certainly be both an event and a process. The abandonment of a fishing village due to the impact of a sudden tsunami is an event but the fall of a building due to the continuous undermining of its foundations by the sea is the result of a process. In all the rest of the disciplines analysed collapse looks like a process but what is confusing is that these processes become part of the explanation of a collapse only once the system has passed a breaking point. This is the case of resilience and the release phase in the adaptive cycle. Release is a phase of the adaptive cycle but it is also an event, one that defines a "before" and an "after" for the system.

After comparing the different fields of research, it is not clear whether collapse always implies an abrupt and rapid change or if it could be linked to slow and long periods of decay. However, the analysis of collapse from different angles and through the study of discrete properties of a system should provide new points of view that hopefully enrich our understanding of collapse.

Why Are These Issues Relevant for Designers?

For designers, trying to learn from the management of ancient settlements and applying this knowledge in modern cities seems like taking swimming lessons in the bath as preparation for a competition in the open sea. However, not acknowledging the history of ancient civilisations will be denying the only evidence and knowledge that we have about collapse.

These historical points of view are helpful for understanding the importance that power relationships, institutional organisations and the behaviour of people have for the future not only of a city but of an entire civilisation. What is relevant for designers is that all these factors are not only linked to the growth, expansion and development of the built environment but simultaneously also to its possible collapse. When the knowledge of collapse is framed within the built environment its study becomes more challenging because the social and environmental factors that sustain a culture cannot be understood in isolation. It is the context that provides meaning to the built environment.

From the historical point of view, the built environment is a passive element wherever the collapse of a society happens. However, from a designers' perspective, the built environment plays a bigger role. If transformations of a habitat make collapse perceivable to the archaeologist, can we say that without a habitat there is no collapse? The hypothesis of Tainter is that we continuously add to the complexity of a society until its decisional systems collapse; however, in this process, we also add to the complexity of its habitat, since people eat, drink and need to sleep somewhere to have comfort and protection from the elements.

Regardless of the specific fields of research, the nature of design is to imagine something before it is made. The act of designing, either as carried out by professionals or by people without any degrees in design, forecasts the complexity of a habitat and in this way impacts its development and growth. Design is a link between the problems and the solutions that have constantly emerged in a built environment. Moreover, the development of a built environment is constrained by the legacy of what is inherited from previous developments.

In this context, one way of looking at the role of designers is as the people responsible for making manifest in physical form the level of complexity of a society. In contemporary cities, design only happens because people with the money to create a building ask for it to happen. The designer is no more than an intermediary in managing this. However, since the result of designing something will be adding something else to the complexity of an environment, perhaps the role of a designer as an intermediary should be creating solutions with the lowest possible investment in complexity and, more problematically, persuading clients to accept these solutions. This is a paradox. What if designers started thinking about minimising the legacy of problems that they will be generating? What if the best thing is not to build? Are designers then redundant when it comes to thinking about a built environment that avoids collapse by addressing the problems outlined in Chap. 1? Any building is not only a solution to the demands of the client, it is also a limitation in the landscape for future designers who will have to meet the new demands in the same landscape but with even more limitations.

Some Final Thoughts About the Collapse

The economic system of capitalism and the political system of democracy have served the world well up to now. They have even worked through world wars, so it could be argued they have managed to deal with crisis on a global scale. However, one major difference between the collapses of the past and those of the future relates to the scale of their implications. Past collapses have been local, even if they affected whole cities, like the eruption of Vesuvius that destroyed Pompeii, or whole civilisations, like the decline of the Western Roman Empire, remembering that the Eastern Roman Empire did not collapse at the same time. When a society collapsed in the past there was always somewhere else to go or somewhere else from where to obtain food, resources and so on. World wars, while global in scale and effect have been limited in time, the assumption of those involved being that the war could be won bringing the conflict to an end.

Future collapses are likely to be global in effect, particularly if they are triggered by the exhaustion of fossil fuel resources, the destruction of the oceans through over-fishing and pollution, and climate change which could render agriculture no longer possible, at least in some parts of the world, and which will cause the sea level to rise. These types of drivers of collapse are more like the events that led to the extinction of the dinosaurs, their effect is on a very wide scale, generally global, so there is nowhere to run to, apart perhaps to Mars. These forms of global collapse are likely to be long term, so unlike war, they will continue and are likely to end with no winners. It may be that the kinds of responses needed in the face of such a large-scale collapse might be totally different from those for a smaller scale collapse and may also mean that we cannot learn from the past because we have never had to deal with this kind of problem before. A global collapse is a large-scale process with large consequences but multiple scales are entangled within it and not all the landscapes might be affected equally. Therefore, it is difficult to forecast the impact of a global collapse and much more difficult to imagine the human responses to radical transformations. However, given the circumstances, it might be good to do everything we can to avoid witnessing a global collapse.

Not only do we need to think about a built environment that can rebound after a collapse but we also need a built environment that does not contribute to causing a collapse. At the same time, it may be that the built environment on its own can do little unless the world can find political and economic systems that do not appear to be working together to bring about existential collapse. The kind of collapse that we may be facing is possibly a different form of collapse from that of previous societies because of its extensive nature. For one thing modern technological development is very different from that of past societies, so might it be possible that technology will come to the rescue in the face of current problems? Technology is discussed in more depth in Chap. 5.

References

Acar, S. (2017). *Natural resources and sustainability*. Palgrave Macmillan.
Anon. (2013). A win-win situation. *Environmental Design and Construction, 16*(6), 30–33.
Anon. (2017). A green office could increase productivity. *NZ Business, 31*(5), 12–13.
Ashley, D. (2017). *Extreme cities: The peril and promise of urban life in the age of climate change*. Verso.
Assadourian, E. (2010). Transforming cultures: From consumerism to sustainability. *Journal of Macromarketing, 30*(2), 186–191.
Associated Press. (2017). New Zealand's ambitious plan to save birds: Kill every rat. Retrieved July 17, 2020, from https://www.telegraph.co.uk/science/2017/05/12/new-zealands-ambitious-plan-save-birds-kill-every-rat/.
Bak, P. (1996). *How nature works: The science of self-organized criticality*. Springer.
Barrie, A. (2014). *Shigeru Ban: Cardboard cathedral*. Auckland University Press.
Bloomfield, R. (2018). Where to find a city cottage; Victorian workers' cottages are just right for first-time buyers. *London Evening Standard*, Jan. 24, 2018, 10.

References

Brunius, S., Cowgill, G. L., & Linné, S. (2003). *Archaeological researches at Teotihucan, Mexico*. University of Alabama Press.

Butzer, K. W., & Endfield, G. H. (2012). Critical perspectives on historical collapse. *Proceedings of the National Academy of Sciences of the United States of America, 109*(10), 3628–3631.

Cambridge University. (2020). *Waterworld: Can we learn to live with flooding?* Retrieved July 26, 2020, from https://www.cam.ac.uk/research/features/waterworld-can-we-learn-to-live-with-flooding.

Celtic Life. (2019). *The highland clearance*. Retrieved July 23, 2020, from https://celticlifeintl.com/the-highland-clearances/.

Chernobyl Tour. (2018). Retrieved August 10, 2020, from https://www.chernobyl-tour.com/english/.

Childe, V. G. (1950). The urban revolution. *The Town Planning Review, 21*(1), 3–17.

Clifton-Taylor, A. (1967). *The cathedrals of England*. Thames and Hudson.

Conzen, M. (1960). Alnwick, Northumberland: As study in town-plan analysis. *Transactions and Papers (institute of British Geographers), 27*, iii–122.

Cook, B. I., Miller, R. L., & Seager, R. (2009). Amplification of the North American '"Dust Bowl"' drought through human-induced land degradation. *Proceedings of the National Academy of Sciences of the United States of America, 106*(13), 4997–5001.

Craven-Todd, E. (2018). WATCH: Horrific moment GIANT tanker ship crashes into historic mansion. Retrieved August 10, 2018, from ps://www.express.co.uk/travel/articles/944799/viral-video-ship-crash-port-mansion-istanbul.

Crawford, J. (2015). *Fallen glory*. Old Street Publishing.

Department of Internal Affairs. (2019). *Section one: Introduction and context*. Retrieved July 25, 2020, from https://www.dia.govt.nz/Government-Inquiry-into-Havelock-North-Drinking-Water-Report---Part-1---Overview.

Dodgshon, R. A. (2017). *No stone unturned: A history of farming, landscape and environment in the Scottish Highlands and Islands*. Edinburgh University Press.

Engle, D. M., Coppedge, B. R., & Fuhlendorf, S. D. (2008). From the Dust Bowl to the Green Glacier: Human activity and environmental change in great plains grasslands. In O. W. Van Auken (Ed.), *Ecological studies, Western North American Juniperus communities: A dynamic vegetation type* (pp. 253–271). Springer.

Evans, G. W., & Ferguson, K. T. (2011). Built environment and mental health. In J. O. Nriagu (Ed.), *Encyclopedia of environmental health* (pp. 446–449). Elsevier.

Garcia, E. J., & Vale, B. (2017). *Unravelling sustainability and resilience in the built environment*. Routledge.

Gillis, A. R. (1977). High-rise housing and psychological strain. *Journal of Health and Social Behaviour, 18*(4), 418–431.

Global Footprint Network. (2018). *Past earth overshoot days*. Retrieved August 8, 2018, from https://www.overshootday.org/newsroom/past-earth-overshoot-days/.

GLOBOsapiens. (2018). Visiting to gateway to the west. Retrieved June 13, 2018, from http://www.globosapiens.net/travel-information/St.+Louis-698.html.

Gray, R. N. (2007). *Residential intensification and the Wellington urban development strategy*. Retrieved July 27, 2020, from https://wellington.govt.nz/~/media/your-council/projects/files/infill-resintens.pdf.

Gunderson, L. H., & Holling, C. S. (2002). *Panarchy: Understanding transformations in human and natural systems*. Island Press.

Hamilton, C. (2010). *Requiem for a species: Why we resist the truth about climate change*. Allen and Unwin.

Hansen, K. (2018). *30 of the most stunning abandoned towns around the world*. Retrieved July 7, 2020, from https://www.architecturaldigest.com/gallery/most-stunning-abandoned-towns-around-world.

Hawken, P., Lovins, A., & Lovins, H. L. (1999). *Natural capitalism*. Little, Brown and Company.

Holling, C. (1973). Resilience and stability of ecological systems. *Annual Review of Ecology and Systematics, 4*, 1–23. Retrieved from http://www.jstor.org.ezproxy.auckland.ac.nz/stable/2096802.
Icehotel. (2020). *Icehotel*. Retrieved July 25, 2020, from https://www.icehotel.com/.
James, P. E., & Martin, G. (1981). *All possible worlds: A history of geographical ideas*. Wiley.
Jencks, C. (1977). *The language of post-modern architecture*. Rizzoli.
Johnstone, I. M. (2001). Energy and mass flows of housing: A model and example. *Building and Environment, 36*(1), 27–41.
Kurtz, D. V., Charlton, T. H., Hopgood, J. F., Kowalewski, S. A., Nicholls, D. L., Santley, R. S., Swartz, M. J., & Trigger, B. G. (1987). The economics of urbanisation and state formation at teotihuacan. *Current Anthropology, 28*(3), 329–353.
McCraw, T. K. (2007). *Prophet of innovation: Joseph Schumpeter and creative destruction*. Belknap Press of Harvard University Press.
Manzanilla, L. R. (2015). Cooperation and tensions in multi-ethnic corporate societies using Teotihuacan, Central Mexico, as a case study. *Proceedings of the National Academy of Sciences of the United States of America, 112*(30), 9210–9215.
Marchese, D., Reynolds, E., Bates, M. E., Morgan, H., Spierre Clark, S., & Linkov, I. (2018). Resilience and sustainability: Similarities and differences in environmental management applications. *Science of the Total Environment, 613–614*, 1275–1283.
Middleton, G. (2017). *Understanding collapse: Ancient history and modern myths*. Cambridge University Press.
Onishi, N. (2011). *Seawalls offered little protection against Tsunami's crushing waves*. Retrieved July 26, 2020, from https://www.nytimes.com/2011/03/14/world/asia/14seawalls.html.
Peace, A. (2015). House prices in boroughs surrounding the traditional west end core are turning increasingly prime as growth radiates outwards across London. *Estates Gazette,* 11 July 2015.
Rapoport, A. (1969). *House, form and culture*. Englewood Cliffs, N.J., Prentice-Hall.
Redman, C. (2005). Resilience theory in archaeology. *American Anthropologist, 107*(1), 70–77.
Richards, E. (2007). *Debating the highland clearances*. Edinburgh University Press.
Rodin, J. (2013). *100 resilient cities*. Retrieved July 25, 2020, from https://www.rockefellerfoundation.org/blog/100-resilient-cities/.
Rosalsky, G. (2014). *Why are Japanese homes disposable?* Retrieved July 25, 2020, from https://freakonomics.com/podcast/why-are-japanese-homes-disposable-a-new-freakonomics-radio-podcast-3/.
Royal Albert Bridge. (n.d.). *Royal Albert Bridge Saltash Cornwall*. Retrieved October 23, 2018 from http://www.royalalbertbridge.co.uk/.
Schwartz, G. M., & Nichols, J. (Eds.). (2012). *After collapse: The regeneration of complex societies*. University of Arizona Press.
Shahid, R., & Bertazzon, S. (2015). Neighbourhood walkability in a major Canadian city. *AIMS Public Health, 2*(4), 616–637.
Soleri, P. (1969). *Arcology: The city in the image of man*. MIT Press.
Stone, R. (2009). Divining Angkor: After rising to sublime heights, the sacred city may have engineered its own downfall. *National Geographic, 126*(1), 26–55.
Szabo, M. J. (2013). Huia. In C.M. Miskelly (Ed.), *New Zealand birds online*. Retrieved August 10, 2018, from www.nzbirdsonline.org.nz.
Tainter, J. A. (1988). *The collapse of complex societies*. Cambridge University Press.
Tainter, J. A. (2006). Archaeology of overshoot and collapse. *Annual Review of Anthropology, 35*, 59–74.
Tainter, J. A. (2014). Collapse and sustainability: Rome, the Maya, and the modern world. *Archeological Papers of the American Anthropological Association, 24*(1), 201–214.
Űrge-Vorstaz, D., Eyre, N., Graham, P., Harvey, D., Hertwich, E., Jiang, Y., Majumdar, M., McMahon, J. E., Mirasgedis, S., Murakami, S., & Novikova, A. (2012). Energy end-use: Buildings. In T. B. Johansson, N. Nakicenovic, A. Patwardhan, & L. Gomez-Echeverri (Eds.), *Global energy assessment (GEA)*. Cambridge University Press.

Vale, B., & Garcia, E. J. (2016). The relationship between resilience and sustainability in the built environment. In *6th International Conference on Building Resilience*, Auckland, New Zealand, 07–09 Sept 2016. N. Domingo, & S. Wilkinson (Eds.), *Proceedings of the 6th International Conference on Building Resilience* (pp. 550–560). Massey University and The University of Auckland.

Walker, B., Holling, C. S., Carpenter, S., & Kinzing, A. P. (2004). Resilience, adaptability and transformability in social-ecological systems. *Ecology and Society, 9*(2), 5.

Whyte, D. (2013). *Environmental history and global change: A dictionary of environmental history.* I B Tauris.

Worthy, T. H. (2015). "Moa" *Te Ara—The encyclopedia of New Zealand.* Retrieved August 10, 2018, from http://www.TeAra.govt.nz/en/moa.

Wright, F. L. (1957). *A testament.* Universe Books.

Wright, M. (2016). *West coasters victims of their own tourist success.* Retrieved July 26, 2020, from https://www.stuff.co.nz/national/81183797/west-coasters-victims-of-their-own-tourism-success.

Chapter 5
Technology and Collapse

Part 1: Technology and Complexity

> *Technology is just a tool. In terms of getting the kids working together and motivating them, the teacher is the most important.*
> Bill Gates

Invention

When technology goes wrong in the built environment, things collapse. The so-called Cheesegrater building in Leadenhall London, named because of its shape, became notorious when two bolts fell off it in 2014. One piece the size of a hand fell to the ground while the other was contained within the building (BBC News, 2014). The steelworks company set aside £6 million to cover the cost of dealing with the problem, which was due to hydrogen embrittlement within the bolts (The Construction Index, 2019). Fortunately, no one was hurt and the building did not fall but the profits of the steelwork firm from the job must have been dented if not come near to collapse. However, in this example as in others, it is not so much the technology that is wrong but the human decision to use a particular type of technology in a particular situation. In Germany, a large hole appeared in Autobahn 20 in 2017 (The Local de, 2017). Where this road passed over an area of peat the decision to use friction piles rather than piling to bedrock ultimately proved the wrong one, leading to the roadbed failure and road closure, which in turn posed a threat to the local tourist industry. The right technology was available but the decision was made to use the wrong one. This shows that technology is to an extent neutral and it is the availability of different technologies, some of which may never have been tried in a particular situation that is the problem. This raises the issue of whether technologies should be limited to what is known to work so as to avoid human error in choosing the wrong one, a course of action that would have those espousing innovation and invention throwing up their hands in horror.

© The Author(s), under exclusive license to Springer Nature Switzerland AG 2021
E. Garcia et al., *Collapsing Gracefully: Making a Built Environment that is Fit for the Future*, https://doi.org/10.1007/978-3-030-77783-8_5

In 1672, Bacon described his ideal society in the book *New Atalantis*. This highly regulated society had at its core a research centre called Salomon's House, which was full of new inventions but it was the scientists and technologists involved who decided which inventions would be offered to society and which would be kept secret (Bacon, 1672:41). This view of technology is that things are invented for their own sake and then uses are found for them (or possibly not, in the case of the control exercised by Bacon's imaginary scientists). A modern example is superglue or cyanoacrylate. The latter was created in 1942 by Harry Coover who was looking to develop a clear plastic that could be used for gunsights. Unfortunately, it stuck to everything and so was abandoned. In 1951, as part of a search for "…stronger, tougher and more heat-resistant acrylate polymers for jet plane canopies," the same cyanoacrylate polymers were the centre of attention. This time when two prisms remained stuck together after an attempt to measure the refractive index of the polymer the light bulb moment occurred and a new family of adhesives was the result (Coover, 2000). The contrasting view of technology is that invention is necessary as it fits a need, and that often this gives rise to the same thing being invented simultaneously in two different places.

Passing over the fact that a third inventor of the powered flying machine, the New Zealander Richard Pearse, was also making trial flights around this time (Ogilvie, 1996), Ogburn and Thomas (1922) argued that the cultural and social factors are extremely important when it comes to inventions. Without the internal combustion engine, powered flight, other than in a steerable airship as developed by Henri Giffard, was not really a possibility. The uptake of aircraft technology began with public demonstrations of flying that often attracted many visitors. Barbara Ganson (2014:1–2) describes how in 1910, 3,500 people turned up in Houston, Texas to watch Frenchman Louis Paulhan fly his bi-plane. The next day there were 6,000 spectators, even without social media. The demonstrations produced a desire for the invention and the rest is aviation history. Even if as a spectator you had no hope of ever flying, you knew it was possible for a man to fly and so could dream of doing so. The fact human society had existed for centuries without the technology of flight, and trade still happened, and buildings were still built suggests the invention was not strictly necessary at that time, even if some saw it as a desirable achievement in itself, and others with more foresight as a future opportunity. The first commercial air service was opened in 1919 not by the Americans but by the Germans with their daily scheduled flights between Berlin and Weimar, although they had not had a hand in inventing heavier than air flight (Dienel & Schielfelbusch, 2000:955). Post First World War Germans felt it was a patriotic duty to partake in air travel "…as a sign of resistance against Versailles and of trust in Germany's resurgent role in the skies" (Dienel & Schielfelbusch, 2000:954), but then the Germans had been the inventors of lighter than air commercial flight with the Zeppelin.

This was the issue Bacon faced in his imaginary world—should there be a limit not so much on the invention of new things and new technologies but rather on which of these things should be filed away and forgotten? Would the world have been a better place without powered flight which made possible the mass bombing of London and Dresden and the dropping of the atomic bombs on Hiroshima and Nagasaki? Flight was not developed as a means of achieving Blitzkrieg bombing but it made it possible. As Aunger (2010) when discussing human history observed, "Science has thus far only been associated with technology at the 'front end'–with innovation, by applying science in engineering–not at the 'back end', when the innovation has an impact on our way of life." The problem is that with some inventions their future effect cannot be foreseen even if it were in some way desirable to pick and choose between them.

Complex systems, like human societies, need to keep solving problems and sometimes redundant inventions supply the answers, as happened when NASA made use of the invention of Velcro, patented in 1955, for securing items in zero gravity. Its inventor, Georges De Mestral, a Swiss engineer, had seen it as the alternative to zips and buttons but the fashion houses had spurned this idea (Eschner, 2017). This is an example of solving a problem that the inventor did not set out to fix, whereas the invention of flight could be seen as introducing both problems and additional complexities to human societies.

What Is Technology?

The previous discussion of invention assumed that everyone understands what is meant by technology. It is a word we use all the time, often meaning "new electronic stuff", but it has more precise definitions. In its simplest form, technology is no more than a way of doing something, either something new that human beings had never done before or a new way of doing something familiar. The invention of the aeroplane is an example of the former and the invention of the water-powered loom an example of the latter. The latter example deserves further scrutiny. Faced with what to wear, the history of human clothing shows that woven cloth came after animal skins in the form of fur and leather, with the weaving of yarns into fabrics occurring some 10,000 years ago in the Neolithic period (Horn & Gurel, 1981:12–15). However, weaving was itself dependent on "…the cultivation of plants [for fibres] and the domestication of animals…" [for wool] (Horn & Gurel, 1981:42). The development of the first loom—the backstrap loom—meant that the technologies of animal and plant husbandry had to be in place, just as before skins could be turned into clothing bone scrapers and needles had to be invented. This suggests that trying to understand what technology is, resembles the issue of which came first, the chicken or the egg? This also means that technological development depends on cumulative change and improvements. It is like a self-rolling snowball, in that every change is built upon the previous change.

Another way of defining technology is once you have people, they will find ways of doing things that make life more pleasant, and these ways are technologies. However, the history of the loom reveals something more about technology. The problem of weaving is to find a way to hold the warp threads under tension so the weft can be woven through. The backstrap loom is made of sticks and ropes with one set tied to a tree or post and the other via a strap to the weaver's waist, with the warp threads running between. The tension is provided by the weaver leaning back. This means the equipment needed is easy to maintain and the loom is portable, meaning the weaver can work inside or out. The disadvantage is only relatively narrow strips of fabric can be woven. This led Yarwood (1952:1) when describing the clothes of Ancient Egypt (3000BC–500BC) to state "The earliest garment for men was a loincloth" made from a long, narrow strip of fabric. Perhaps more importantly for the history of technology, Yarwood goes on to state "In the early centuries everyone wore this garment" but later "the loin-cloth remained the attire of slaves and the working classes, being worn only as an undergarment by wealthier men". Narrow strips can be sewn together to make clothing, but the extra effort required to do this in Ancient Egypt meant that some people could afford clothes but not all. A simple technological advancement produced a visual inequality because dress then established a person's standing in society. This suggests a technology also has a social impact that goes beyond the technological purpose. In this way, technologies contribute to making a society more complex by creating differences between people of different socio-economic standing. Technologies thus arise through problem-solving and adding to social complexity, but their ability to create social differences also increases that complexity.

The development of the warp-weighted wide loom, which was used in much of Europe and all around the Mediterranean Sea (Wild, 2003:13), meant both wider and longer lengths of cloth could be produced. "The speed at which fabrics could be produced was of primary importance. Not only was the weaving process itself faster, but the tedious task of sewing strips (or animal skins) together was no longer necessary" (Horn & Gurel, 1981:43). The wider cloth was to have an obvious influence on the type of clothes that could be created, but this simple development also reveals something about technology. The new loom allowed for cloth to be produced faster, and if more cloth can be made in a given time then more can be sold for the same input of labour. The development of the hand loom in the thirteenth century speeded the process as did the flying shuttle in 1733 (Park & Jayaraman, 2017), a mechanism which was capable of being operated by machine rather than hand, leading to Cartwright's power loom of 1785 (Strickland, 1843:59). The latter meant cloth could be produced much faster. The Jacquard loom of 1801 used punched cards as a means of producing patterns in the woven cloth, which again speeded the process of producing patterned cloth and this changed clothing into fashion "…for it was in the second half of the eighteenth century that fashion really began to move" (Laver, 1966:9).

The development of the loom also illustrates the economic side of technology. Technological development helps people to do more with less effort and also to do it faster. The efficiency and speed of technological advances have been used to increase human productivity. When these factors are mixed with the social impact of technology, something new emerges—consumerism. This in turn can create a new problem if the resources used to produce fashionable consumer artefacts either start to become exhausted or lead to the need to discover new technologies, such as the development of Bakelite as a means of making billiard balls from plastic rather than from the ivory tusks of elephants, which were in short supply (Kaufmann, 1963:26–27). The technological spiral of change can have intended and unintended consequences which contribute to the complexity of the social behaviour and this, in turn, can affect the built environment. To continue with the history of the loom, from weaving on a fine day in the open air with the simple backstrap loom attached to a tree, the invention of the hand loom brought the machine inside to protect it from the weather, since considerable resources were now invested in the machine as well as in training the weaver to operate it. This led to a change in houses in areas where weaving occurred, as the top floor was now devoted to the craft, being nearest the light, and large windows were inserted (Fig. 5.1). The development of the powered loom meant the machines and the workers had to be near a source of power, usually water at the start, and so the factory was developed as a new type of building with the workers now having to move some distance between home and factory rather than just going upstairs. The technological developments that allowed faster production of cloth in terms of both spinning (Fig. 5.2) and weaving, both changed the built environment and added to the complexity of using it.

In the history of the loom, each new technological advance affected not only the way cloth could be produced but also what could be produced for a particular input of effort, and hence what could be sold for that input of effort. This series of inventions thus turned clothing from something necessary to the modern fashion industry, where clothes are discarded not because they are worn out but because they no longer represent the latest fashion trends, provided you have the money to take part. This has perhaps unintended consequences as the UN states that "Nearly 20 percent of global waste water is produced by the fashion industry, which also emits about ten percent of global carbon emissions" (UNECE, 2018). Not all can take part in the fashion industry. Those who do not have the money have to wear what they can find (Byrde, 2003:891), just as the working people and slaves of Ancient Egypt wore only the loin-cloth.

Fig. 5.1 Weavers' cottages, Wardle, UK (adapted from https://en.wikipedia.org/wiki/Weavers%27_cottage)

Fig. 5.2 1771 Cromford mill, water-powered spinning (adapted from https://historystack.com/Cromford)

This extended example has been offered as a way of describing what technology is. It is not just the invention but also what the invention allows us to do that is different from what we did before. The other aspect is that one technology builds on another through modification of what is already there—the backstrap loom becoming the hand loom—and through an invention of something new—the flying shuttle, which in turn led to the loom being mechanised. However, the loom illustrates another aspect of technology, whereby an existing process is transferred into a new domain. Park and Jayaraman (2017) note "The Jacquard loom proved to be the inspiration for Charles Babbage's Analytical Engine and then Hollerith's punched card," going on to argue that these developments in weaving ultimately led to the computer revolution. Again, however, the point of the computer is to do what humanity had done before but faster.

The example of the loom shows that every development in a particular technology has an effect, whether foreseen, as being able to make wider cloth, or unforeseen as in the development of the fashion industry from what had formerly been the need for clothing. This challenges the idea put forward by Heidegger (1977:6) that:

> …modern technology is a challenging [Herausfordern], which puts to nature the unreasonable demand that it supply energy that can be extracted and stored as such. But does this not hold true for the old windmill as well? No. Its sails do indeed turn in the wind; they are left entirely to the wind's blowing. But the windmill does not unlock energy from the air currents in order to store it.

The earliest post mill had to be turned by the miller so that the sails would take the maximum energy from the wind, and windmills were also spaced apart so that the wind had time to regain its speed before hitting the next set of sails. The wind energy was also "stored" in the form of the mill's product, whether this was as milled grain or water that was pumped up for irrigation under gravity. All technology is simply a way of trying to do something faster for the same effort, the starting point being what one person can achieve. This has a link back to the origins of the word technology, which appeared in 1765–1783. The word is etymologically linked to technique (*tekhnikos*, relative to an art). The Greek word *tekhne* means art, industry and also the ability to do something (Corominas, 2003:560). The point is that the etymology of the word seems to highlight the human ability to do something, more than the tool to do it with. This is the opposite of the current understanding of the word technology that prioritises the novelty of the tool over the capacity to do something well.

Returning to the example of the miller, obviously, a windmill can grind far more grain in an hour than one person with a stone pestle and mortar. However, the windmill will draw on natural systems in terms of the wood that has to be felled and processed to make it and the stone that has to be extracted for the grinding stones. The impact of a mill on natural systems is greater than that of one person with their simple grinding equipment. Simply, if you want to avoid technology you have to get rid of people and their desire to do more for less, or conversely, where there are people there will be technology. In turn, the development of technology has an impact on the complexity of society and its built environment. The question is whether the investment in technology and therefore complexity is justifiable.

Energy Return On Investment (EROI)

Energy return on investment has been introduced in Chap. 1. The purpose here is to link it not just to fuels but to technology. The example of the loom also shows that greater investment in terms of materials needs to be put into the technology for an increase in output. When it comes to the power loom this input extended to the energy needed to power the machine. Where this was water power the investment was in the materials for the water wheel to extract potential energy from the stream or river and convert it to rotational energy. The energy in the water came from the hydrological cycle. This is what modern hydropower schemes do on a much larger scale, though for these more materials are required to make the dam and the turbines that together convert the water flow (potential energy) to electrical energy. The energy needed to make the materials and bring them to the site also needs to be taken into account. The bigger the technology, the more complicated this becomes. Thus the energy it takes to make energy becomes important as the more energy that comes out for the energy put in the better. The ratio between output and input is termed energy return on investment (EROI). "Energy return on investment…is the ratio of energy returned from an energy-gathering activity compared to the energy invested in that process" (Hall & Klitgaard, 2012:310). Values of EROI can range from 100:1 for hydropower to 0.8–1.7:1 for ethanol from sugarcane, while solar energy is also low at 2–8:1 for a flat plate collector to 6.8:1 for photovoltaics, although passive solar design in buildings would probably be high (Hall & Klitgaard, 2012:313). Is there an equivalent to EROI for technologies? Technologies have certainly moved from being human-powered (the loom) to being powered by renewable energy (the powered loom using water power), and being powered by fossil fuel energy, first as steam and then as electricity. Each step has made the technology more complicated and harder for the user to understand. This, in turn, has moved the user from being a crafts person who would know not only how to weave but how to maintain the loom, to being a machine operator with people with different skills needed to maintain the machine. At the same time, the built environment has also had to change to accommodate the changes in technology, moving spinning and weaving from the home to the factory. This is an example of what would be considered progress in technology—moving from the human-powered to the energy powered loom—because it allowed for producing more cloth at the same time, leading to an increase in complexity and, of course, an increase in profit. The two variables here are time and energy, and to save time other sources of energy have to be exploited. However, this move to maximise time at the expense of energy leads to other problems. If the loom breaks down it cannot be mended immediately as the mending depends on someone else, who may be engaged elsewhere doing something different. If the power supply breaks down, then another chain of people has to be brought in far removed from the site of the breakdown to restore normal working. This increase in complexity was the basis of Tainter's argument that it was not environmental conditions that led to the collapse of earlier societies but increasing complexity in the way societies were organised. Part of this increase in complexity comes from developments in technology.

However, this does not answer the question as to whether there is an equivalent EROI for technology. Is there a return from new technology and if so, what does return mean in this context? One of Heidegger's arguments in *A Question of Technology* (Heidegger, 1977:16) is that effectively we all view the world through technological eyes. Only in the Garden of Eden before encountering the Tree of Knowledge do we have an image of living without technology or tools. Perhaps another way of thinking about this is we see the world in terms of tools, or perhaps as a first step, what can we use (or invent) to help us do a particular task. This leads on to the next step of what can we do or invent to do a particular task faster or with less effort, and finally what can we do or invent to gain as much monetary profit as possible from doing a particular task. However, at no point is there a questioning as to whether the task needs doing, especially once money comes into it. Instead, such development of a technology is seen as progress and progress is always viewed as something positive—a going forwards. What might the world look like if not viewed through the eyes of technology? Since human beings, along with some of our primate relatives, are tool users this is not possible.

What we can do is to try to look at the world but not through the eyes of modern technology, which could be loosely defined as technology during and after the industrial revolution. The Aboriginal people survived in Australia, a country known for its harsh climate except along some of its coastal fringes, and to do this they needed to invent sophisticated tools such as the boomerang and woomera (spear thrower) as well as creating a thermoplastic resin from natural sources that could bind a stone to wood (Delacey, 2015). Dr. Ormond-Parker of the University of Melbourne is quoted as saying "Certainly, when it came to social complexity, Australia was just as diverse as Europe. This was a constantly changing, fifty-thousand-year-old civilisation with over 500 different cultures and languages" (Delacey, 2015). From this it again seems that to have technologies is to be human, so the issue is not so much having technologies, but as Bacon observed, whether there should be a limit on the technologies we have. Here it is worth discussing an example from the built environment, that of prefabrication.

Prefabrication

Building, the need to create shelter, is as old as the need to find tools. Building depended in the past on the resources available in sufficient quantity that they could be used to create something relatively large. This tended to lead to vernacular building styles that were specific to certain locations, such as the bamboo houses of Indonesia and the Philippines (Polkinghorne & Polkinghorne, 1945:81–85) or the cob cottages of SW England (Williams-Ellis et al., 1947:82–89). Within this vernacular were the designs for dwellings that could be moved because of the nomadic needs of those using them. Examples are found in many climates, from the Bedouin tents of the north Sahara woven from the hair of their livestock of goats, camels and sheep (Polkinghorne & Polkinghorne, 1945:14–16) to the felt tents or yurts of Mongolia, made from trellis work of willow, with willow stems for the roof framing, the whole being covered with large pieces of woollen felt that are stretched and tied down (Polkinghorne & Polkinghorne, 1945:24–26). These tents are of necessity prefabricated, and unlike modern prefabricated buildings, prefabrication is the necessity driven by the need to assemble, disassemble and move the building as part of a nomadic way of life.

The modern interest in prefabrication stems from two different sources. The first is the interest of modernist architects like Gropius and Le Corbusier in standardisation and the second the need to build numbers of dwellings in a short time, such as after both world wars. Both sources had an impact on the technologies involved in prefabricating dwellings. For modernist architects prefabrication was seen as a more efficient way of producing buildings on the grounds that they could then be produced like factory-made cars in a limited number of types and styles. This led Le Corbusier to the Dom-ino House, which was a structure in reinforced concrete that allowed for walls to be placed anywhere within the structure so as to create an interior. Gropius, in contrast, developed the Copper-Plate house, which was a modular construction where prefabricated components, such as a variety of wall types, could be assembled in different ways (Gropius, 1935:59). Neither approach was instantly successful. When Le Corbusier designed a series of concrete houses with a structure detached from the building envelope at Pessac, the users immediately set about infilling and changing the design because of the freedom the structural system allowed, leading Le Corbusier to note that "…it is life that is right and the architect who is wrong" (Boudon, 1972:2). The approach of Gropius in partnership with Konrad Wachsmann also failed because of the high cost of tooling up to produce the Copper-Plate house (Vale, 1995: 84). After the world wars that had led to a housing shortage because of the cessation in house building, it was noted that there was insufficient skilled labour to build the required dwellings, and so the proposal was that building work would be done in the factory on a production line basis, with assembly on-site, thus breaking down the required skills, along with the dwelling, into component parts. This led to the search for suitable techniques that could be factory-based, exacerbated by shortages of conventional building materials. Outcomes were steel houses produced using ship building technology post First World War and light steel frame and even aluminium houses post Second World War (Ministry of Works, 1944; Anon, 1945). This was fine whilst industry was in disarray but as soon as metals were required for making other products in factories, the prefabrication of houses using unusual (at least for houses) materials stopped.

The point of this anecdotal history of the prefabrication of houses is to show that a change in technology—moving from conventional to prefabricated building methods—is a means to achieve an end, whether that end is an abstract ideal in terms of Modernism or a way of producing houses at a time of conventional material and labour shortages, or simply a way of making dwellings that can be moved regularly.

Is modern talk of prefabrication a means to an end or a way of selling something in a competitive market? Many components in modern buildings are prefabricated off-site, such as pre-stressed concrete beams, staircases and even the humble roof truss. Such items are small enough to be easily moved and the tooling up to make one type of building component is manageable. The problem comes with the idea of prefabricating the whole building—normally a house—off-site. The claims made for doing this are that time will be saved, quality will be better and there will be less waste, thus adding to a building's sustainability credentials (Gorgolewski, 2005). However, when these claims are examined (Moradibistouni et al., 2018) it seems prefabrication has less to offer than as seems at first. It is applying a technology to a problem that we already know how to solve since most buildings are erected without the benefit of whole building prefabrication. In fact, prefabrication adds to the complexity of the building process since it is done off-site, so cannot adapt to local conditions and the problems that emerge from them. Transport is another added complexity as now rather than timber coming from the sawmill to the local stockist to the site, timber comes from the sawmill either to the factory, where the room has to be found to store it, or to the stockist and then to the factory and finally to site. De-skilling is another potential complexity. A builder building a house on-site does most of the jobs involved in its construction. A factory worker building a prefabricated house may spend their entire working career doing only one task, meaning they have less ability to seek alternative employment if the prefabrication company goes out of business. Finally, it is worth pointing out that if prefabrication really were a profitable way to build houses, it would have already replaced on-site construction.

Prefabrication is again solving a non-problem in the name of doing something differently through introducing a new technology, which in turn brings added complexity, except prefabrication is not a new technology and its past failures should suggest that maybe it is not worth reviving. As philosopher George Santayana said, "Those who cannot remember the past are condemned to repeat it" (Big Think, n.d.).

Technology and Complexity

All these findings of technology can be partially linked to the problem-solving mechanism (see Chap. 3), where societal problems and solutions are tightly bound to escalate investments in resources that produce marginal returns until the system collapses. Prefabrication offers such an example since in solving one problem—how to build more quickly—it introduces others, such as having to build a factory in which to prefabricate. At a society level, according to Kohler et al. (2018:289–318), the growth and development from smaller scale foragers and horticultural groups to larger agricultural societies implied an increase in complexity through societies being more hierarchical, with an increase in both political structures and the use of new technologies. The marginal returns from this investment in agriculture were also made visible through the inequalities generated by this change in the way of exploiting natural resources. These inequalities became visible in the built environment because those at the top of the hierarchy could afford to live in larger dwellings than those at the bottom, who often struggled for survival. These issues are explored in detail in Chap. 6.

The point here is that it is the use of new technologies, such as the invention of the horse collar, that allowed fewer people to be involved in agriculture, and hence introduced the idea of a more hierarchical society where only some worked on the land, freeing others to engage in different trades and businesses. This raises the issue of whether adopting new technologies inevitably leads to greater complexity in a society. At first sight, this seems unreasonable. Why should something like the move from large desktop computers to small laptops and even smaller tablets increase complexity? After all, a computer is a computer whatever its size. Working at a desktop ties the user to the desk and although they can stay at work late this is not the same as bringing work home because you can use a laptop anywhere. However, this establishes a new pattern of working at home not just at work and computers, as everyone experiences, do not lead to a paperless office. It is claimed the average document gets physically copied 19 times (HalFILE, n.d.). Such home working often means that documents have to be kept at home as well as work, and these documents can also escalate as the rate of working goes up because of the use of computers. However, these documents have to be kept somewhere so that dwellings now need to have a home office space, houses need more floor area to accommodate this, but this makes them more expensive. Since now both partners go out to work to earn the money to pay for the larger dwelling, both partners need a home office, so dwellings become larger again. This is a simple example of the problem of diminishing returns as a result of a change in technology.

Changes in technology, like the horse collar, led to societies becoming more hierarchical and hence more complex. Aunger (2010) defined hierarchy in technology as "Production within a social system in which certain individuals have the power to enforce others to help produce artefacts", and gives the pyramids as an example for this. To an extent, most societies have moved on from the simple division into owners—with the power of enforcement—and slaves—with no powers to resist being enforced. However, in any large-scale project, there are always winners and losers. The construction of a new motorway is a typical example in the modern built environment. The land required for the motorway will already be in use, whether as a farm, forest or a reserve for bio-diversity but these uses, which might benefit the few, will be taken away from them and given to the new road, with the idea that the latter will be of benefit to the many—at least the many who drive vehicles, including those businesses who make money from driving vehicles on public roads. New motorways also serve people and this means they have to be connected into areas where people live. This in turn can see people lose their houses to feeder roads, or finish with a stream of noisy traffic behind a fence that is supposed to screen them from the disturbance. They lose so that the many can gain. Enforcement here comes from an elected government, which is supposed to somehow represent the wishes of disparate people, each with their own personal goals—or at least the wishes of a majority of the population. Unfortunately, many democracies do not work like this. As *The Economist* (2019) noted at the time the President of America was being threatened with impeachment, "A plurality of Americans—but not of states—want Donald Trump impeached", further noting that the senate had more than its fair share of rural and under populated states that were against impeachment. However, this is perhaps moving away from technology.

The motorway raises another issue about technology, which is the number of people involved in making a project happen. Unlike the village chairmaker who did everything in his business, from consulting with the client, finding the materials, making the chair and banking the money from a grateful client in a pot on the dresser, a motorway project involves not only those directly affected by it but also the drivers who are held up because of the new roadworks and then come the battery of people involved in its design, construction and financing, all of whom have to be in communication with each through some kind of hierarchy. Big projects lead to complexity. This was what was addressed by the alternative technology (AT) movement of the 1970s. Smith (2005) noted, "AT activists in a variety of industrialized countries called for technologies that would facilitate the radical transformation of industrial society: a transition to a more ecologically harmonious, socially convivial, and economically steady-state society." An alternative technology was thus much more than just the use of renewable energy systems, although these were part of it. It was also about simplifying the hierarchy that comes with developments in technology. This was an era where if you used technology you should be able to look after it yourself. This is perhaps best expressed in the foreword to a 1970 version of the *Whole Earth Catalogue* (Portola Institute, Inc., 1970: front cover).

> ...So far remotely done power and glory—as via government and big business, formal education, church—has succeeded to the point where gross defects obscure actual gains. In response of this dilemma and to these gains a realm of intimate, personal power is developing—power of the individual to conduct his own education, find his own inspiration, shape his own environment, and share his adventure with whoever is interested.

This in turn led to theories about self-sufficiency, whether for food or power. However, it did not take long to realise that people could not knit their own light bulbs and that the solar panels on the roof had to be made somewhere. Despite a situation that would have required a full revolution to change the means of production, the simple idea that what you had in your house you should at least be able to maintain and probably repair was an attempt to reduce complexity and at the same time increase resilience through spreading skills through a community rather than concentrating them in groups of experts. The book *Radical Technology* argued for "...the growth of small-scale techniques suitable for use by individuals and communities, in a wider social context of humanised production and workers' and consumers' control" (Boyle & Harper, 1976:5). This suggests that those writing about alternative technology understood it would have an effect on how communities were organised and hence on the built environment. This was not to be a world of mega cities, but rather one of those living on self-sufficient small holdings in the country to help repopulate rural areas (Harper, 1976:164), and autonomous terraces of modest houses in the town, where not only was solar energy collected but facilities like a laundry, bakehouse, library and pool were shared and food was grown in the collective gardens (Harper, 1976:168–169). Since the 1970s, there has been a huge change in the human population from 3.7 billion of which 37% were urbanised in 1970 to 7.8 billion with 56% urbanised in 2020 (Worldometer, n.d.), which makes the ideas of the AT movement now seem impractical. An increased urban population means trying to increase population density in the inner urban areas by building large, which in turn requires a more complex building industry. Building large increases construction costs (see Chap. 8) and reduces affordability. Displaced people move to the periphery of the city to live in smaller and cheaper buildings, very similar to those they were living in before these were destroyed to build the new large buildings. As a consequence, the urban area keeps growing at the expense of the rural periphery.

The 1970s and the AT movement were also a period that promoted self-building as a means to provide affordable housing, with examples like that of Lewisham Council who worked with architect Walter Segal in the 1980s on a self-build housing project (Grahame, 2015). Even then, self-building was not a new route to affordable housing. From the 1920s in Sweden people could build their own timber houses using pre-cut timber, prefabricated wall panels or by using a press to make concrete blocks so that "With this set up, an able bodied workman putting in his weekends, holidays and daily after-work hours in the long northern summer days, with the aid of family and friends, can build the greater part of his own house" (Gray, 1946:88). Given the earlier discussion of prefabrication in this chapter, this is a coming together of technology for a social purpose that again reveals it is not the technology that causes problems but the context in which it is used. At that time in Sweden standardisation and a degree of prefabrication reduced the cost of materials to the point where a family could afford a house by building it themselves. This situation is no longer possible in many developed countries because of the complexity of negotiating building and planning regulations and also the way in which money is lent to finance the buying of a dwelling. Building a house will require certain guarantees that self-builders, unless they are already part of the construction industry, may find hard to provide.

Progress

The essence of progress is that it implies a goal towards which one works. This implies that progress can be both good and bad, depending on the goal. Progress towards collapse might not be perceived as a good thing. The problem is that modern progress is generally viewed as something positive but without defining the goal that we are all progressing towards. In these terms moving towards a simpler life with less technology would be viewed as retrogressive. However, this can be turned on its head so that a future that was simpler in terms of the technology used, as advocated by those interested in alternative technology, could become a goal that we could all progress towards. This is different from the modern idea of progress where the undefined goal is in all probability the idea of everyone having more stuff or more money. However, in a finite system, this eventually leads to some having more at the expense of others who perforce have to have less. This is made visible by an examination of environmental indicators, such as GHG emissions and ecological footprint (see Chap. 8), which show that some people on the planet are emitting carbon or using resources far above their fair share for a finite world, while others are living equally far below their fair share of these things. Inequality and its effect on the built environment are also discussed at length in Chap. 6.

Technology is also bound up with progress since it is the technology that is held up as the hope for rescuing us from present problems. The Wyss Institute in the USA (Gordon, 2020) was set up to solve environmental problems through the development of new technologies. To tackle air pollution in urban areas, they have developed a new catalytic converter for petrol-fuelled vehicles. "Inspired by the nanoscale structure of a butterfly's wing, the researchers developed a honeycomb-like scaffold with catalyst nanoparticles precisely placed along the structure to maximize the area of catalyst exposed to exhaust." This begs the question of whether oil-powered vehicles are the best way of moving people around cities. The following case study explores the use of technology to tackle another effect of climate change—more water from floods and sea-level rise.

Part 2: A Case Study of Technology and Climate Change

Water, water, everywhere...
Samuel Taylor Coleridge: *The Rime of the Ancient Mariner*

Introduction

As we saw in Chap. 1, climate change is a major problem. Since problems in the past have been solved with the development of technologies, then it seems reasonable to assume that technology might again be the means through which climate change mitigation is achieved. Part 2 sets out to explore whether and how technology is being called on to deal with sea-level rise, increased heavy rain events and subsequent flooding. As seen in the discussion of floating cities in Chap. 2, one of the issues of dealing with more water is that both rich and poor are affected—the rich because they want to live on prime waterfront land for the view, and the poor because often the only land on which they can build for themselves is land that no one else wants because it floods. The newly wealthy families that formed as a result of the UK industrial revolution could afford to migrate to suburbs in the south-west of the cities to avoid pollution from the factory chimneys blowing over their heads, so it is the poor who have to put up with living in less than ideal circumstances (see also Chap. 6). If you have little money you have to live on the cheapest areas of land and this is often land that is polluted or prone to natural disasters, such as flooding.

A World Bank Report (Hallegatte et al., 2016:7) found that although river floods as a result of changes in the climate would affect both rich and poor, in general, urban dwellers had higher exposure to the risk of flooding. The Intergovernmental Panel on Climate Change (IPCC) (2001:8) recognise that poorer nations will be affected more by climate change stating, "...the least developed countries, are generally poorest in this regard. As a result, they have lesser capacity to adapt and are more vulnerable to climate change damages, just as they are more vulnerable to other stresses. This condition is most extreme among the poorest people." The same report estimates annually 200 million people would be affected by flooding from coastal storm surges for the mid-range scenario of a 400 mm rise in sea level by the 2080s compared with no sea-level rise (IPCC, 2001:38). A more recent estimation by Mcgranahan et al. (2007) states that although what they define as the low elevation coastal zone (the land at risk of flooding from climate change) forms only 2% of global land area, it contains 10% of the global population, or in 2007 some 660 million people, of which 75% are in Asia. Flooding damages the physical environment and increases the risk of respiratory diseases and those linked to diarrhoea for everyone, while in developing countries it also leads to hunger and malnutrition (IPCC, 2001:12). It is worth remembering that reports from the IPCC represent what everyone can agree on and could, therefore, be considered conservative (Herrando-Pérez et al., 2019).

Climate change thus brings potential disasters nearer to everyone, whether rich or poor. It is no accident that some of the most valuable real estate in the developed world can be found along both coastal and river margins, for the sake of both the views and access to a watery playground. An old study (Benson et al., 1998) suggested a high-quality ocean view could add nearly 60% to the value of comparable houses. A later study of Auckland residential properties (Bourassa et al., 2004) also found that great water views could add 59% to the value of the property but the advantages decreased with distance from the coast. Climate change could undermine this but as McNamara et al. (2015) note, "Rising sea-levels and increased storminess threaten to accelerate coastal erosion, while growing demand for coastal real estate encourages more spending to hold back the sea in spite of the shrinking federal budget for beach nourishment." Thus, rather than retreating from the coast, there is a sense that technology should somehow protect people from sea-level rise so that nothing changes, and coastal properties continue to enjoy a real estate premium.

Risk

The issue, however, is not as straightforward as this might sound. Risk, or rather perceived risk, depends on where you live and what you believe. The district known as the Kapiti Coast lies near the bottom of the west side of the North Island of New Zealand. Most of this district is a coastal plain (Figs. 5.3 and 5.4). The website of the Kapiti Coast District Council (2019) in New Zealand states that "Floods are a common hazard in Kapiti." The updated website has a series of maps showing where flooding might occur around the water courses that drain into the sea, the location of low-lying areas of land, and which properties would be affected (Kapiti Coast District Council, 2020a). However, the question of what might be affected by sea-level rise remains a matter of debate, with an interested person having to read the long list of reports and submissions available on the website (Kapiti Coast District Council, 2020b). This may be because of the contested 2012 report on coastal hazards in the area (Kapiti Coast District Council, 2012) which stated 1000 dwellings would be at risk in 50 years from storm erosion and sea-level rise due to climate change, with the figure rising to 1800 in 100 years. This represents nearly 5% of all occupied dwellings in the Kapiti Coast District (Statistics New Zealand, n.d.). This Kapiti council report was later questioned on the basis of its science (De Lange, 2019).

> The study includes an analysis of historical shoreline trends, and found that the Kapiti Coast has undergone long-term accretion with episodic erosion associated with storm events; a finding consistent with several earlier studies and the evidence provided by coastal landforms. It is clear from the available evidence that this long-term accretion occurred while sea level was rising.

The story may not be that simple, however, as it depends on which part of the coast is under consideration. A thesis from 1972 (Gibbard, 1972:58) states that the shallowest gradient beach profiles backed by sand dunes were found between the

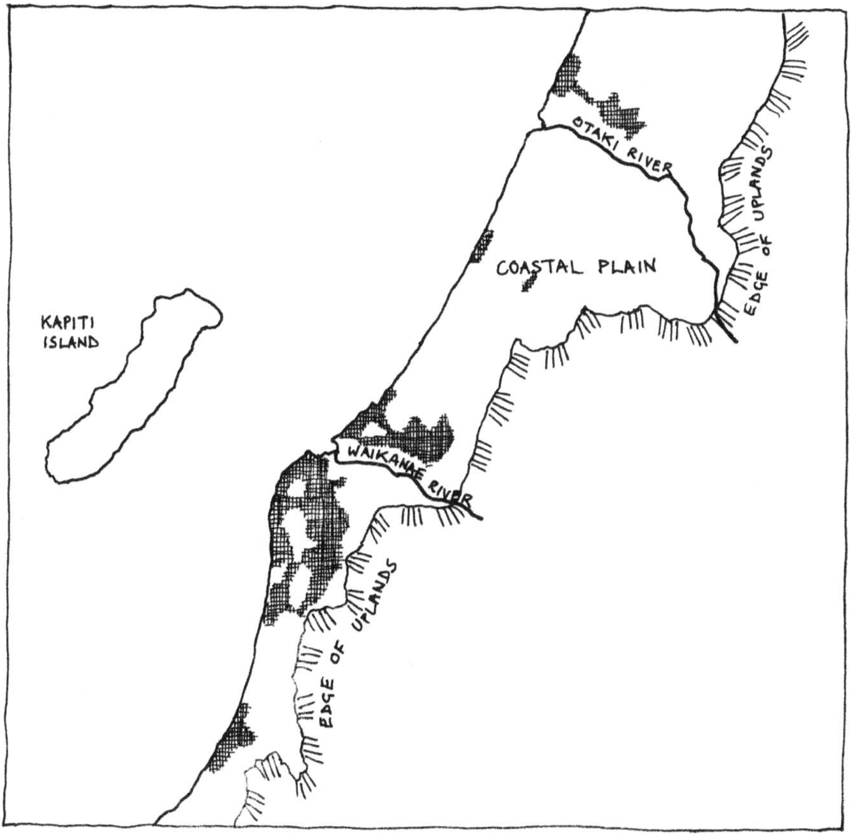

Fig. 5.3 Kapiti Coast map. Hatched areas indicate major settlement (adapted from https://www.gns.cri.nz/Home/Our-Science/Land-and-Marine-Geoscience/Regional-Geology/Urban-Geological-Mapping2/Kapiti-Coast)

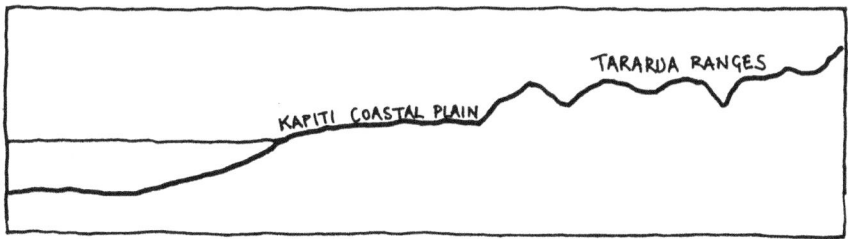

Fig. 5.4 Kapiti coast section (adapted from Nolan, 2017:11)

mouths of the Manawatu and Ohau rivers, in the north of the Kapiti Coast District. These beaches were then undergoing accretion and so increase in size, but north and south of this zone gradients were steeper and the base of the dunes showed signs of cliffing, leading to shore line erosion, while those beaches facing the southern end of Kaptiti Island at Paekakariki had almost no foreshore and the dune cliffs were of significant height, again suggesting the coastline was being eroded. Obviously, this thesis is historic but does demonstrate the problem that while we argue over the details of what and what might not happen, climate change and its consequences are not going away.

In the event, having a thousand dwellings at risk in 50 years' time does not seem a big issue given that under the New Zealand building code houses are assumed to have a 50-year life. All this would suggest is that no further housing should be built in coastal areas that are perceived to be at risk. However, once the spectre of risk is raised it seems that property values could fall and insurance becomes harder to get. The unforeseen risk is that legal battles and more stress for the victims of an extreme event may ensue over what is and what is not insured in the face of a natural disaster. After Hurricane Katrina there was an argument over wind damage, which would have been covered by homeowners' insurance, and damage caused by surface water flooding, which would not (Kunreuther & Michel-Kerjan, 2009:26). In March 2017 in New Zealand, the then Minister Nick Smith, in relation to the delay in releasing a National Ministry of the Environment report on coastal hazards, was said to be "…worried about the economic implications, including costs to property owners if insurance or value was affected. He decided, over strenuous objections from his ministry, that the guidance needed to go through Cabinet" (Gibson, 2017). The report was eventually published in December 2017 after there had been a change in government (from centre-right to centre-left with the support of the supposedly environmentally focused Green Party). The report was commissioned to offer guidance to local councils and so circumvent the problems the Kapiti Coast District Council had encountered in issuing its own report. The report contains a number of fact sheets including those on the effect of climate change on coastal erosion, on coastal flooding as a result of storms and sea-level rise, and on sea-level rise alone (Ministry for the Environment, 2017a).

When it comes to sea-level rise generally, the IPCC scenario predictions are given as 0.2–0.4 m by 2060 with the note that local conditions will vary. A warning note is given that "In the more distant future, it is virtually certain that sea-level rise will continue for many centuries, well beyond 2100, as rising temperatures warm the oceans and make them expand" (Ministry for the Environment, 2017b). Thus, any move to cope with sea-level rise in a "graceful" manner is something for the future, despite the fact alterations to coastlines through development are acknowledged as having an effect on coastal flooding as a result of storms (Ministry for the Environment, 2017c). The same document also suggests not allowing development in coastal areas prone to flooding, which is exactly what local authorities like Kapiti Coast District Council were trying to do when outlining the perceived danger areas in their 2012 report. The Greater Wellington Regional Council (GWRC), have now voted to impose a buffer zone along the edge of a coastal park in Kapiti, a proposal which has gone out for public consultation. The GWRC stated, "Our policy of managed coastal retreat is a taste of things to come as we consider the implications of regional coastal adaptation in the face of climate change" (GWRC, 2019). At least this is a toe in (or maybe out of) the water. Unfortunately, accurate predictions rely on good science, which takes time, and storms sadly do not wait. Dwellings in unidentified flood-prone areas that experience an extreme event will then suffer the probable problem of increased insurance costs and loss of property values, but owners do not wish to confront this by having their dwelling identified as being at risk, and so the cycle starts again.

Ultimately, the issue is one of how risks are perceived, so you take a chance when deciding what you feel about risk. A coastal study of the Hawkes Bay in New Zealand felt the main risk to property and loss of life was a tsunami as the result of an earthquake, something that is unpredictable; "In terms of relative scale the tsunami hazard has a significantly greater exposure than erosion and coastal inundation" (Napier City Council et al., 2016:11). However, this emphasis on the local is supported by research that shows climate change initiatives need to be related to local circumstances. In a study of data from 2007 to 2008, Lee et al. (2015) found that awareness of climate change was variable, with for example 65% of those sampled in India not being aware of it. They also found that in Europe and Latin American understanding how people cause climate change was the strongest predictor of people perceiving the risks from climate change, whereas in Africa the strongest predictor was understanding its link to local temperature changes. They also speculated that, based on their analysis, education was the key to being aware of climate change and understanding that climate change is human-induced was key to perceiving the potential risks it poses. A similar situation has arisen with the COVID-19 pandemic (see Chap. 9) where not everyone is taking the risk seriously, in part perhaps due to optimism bias, which is "…a general tendency that we all have whereby we tend to underestimate personal risks" (Pierre, 2020).

However, the Lee et al. (2015) study of attitudes to climate change was based on old data, and a more recent longitudinal study in New Zealand found that over six years there was an "…observed steady increase in climate change belief," although more people believed that climate change was real than that it was caused by human actions (Milfont et al., 2017). When it comes to flooding another New Zealand study found that proximity to the coast led to increased concern about the effects of climate change and a greater willingness that government should do something about it (Milfont et al., 2014). From this, a paradox emerges. If you live near the coast it seems you are aware you could be affected by climate change. However, if your house is identified on a map as being at risk this leads to resistance to accepting this knowledge because of the potential effect on the value of most people's greatest asset—their house—and the cost of insuring it. Although most people in developed countries have lifestyles that contribute to climate change, there seems to be a reluctance to grasp that it will affect you directly. There is evidence that repeated flooding does lead to relocation, though people are more ready to do this if there are supportive schemes to make the transition relatively easy (Bukvic et al., 2018). The only other options to relocate in the face of the risk of flooding are taking measures to prevent it and learning to live with its effects. The remainder of this chapter will look at each of these issues in turn before considering the relationship of each to collapse.

Flood Prevention

The obvious thing to do if you are under threat of flooding or sea surges is to take steps to prevent it. Historically, towns and fortifications have built walls to keep those inside safe from threats from outside. The history of building walls to keep out the sea is also long. Some 2000 years ago the Romans built sea walls to form harbours, and a recent discovery has found that rather than being damaged by the waves over the years the type of concrete used, by interacting with the waves, has made the walls harder as they age (Jackson et al., 2017).

The downside of building walls to keep out water is that they cost money and more money has to be spent on their regular maintenance to repair wave damage unless, like the Romans, a technology can be found that makes sea walls stronger with age. Even when such steps are taken, safety from flooding is not guaranteed. Brisbane in Queensland, Australia was a city with a history of flooding from the Brisbane River. In 1974, cyclone Wanda caused severe inundation in the north of the city. This led to the construction of the Wivenhoe Dam upstream of the city. However, this investment ameliorated but did not prevent, the January 2011 floods (Smith & McAlpine, 2014). There is thus no guarantee that spending money on flood prevention will always keep you dry. The Thames Barrier in London, built as a result of the east coast and Thames estuary floods in 1953 (Met Office, 2017), has still to be tested.

In the face of this, the Dutch have a long tradition of keeping water out, which is essential given approximately one-third of their country lies below sea level (Netherlands Tourism, 2019). Early dikes were no more than piles of peat but from 1500 onwards the Netherlands witnessed a period of prosperity and many dikes of wood were constructed, only in 1730 to fall foul of the teredo worm which ate them away. From then on dikes were reinforced with stone and later with concrete (Pleijster et al., 2014:106–131). Modern Dutch sea defences are far more sophisticated, such as the Oosterscheldekering storm surge barrier (Fagan, 2013:219), which after a disastrous flood in 1953, was built to protect the delta region from flooding from the North Sea. This dam is part of a long defensive wall, parts of which can be opened to allow for salt water fishing or closed when floods threaten. The original plan had been for a solid dam with the lake behind reverting to fresh water. Public pressure to maintain the salt water estuary meant that a far more complex and expensive dam had to be devised (Goemans & Visser, 1987). Even when flood defences seem an obvious solution, they are not always welcome.

Tai O on the eastern side of Lantau Island, Hong Kong offers a smaller example of how sea defences are not always all good. This is a traditional fishing village where houses were built on stilts above the sea water which has become a tourist attraction (Hong Kong Tourism Board, 2017) (Fig. 5.5).

Fig. 5.5 Tai O stilt houses (adapted from https://en.wikipedia.org/wiki/Wikipedia:Featured_picture_candidates/Tai_O#/media/File:1_tai_o_hong_kong_2013.jpg)

The old sea wall that protected the village had earlier fallen into disrepair and in 1989 a flood destroyed half of the village (Loh and Civic Exchange, 2002:3–4). Since then various storm surges have flooded the small town. A new sea wall has been constructed to preserve the town centre, but only from the annual flooding rather than storm surges, and these may well be more severe because of climate change (Webster et al., 2005; Zhang et al., 2011). The wall was to have been taller but residents objected because they felt it would interfere with tourism, an important source of livelihood now fishing is almost non-existent (Chan et al., 2013; Dryland & Syed, 2010). However, the village remains under threat from climate change linked sea-level rise and it seems inevitable that Tai O, with its elderly population and very high insurance premiums following recent storm surge floods (Chan et al., 2013), will not survive in its present vernacular form.

Sea walls are expensive items and there is always the problem of what happens when a sea wall ends and the natural coastline takes over, as erosion can happen at different rates. Using sea walls as a defence against sea-level rise will depend on economies generating sufficient surplus to afford them, while concurrently it is the growth in economies that is leading to environmental degradation and climate change (Clement, 2010). This problem of available resources also applies to river and estuarine defences and it is not a new issue. The Thames estuary has always been subject to flooding and its sea walls and other defences date back to the Middle Ages. However, the 1349 Black Death (for more discussion, see Chap. 9) produced labour shortages leading to the abandonment of some of the defences and people relocating further inland, in what would now be called a managed retreat (Galloway & Potts, 2007).

In part actions such as these have led to the idea that rather than keeping water out it might be possible to change the landscape so that flooding is controlled. Historically, earth banks were used to keep the river in its channel and often similar banks were also used around the adjacent agricultural land, so if the river banks were overtopped the flood water would be contained, preventing damage further downstream (Rose, 2014). However, these past lessons are often ignored. In a study of the Vistula Basin in Poland, Wyzga et al. (2018) point out that channelling and cutting of the meanders in the river in an effort to move the flood water downstream fast, thus reducing the flood plain, could produce catastrophic results downstream through embankment failures. Rather than reducing the flood plain, one modern idea is to lower the flood plain so as to make room for more water when rivers flood (Rijkswaterstaat, n.d.). Following the 1993 and 1995 floods in the Netherlands, the dikes were again reinforced but a new policy document appeared entitled *Ruimte voor de Rivier* (Space for the River) which suggested a number of ways of widening the riverbed from lowering the flood plain to removing obstacles, such as bridges (Ruimte voor de Rivier, n.d.).

This seems like a return to the idea of keeping space by the river that will flood at times, as exemplified by the traditional flooding of water meadows, or irrigated grass pasture, that has been around in southern England since 1600 AD (Cook et al., 2015). Water meadows are not the same as flood plains. They were part of farms and were deliberately flooded with a movement of water across them as the aim was "…to force early growth of grass in the spring, to improve the quality of the grass sward and to increase the summer hay crop" although they can also "contain flood water, trap silt and help to reduce the nutrient load in water returned to rivers" (Historic England, 2017:2, 3). A flood plain is a flat area of land next to a river that will flood during periods of high rainfall as river levels rise. Emergency flood plains also occur where land is deliberately flooded to avoid floods downstream. This was recently the case in New Zealand, where in July 2017 dairy farms were flooded to avoid the water reaching settlements. This is in an area where the danger of flooding has been "…modified (but not eliminated) through engineering works over the past 150 years" (Otago Regional Council, 2012). The problem is that the necessarily flat flood plains make very attractive flat areas for building and as populations increase it is conveniently easy to forget why this land was originally set aside and not built on. Despite the UK plans to limit building on flood plains "…each year 1,500 new homes are built in areas of high flood risk and 3,100 homes per year in areas of medium flood risk" (Committee on Climate Change, 2015:10).

Creating flood defences also means the land is then open to development. This is a current problem in the city of Jakarta, which is sinking, because the lack of a proper water supply means people are extracting water from beneath the city. Although the latest idea is to move the capital city to Borneo, an earlier plan was to build a new sea wall, named the Giant Garuda, with Dutch help to protect the low-lying land. However, as Dawson (2017:47) comments the outcome is not good for all as the city is "…pursued by a consortium of global experts and local developers who use the extreme plight of sinking cities like Jakarta to justify violent land grabs, further displacing the poor majority of urban citizens." If you are wealthy then sea and river defences seem like a good idea but great care needs to be taken so that they do not displace and disadvantage the poor, and also to ensure that the economic benefits that come with them are channelled into combating climate change and not adding to it.

Living with the Effects of Coastal Erosion and Flooding

The village of Happisburgh (perversely pronounced "Hays Borough") in Norfolk in the UK (Fig. 5.6) is being left to fall into the sea as it is not cost-effective to put in defences against the erosion of the soft glacial till cliffs. There were coastal defences in the past but these were not replaced after a major event in the 1950s. As a consequence, houses and other structures, including the lifeboat station, have been lost as the cliffs battered by storms give way. The local pressure group would like to see these defences replaced and also blame off-shore dredging as a contributing cause of the recent rapid erosion. In opposition are the national and local government who take what Tebboth (2014) calls a hierarchical worldview, in that the erosion should be left to take care of itself, and that sea defences would not be cost-effective. This has led to a policy of managed realignment, which means some attempts have been made to slow erosion but ultimately it will be a case of relocation as the sea claims the buildings (Beeler, 2018). This is an example of an inevitable but slow collapse at the local scale. Ultimately this village will be lost, as others have before. The reasons for the collapse are not that it is impossible to avoid but that it is not cost-effective to avoid it, so although the erosion caused by storms coming off the North Sea is the environmental reason for the predicament, the collapse will come about because of socio-political decisions. This illustrates the point that collapse is a socio-political process. It also echoes the approach of Pascal's wager (see Chap. 1) in that if people

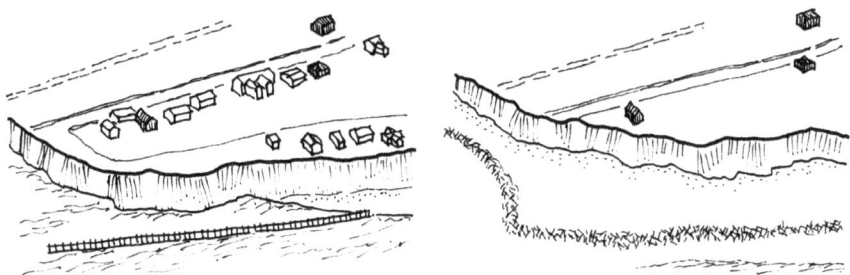

Fig. 5.6 Happisburgh in 2001 (left) and 2014 (right). The shaded buildings are the same in both images. The groin (left) has gone to be replaced by a different sea defence (right) (adapted from Mike Page Aerial Photography) (https://www.pri.org/stories/2018-04-05/british-village-crumbles-sea-family-holds-home-cant-be-saved)

stay living close to the danger area it will cost money and there will still be a risk. Moving people inland will ultimately cost less and save lives, so it is the path for a graceful collapse in that it is buildings that will be lost, not people.

Collapse is also about opportunity as the lost Suffolk town of Dunwich demonstrates (Fig. 5.7). This medieval port, once the largest in East Anglia, has all but been lost, a process that started with two severe storms, the first in 1286 and the second in 1328. The first filled the harbour with silt and the second blocked it, which affected the economy of a town that relied on trading. Coastal erosion was the other problem that was evident even before these storms and that eventually claimed most of the original settlement.

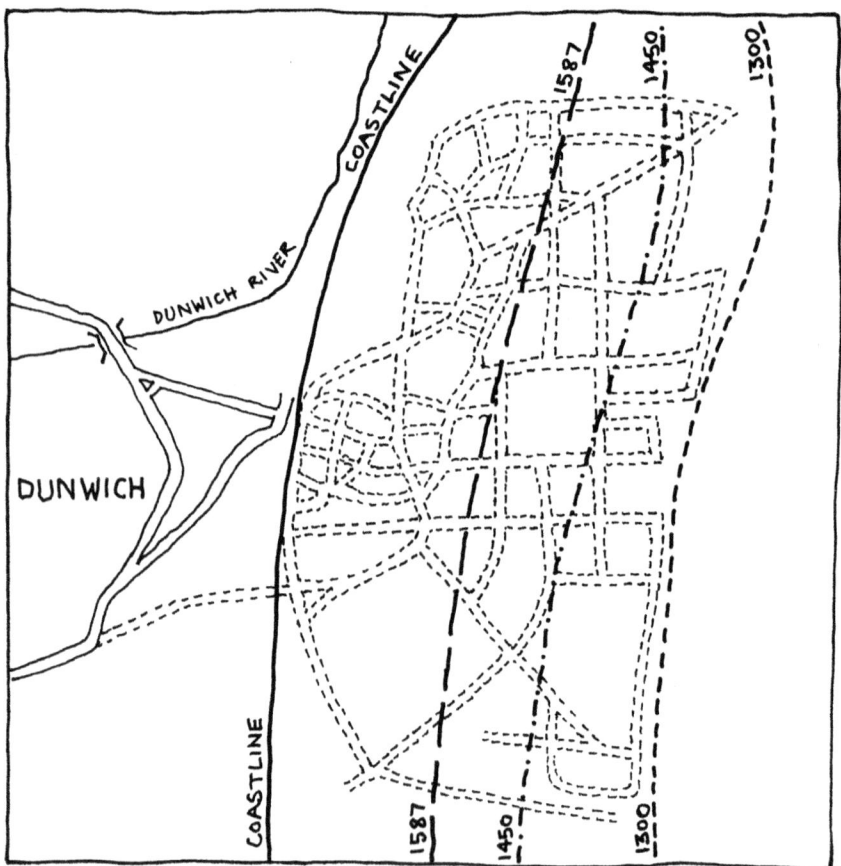

Fig. 5.7 The receding coastline at Dunwich: The dashed road layout shows what has been lost: The large central square was the marketplace (adapted from https://flickeringlamps.com/2016/06/12/the-last-ruins-of-dunwich-suffolks-lost-medieval-town/)

The village now has a very small permanent population of 100, mostly older people (Morris, 2014). It has also made itself into a tourist attraction with a museum about the much larger lost town, just because it is possible to walk along the beach and pick up bones and other remnants from the cemetery and buildings under the sea (Atlas Obscura, 2019).

However, in the UK for those whose coastal houses and farmlands are lost to the sea under a policy of "managed realignment" (Environment Agency, 2010) life will change. The question then becomes one of when this will happen and it seems nobody wants it to happen immediately. In another UK example, that of the harbour at Mullion Cove in Cornwall which is threatened by storms and sea-level rise, a compromise position was reached whereby the harbour would be repaired in the immediate future but would eventually be left at the mercy of the sea so that "…at an unpredictable date in the near or distant future, the cove will once again look like it did in 1890 [before the harbour had been built]" (DeSilvey, 2012). This, however, is delaying the problem rather than collapsing gracefully, the essence of which is to plan what to do and when in the face of erosion caused by the sea. Relocation in the face of such a threat would be the carefully planned retreat to a place of safety, through accepting that change must happen. The issue is whether waiting will mean the retreat will not be as carefully planned as it should be.

This is what happened to communities after the 2004 Boxing Day tsunami. After the tsunami, traditional fishing communities in Sri Lanka were forced to relocate because the government initiated a coastal no-build zone, and "Some 50,000 families were relocated inland, sometimes by as much as 10 km, and settled into new housing developments with no public amenities. Isolated by distance and lack of private and public transport not only from livelihood and food sources but also from traditional social environments" (de Silva et al., 2017). Formerly much housing in Sri Lanka had been built with local community participation through the National Housing Authority, thus ensuring the housing suited people's traditions and needs, but after the tsunami the government stepped in to build 120,000 new houses (Schilderman & Lyons, 2011). This was because donor support favoured supplying complete houses rather than materials and labour, with the consequence that the houses did not match the cultural needs of the displaced communities, at the same time as these communities had lost their livelihood because of being relocated far from the coast (de Silva et al., 2017). This could be viewed as a cure as bad as, if not worse, than the disease.

Living with Regular Inundation

The River Nile allowed human habitation in a desert and the annual floods that came from the monsoon rain in its upper reaches were the means of adding both river silt and water to the land to maintain its fertility. In their turn, the Ancient Egyptians learned how to channel this water for irrigation so as to make the best use of it. Floods can be both disruptive and essential. In contrast, Pliny the Elder describes the poor people of Friesland dwelling on the mounds of earth (terps) they had built themselves, with sea and river water lapping around them and wondering why a Frieslander would prefer that life to being a Roman slave (Pleijster et al., 2014:107). The Friesland dwellers of that time had learned to live with water and preferred to do this as a free people rather than live on dry land in slavery. The Ancient Egyptians are not the only society to develop a life around regular flooding. People, despite what Pliny the Elder might think, adapt to their conditions if they have no alternatives. However, if and when the systems that societies have built themselves around are disrupted by unwanted water, then problems ensue.

The Amazon River, as might be expected, has the world's largest floodplain along its edges, forming areas that were, and still are, important for agriculture. "Cultivated since pre-history, the active bars, levees, back swamps and terraces of the Andean tributaries and main stem of the Amazon river are sown in commercial and subsistence crops by farmers living in floodplain and river bluff communities" (Coomes et al., 2015). Crops are planted as the flood recedes but there are sudden natural reversals in flood water which can ruin newly planted crops. During the flood season farmers will watch the river levels looking for signs. However, the Amazon is now flooding more frequently due to "anthropogenic and natural factors" (Barichivich et al., 2018), which makes precarious living even more precarious. These same subsistence farmers are not the ones who are responsible for the anthropogenic changes to the earth's climate which are causing the increased flooding. The situation may not be helped because, as a result of democratic elections, "Brazil's president-elect Jair Bolsonaro has chosen a new foreign minister who believes climate change is part of a plot by "cultural Marxists" to stifle western economies and promote the growth of China" (Watts, 2018).

Reliance on regular flooding can also be found in the Tonle Sap community in Cambodia (Fig. 5.8). Tonle Sap is a lake that expands annually with the flow of the Mekong River, which has its origins in Northern China, and flows through Laos, Thailand, Cambodia and Vietnam before reaching the sea (Kummu & Sarkkula, 2008). The water levels of the lake rise from a low of around 1.5 m to a high of 9 m, with a consequent expansion in the lake area from around 2600 km^2 to 15,000 km^2 (Arias et al., 2014). This annual flood is essential for livelihoods in regions around the Tonle Sap and the lower reaches of the Mekong River. The flooding of the Tonle Sap Lake is caused by a phenomenon known as flow reversal (Arias et al., 2014). As the water in the Mekong increases substantially from the annual monsoon rains and ice melting in the Himalayas, rather than water flowing from the lake into the river, the flow reverses bringing the lake water back together with some 60% of the water flowing down the Mekong (Kummu & Sarkkula, 2008). These waters are rich in sediment which is deposited as the water retreats, and this is important both for biodiversity (Kummu & Sarkkula, 2008) and Cambodia's fish production (Evans et al., 2004:iii).

The settlement around this fertile area happened centuries ago and gave rise to both floating and elevated buildings (Evans et al., 2004:7; Grundy-Warr & Sithirith, 2016) (Figs. 5.9 and 5.10). The floating architecture consists of houses, shop boats and other structures on rafts, including aquaculture pens under floating homes and sheltered pens for livestock built on rafts. All structures move with the waters of the

Flood Prevention

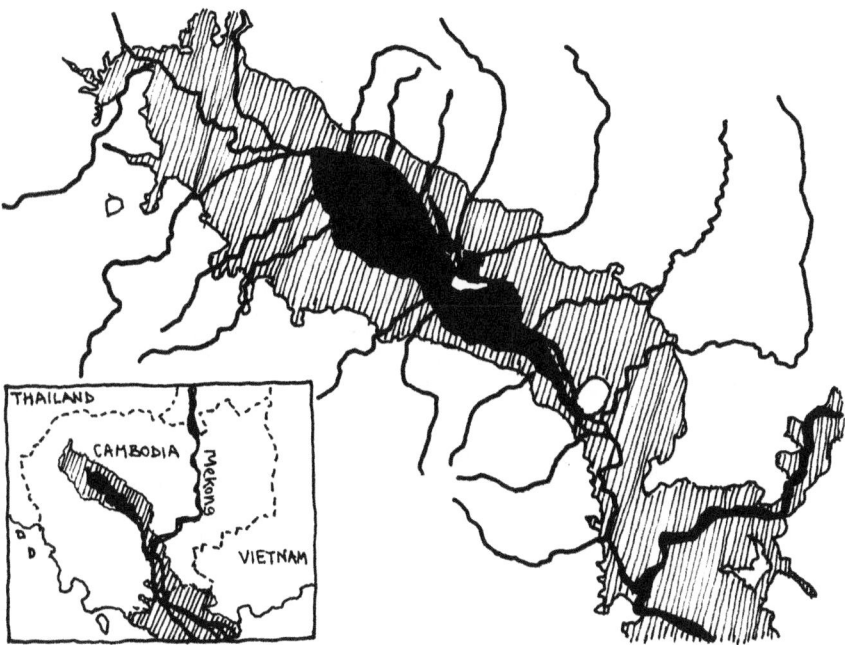

Fig. 5.8 Tonle Sap lake; the lake is black and the shaded area is the flood plain (adapted from https://en.wikipedia.org/wiki/Tonl%C3%A9_Sap_Biosphere_Reserve)

Fig. 5.9 House on stilts at Tonle Sap lake (adapted from author's photograph)

Fig. 5.10 Floating house on Tonle Sap lake (adapted from author's photograph)

lake, whereas the stilt structures rise 6–8 m above ground level and are found along the feeder rivers (Au Morris, 2014). The community is thus designed to deal with differing water levels, and as the levels rise boats become the means of transport. The structures are made of timber, bamboo, palm thatch and other materials, and are designed not only to deal with the lake but also to cope with tropical storms.

However, since the 1990s this traditional way of living with flooding has become increasingly under threat from modernisation, industrialisation and growth in settlements within the Mekong river basin (Dugan et al., 2010; Evans et al., 2004:iii; Kummu & Sarkkula, 2008). The largest threats that have emerged in the last 30 years stem from the development activities taking place in the upper reaches of the Mekong, particularly in China. The numerous large-scale hydropower dams, reservoirs and irrigation schemes along the upper reaches of the Mekong not only cause flow alterations to the river and increase the silt it carries but also affect the flood pulse of the Tonle Sap. This increases the Tonle Sap Lake's dry season water levels (Kummu & Sarkkula, 2008) resulting in shrinking of the land-based dry season food production areas and thereby loss of food for the lake communities (Evans et al., 2004:iii). In addition to the hydropower projects, industrial and agricultural pollutants and human waste enter Tonle Sap and pollute its waters, causing damage to fish populations and hence the livelihoods of the communities (Dugan et al., 2010; Kummu & Sarkkula, 2008). Although the traditional communities of the Tonle Sap were able to devise a way of living with flooding, the interventions of modern societies in the search for cleaner energy through investment in hydropower now threaten the resilience of the community and what had hitherto been the sustainable, if watery, way of life. Unlike the Amazon example where cause and effect are widely separated, here the cause and effect are much more visible.

Developed world solutions to living with flooding are somewhat different from the communities of Tonle Sap, with a focus on keeping the water out at a small scale or using materials that will not be damaged by the floodwater. Examples range from the installation of concrete walls and rubber-sealed flood gates to houses along the banks of the River Ouse in York in the UK to bolting down manhole covers (Dhonau et al., 2017). Other advice is to adapt to flooding by not storing valuables or having expensive furnishings at a low level (Osberghaus, 2015), and this behavioural approach comes closest to that of less well-off communities that have traditionally learned to put up with flooding that does not bring the benefits of soil enrichment as in the examples described above. If you are a migrant or part of the urban poor you will look for cheap accommodation and this is generally to be found on marginal land that is not thought good enough for permanent settlement. An investigation of slum areas in Mumbai found that after a flood in 2005, 72% of households surveyed had constructed some sort of elevated platform in the house to protect family treasures, approximately 53% raised their floor level by filling it with broken bricks and other materials, and a surprising 22% constructed a second floor (some had done more than one thing) (Chatterjee, 2009). Obviously, what was done depended on what the family could afford, but such actions also demonstrate that those living in these areas of Mumbai anticipated further flooding and were finding ways to live with this idea, however unpleasant the experience might be.

In the Indonesian port city of Semarang living with flooding has become a way of life. Since 1957 it has suffered from tidal flooding caused by land subsidence that has been made worse by sea-level rise as a result of climate change. In 2015, the land was estimated to be sinking annually by 130 mm each year (Semarang City Government, 2016:37, 56), as a result of soil consolidation because of the weight of the roads and buildings being constructed with continuing urbanisation, together with increasing extraction of groundwater (Abidin et al., 2012, p. 71). This local seawater flood, called the rob, affects nearly half of the city, or nearly four million hectares (Hadi, 2017:1). Apart from the physical damage caused by the water the rob affects the septic tanks used by households for sewage treatment with consequences for health (Hadi, 2017:2). Research carried out by Harwitasari and Van Ast (2011) in some of the most flood-prone areas of Semarang found 50% of the residents of these areas experienced floods of 500 mm to 1 m between four and nine times a month, with 10% suffering daily. In spite of this discomfort, people stayed not only because they could not afford to move but also because of being near to local services, being near the city centre, employment and ancestral property ownership. Local social contacts were also found to be very important (Brus, 2012:37). This means people have had to find ways of living with water. A study of Genuk District, one of the areas regularly flooded, found that the economic benefits of staying outweighed the problems caused by flooding. People built up the level of the ground around the house then raised the floor of the living room and ultimately built a second storey on the house (Khadiyanta & Dewantari, 2016). Another study of six flood-prone districts found that the most common strategy was to raise the house higher above the ground, and that at the neighbourhood scale it was common to increase the height of the streets (Harwitasari, 2009).

The example of the Semarang rob shows how humanity and the quest for urban development have exacerbated floods. On a smaller scale, in the developed world, impermeable surfaces are also causing more people to live in small temporarily flooded areas as rainwater cannot drain away at a sufficient rate. There was a 22% increase in impermeable surfaces in the UK between 2001 and 2019 (Committee on Climate Change, 2019:13). Another study (Kelly, 2016) reported, "Large areas of front gardens have already been lost…nearly half of all front gardens in North-East England, and just under a third in Scotland, South-West England, and Eastern England, are at least 75% paved over." The paving is only done to provide off-street parking for cars and the reason people have cars is that they cannot, or more likely choose not to, walk, cycle or use public transport to get to work and school, as would have happened in the middle of the twentieth century. This is an example of solving one problem—where to put the car—that leads to another problem—surface water flooding. The real issue is that surface water flooding may not affect those who park their cars on their paved front gardens. A study of Manchester found that yet again "…for poor and diverse communities vulnerability coincides with the occurrence of the hazard of surface water flooding" (Kaźmierczak & Canvan, 2011).

Pipe Dreams

If one problem always leads to another, will some of the proposals put forward for dealing with the increased risk of flooding be true solutions? From floating houses to floating cities the simple solution would seem to be taking to the water rather than trying to keep it out. Rather like air, water seems something limitless that is there to be exploited. Who owns the sea? At the moment it could be the person whose house is floating on it. In a discussion of houses built on a floating concrete raft filled with Styrofoam, the president of International Flotation Systems Inc. stated, "Mortgage companies now see the surface of water as real-estate and insurance companies will now cover floating homes because of the advances in floating platforms" (Build LLC, 2008).

In her book on the changing cultural attitudes to the canals of Amsterdam, Kinder (2015:14) described how in the 1960s and 1970s water was perceived as a "…residual space available for informal reappropriation", in this instance being appropriated by "…squatters [who] built homes on outdated industrial barges and, turning regulatory structures to new ends, transformed mooring spaces into residential addresses." However, once the value of water is appreciated, whether for tourism purposes or as a chance to enhance real estate prices, squatters are moved out and developers move in. This links back to the point that such development and consequent boosts to the local economy inevitably, at least at the moment in the paradigm of economic growth, lead to increased GHG emissions, and hence increased risk of flooding from storm surge and sea-level rise, which could then imperil these very developments.

The counter-argument to this situation is to avoid the land on the coastline and float the buildings on the water. In Maasbommel, again in the Netherlands, along the banks of the River Meuse houses have been built with hollow concrete bases with a foam core. Most of the time they sit on piles but if and when the water levels rise they can float, while still being anchored to the piles and linked to the bank (Beizer, 2006). However, the developer DuraVermeer notes these "…houses are not exactly cheap" (Inhabitat, n.d.). Floating houses still need infrastructure and in this case services to the houses need to be in flexible linkages to allow for the rise and fall. The houses also need an access road and are thus conventional in everything except the technology of their foundations, service connections and drains. As Steinberg (2010) points out, as a result, their social impact has been minimal. Steinberg also discusses *The World*, the only residential cruise ship, that is home to 165 people (The World Residence Holdings Ltd., 2019). Such a ship cannot exist, however, without the support structure to provision it and dispose of its wastes, and for this, it has to make use of the infrastructure of the countries it visits. The same problems of food supply and waste disposal will exist for floating city projects and the problems will be more difficult for a city on water.

Despite such obvious drawbacks, the United Nations has given its support to the concept (Pollock, 2019), with the idea of building a water-based city close to the shore that would house 10,000 residents, and which would be autonomous for energy and water with food being farmed in cages under the sea. Assuming the residents are prepared to eat only fish and seaweed as it might be difficult to grow fruit, grain, vegetables and dairy produce under water, such a city would still not be cheap and 10,000 residents look more like a large village than a modern city. Would such a concept ease housing shortages, especially for the urban poor and disadvantaged, or are they just dreams for the rich? An example from Nigeria illustrates this problem, where on the edge of a lagoon "Several hundred thousand people live in a maze of tethered boats and rafts in the Makoko section of Lagos, Nigeria" (Revkin, 2019), described as "…a settlement where most inhabitants live in unhygienic and blighted conditions" (Olumuyiwa et al., 2019) (see also Chap. 2). However, the settlement has developed its own economy based on "…fishing, trading, sand mining, and low-rank civil service employment". The government wants to displace the current citizens on the grounds that the area is unhygienic, and develop the waterfront for recreation and better navigation (Olumuyiwa et al., 2019). As the current residents do not own the water and have no rights to use it, they can simply be evicted but with no plans as to where else they might go, or how a local economy might survive such an eviction. Floating cities exist but not in the form that would make them acceptable to those in the developed world. However, where people build their own versions of a floating city because this is all they can do, this attracts the attention of others who see that there might be a development opportunity if they evict the residents. The development will again lead to economic benefits for some that will subsequently fuel the growth in GHG emissions. Solving one problem just leads to another.

To link these ideas with theories of collapse (see Chap. 3), it seems that segregating poorer people in these vulnerable areas may increase the complexity of the city, not least because more bureaucracy is created through new government departments to deal with the issues that are raised by informal settlements (Florian, 2012). Allowing such informal settlements to develop creates new problems by generating "soft" spots in the landscape where hazards will hit harder because when they do so much is destroyed. In the case of Latin America, disasters are blank cheques with which to borrow money, increase corruption and for those governing to appear on the media like a hero; therefore, nobody is interested in fixing the problems, the latest COVID-19 pandemic being a sad example (Goodman, 2020).

Flooding and Collapse

Theory says that societal collapse comes about because of increasing complexity as each solved problem leads to another. This is illustrated in the discussion of flooding above. It is not the sea-level rise and the increased flooding that will lead to societal collapse but the decisions taken around how to cope with issues such as maintenance of sea defences or the displacement of people. The obvious solution of drastically decarbonising society immediately, in order to reduce the effects of climate change might, if taken seriously, lead to a simplification in the organisation of societies, at least in the developed world. There would be no or many fewer private cars. People would have to walk or those with bicycles might use them. There would be no economic growth in the sense the term is now accepted and the economy would be run on different lines, such as those proposed by Schumacher (1973). People would have to learn to live with the basics. Securing food and water would be the priority. Everyone would have to live with much less.

Currently, the ways of combatting flooding as a result of climate change are not focused on this but rather on keeping the water out or learning to live with it. The former is expensive in money and resources and can lead to the displacement of those already living near the water, as land values rise once the land is protected. Does this just displace the poor to areas with no such protection and is this an ethical thing to do? Living on the water brings its own complexities in terms of the need for technological advancements to make this happen. No one has yet grown food in underwater environments and the closest water farming has come is the 40 cow Dutch "farm" on a floating concrete raft in Rotterdam harbour, which cost three million euros (Stuff, 2019a) (Fig. 5.11).

Part of this cost comes from the fact the processes on the floating farm are fully automated, powered by PV panels situated on an adjacent floating structure. The manure from the cows is processed by machine for fertiliser, and cows are milked by a robot, and there is also an automated feeding system, and "80 per cent of the cow's fodder will come from city waste, including by-product from a local brewery, grass from sports fields, bran from bakeries and potato peel from French fry companies" (Constable, 2019). However, the energy to collect this material is presumably not

Fig. 5.11 Cow in Rotterdam harbour (adapted from https://apnews.com/article/9d1f901a48b04843a06052d652b1050)

included in the idea of the farm generating all the energy it needs for operation from renewables. This also begs the question as to why these food wastes are produced in the first place. Just taking brewing waste, in the past the mash would have been rebrewed until there were no sugars left. The first brew was known as strong beer and the following as small beer, and a brewery in London still uses this approach (Alworth, 2015). This is another example perhaps of not solving the real problem, which is the avoidance of food wastes. In turn, this is an issue of scale. In subsistence farming there is no food waste as what people do not consume goes to the animals or the compost heap, to be returned via either path as nutrients to the soil.

The cost of creating this additional, artificial land is also an issue. To investigate the costs further, Table 5.1 gives the 2016 breakdown of the costs of producing a kilogram of milk from a conventional dairy farm in the Netherlands.

The point of the table is to show that the costs related to land, at 2.19 cents, account for only 4.8% of the total cost of producing a kilogram of milk. In the floating farm, land is not required but there has to be an investment in making artificial land in the form of the concrete raft for the cows to stand on. The cost breakdown of the floating farm is not available, but we can get some idea by looking at the cost of a house. The foundations of a conventional house normally account for around 10% of the total costs, though a more expensive floating foundation might push this up quite a bit. Obviously, the floating farm is somewhat different from a house, but these

Table 5.1 Costs in 2016 of producing one kilogram of milk in the Netherlands (European Milk Board, 2016:6–8)

Process	Cost in cents
Inputs (seeds, fertilizers, plant protection products, bought-in feed, cost of equipment and machine maintenance and energy)	16.46
General operating costs (other specific costs for plant and animal production and all non-specific costs like labour, other general costs, paid wages, rent, interest and taxes)	21.59
Equivalent cost of labour contribution from family members	9.31
Beef production	2.53 (negative cost)
Subsidies	2.17 (negative cost)
Land (rent)	2.19
Interest on capital outlay	0.57
Total	45.42

figures suggest the milk from a floating farm will be more expensive than that from land, since the cost of the buoyant structure will form a higher proportion of the total cost of the farm, all other parameters being equal. In this case they are not, as on the robotic floating farm energy has been substituted for labour, which seems a pity as the one thing the world has is plenty of people even if renewable energy is in short supply. By 2023 renewables are predicted to form only 12.4% of global energy demand (IEA, 2019).

Perhaps a better sense of the cost of floating farms comes from a comparison with Dutch barn milk production. A barn for 190 cows with automatic feeding and three robots costs approximately €1.4 million (Kelleher, 2016), or around €7,400 per cow, compared to €75,000 per floating cow. Growing urban food in this way will increase its cost and a warning should have come from history. Cow keeping in London was common until the 1860s when the railways meant milk could be brought in fresh from rural areas. As the pressures of urbanism increased and fields were lost to buildings cows were increasingly kept indoors all year round, with their food being brought in, supplemented by waste from the brewing industry (Atkins, 1977). However, when diseases arrived with imported cattle, ironically imported Dutch cows which gave higher milk yields, the price of milk rose until it was "…beyond the price range of a large section of society…" (Atkins, 1977). Around this time condensed milk was a cheaper alternative and was used for feeding infants (Stevens et al., 2009), as it was cheaper than cow's milk and lasted longer. However, Atkins (2003) points out condensed milk was often made from skimmed milk and so for the very young was "…nutritionally of dubious benefit…" Thus, history shows that if local urban milk costs more, then it is only the rich who will be able to afford it, not solving the problem of supplying food to those who live in urban areas. Indeed, it seems it is the wrong problem that is being solved here. Rather than creating artificial land, it is the way of farming on conventional land that needs to change with far more arable and far less livestock farming than at present. The floating farm is an example of the solution being more complex than the problem by creating a more complicated situation without really changing anything.

> Transformation to healthy diets by 2050 will require substantial dietary shifts. Global consumption of fruits, vegetables, nuts and legumes will have to double, and consumption of foods such as red meat and sugar will have to be reduced by more than 50 per cent. A diet rich in plant-based foods and with fewer animal source foods confers both improved health and environmental benefits (EAT-Lancet Commission, n.d.:12).

Growing what needs to be grown for sustainable food production would suggest decreasing the number of cows in the world rather than building additional "land" for them in the water. The floating farm seems to be tackling the wrong problem.

Conclusion

The example of the floating farm also shows that solving one problem—how to bring food closer to the city—involves additional complexities, in terms of having to devise new technologies to create systems for disposing of the manure and floating the energy production system to power these systems. Just because such technology makes such actions possible, should they be done? This comes back to the argument that solving one problem leads to another and it is this increase in complexity that has led to the societal collapse in the past. If food needs to be grown closer to the people then maybe the problem is one of how to get people to grow food, which may mean a change in the way cities are designed.

The reason that flooding is a global problem is that there are too many people on the earth, forcing those who have no other options to live on the land which should be left as part of the hydrological systems. At the same time, a wealthy part of the population is failing to deal with climate change even though the issue was accepted as something that has to be addressed by all the countries of the world bar one (the United States). This is the problem that has to be tackled and all the sea defences and floating houses that can be created by using old and inventing new technologies are not the answer to the problem. In fact, they add to the problem. Building sea defences, as proposed in Jakarta, will displace people, and they then have to go somewhere. Such displaced people also often lose their livelihoods because they have lost access to the water. When this happens, there are now two problems, finding the money to build new homes for the displaced and creating employment for them, whereas before there was only one problem. To this we could add a third problem of displaced people returning to the site of their former homes through existing property rights, which has led to a whole new layer of bureaucracy (UN, 2007). This escalation of problems through solving one only to create others is reminiscent of the traditional American folk song "Momma buy me a china doll". The little girl suggests they could afford the doll if they sold Father's feather bed. So where will father sleep? He sleeps in Horsey's bed, so where will Horsey sleep, and so on, until in my version, Granny finishes up sleeping in the piggy pen.

Floating structures come with their own problems, not least being that they are more expensive than building on land (Louisiana Resiliency Assistance Program, 2017). If buildings cost more, then economies will have to grow more, with all the problems that this could bring for the environment. This is using technology to try to maintain business as usual. The only way for flooding not to cause a problem is to learn to live with periodic inundation, as many of the world's poor do. This is the uncomfortable equivalent of collapsing gracefully—learning to live with the problem without adding to its complexity. However, there comes a point when flooding brought about by climate change could become unpredictable in terms of its severity, and retreat is the only feasible option, leading to the same issue of having to rebuild and create an economy to support this, which in turn could feed climate change if tackled in the wrong way. The only way to tackle the problem is at its source.

It seems that through trying to solve the problem of more water through using technology you might be using money and energy to try to stem a problem but are creating a problem of constant maintenance. Ultimately if the technology fails that money and energy are lost and you are back where you started. It might be better to invest that money in the resilience of the system or finding ways to cope with coastal erosion, such as withdrawal. In this case you are losing something, buildings on the coast, to allow for persistence further inland, as in the example of Happisburgh. You have to lose to win.

However, resilience theory shows that with collapse comes reorganisation and opportunities. When the bridge was swept away by the river in full spate in a March storm in 2019 at Fox Glacier on the west coast of New Zealand, a local builder used a dump truck to ferry cars and people across the River Waiho (Hayward, 2019; Stuff, 2019b) until the bridge repairs were finished. Although the service was closed as it was deemed unsafe, it demonstrates that there are always alternatives but they might not fit the current regulatory system, as many poor people who live with floods already know. The inequality highlighted by the latter forms the focus of the next chapter.

References

Abidin, H. Z., Andreas, H., Gumilar, I., Fukuda, Y., Nurmaulia, S. L., Riawan, E., Murdohardono, D., Supriyadi. (2012). The impacts of coastal subsidence and sea level rise in coastal city of Semarang (Indonesia). In L. Ching-Hua (Ed.), *Advances in geosciences, solid earth science (SE)* (Vol. 31, pp. 59–76). World Scientific.

Alworth, J. (2015). Quirks of brewing: Parti-gyle brewing. *All About Beer Magazine*. Retrieved November 7, 2019, from http://allaboutbeer.com/quirks-of-brewing-parti-gyle-brewing/.

Anon. (1945, June 8). The aluminium from war to peace exhibition. *The Architect and Building News*, 151–152.

Arias, M. E., Cochrane, T. A., & Elliott, V. (2014). Modelling future changes of habitat and fauna in the Tonle Sap wetland of the Mekong. *Environmental Conservation, 41*(2), 165–175.

Atkins, P. J. (1977). London's intra-urban milk supply, circa 1790–1914. *Transactions of the Institute of British Geographers, 2*(3), 383–399.

References

Atkins, P. J. (2003). Mother's milk and infant death in Britain, circa 1900–1940. *Anthropology of Food, 2*(September). Retrieved November 7, 2019, from https://journals.openedition.org/aof/310.

Atlas Obscura. (2019). *The lost town of Dunwich*. Retrieved September 24, 2019, from https://www.atlasobscura.com/places/lost-town-dunwich.

Au Morris, J. (2014). Adaptive landscape architecture: Embracing amphibious environments and empowering community sustenance, Master of Landscape Architecture Thesis. Victoria University of Wellington.

Aunger, R. (2010). Types of technology. *Technological Forecasting and Social Change, 77*(5), 762–782.

Barichivich, J., Gloor, E., Peylin, P., Brienen, R. J. W., Schöngart, J., Espinoza, J. C., & Pattnayak, K. C. (2018). Recent intensification of Amazon flooding extremes driven by strengthened Walker circulation. *Science Advances, 4*(9), 7.

Bacon, F. (1672). *New Atlantis*. In H. Osborne (n.d.). (Ed.). University Tutorial Press.

BBC News. (2014). *Bolt part falls off Cheesegrater skyscraper in the City of London*. Retrieved November 15, 2019, from https://www.bbc.com/news/uk-england-london-29929761.

Beeler, C. (2018). *As a British village crumbles into the sea, a family holds onto a home that can't be saved*. Retrieved September 22, 2019, from https://www.pri.org/stories/2018-04-05/british-village-crumbles-sea-family-holds-home-cant-be-saved.

Beizer, D. (2006, September 18). Survival guide: Perspectives from the field: Koen Olthuis, architect and designer of the Amphibious House. *Washington Technology*, 50.

Benson, E. D., Hansen, J. L., Schwartz, A. L., Jr., & Smersh, G. T. (1998). Pricing residential amenities: The value of a view. *The Journal of Real Estate Finance and Economics, 16*(1), 55–73.

Big Think. (n.d.). *Those who do not learn history are doomed to repeat it. Really?* Retrieved December 26, 2019, from https://bigthink.com/the-proverbial-skeptic/those-who-do-not-learn-history-doomed-to-repeat-it-really.

Boudon, P. (1972). *Lived-in architecture: Pessac revisited*. Lund Humphries.

Bourassa, S. C., Hoesli, M., & Sun, J. (2004). What's in a view? *Environment and Planning A: Economy and Space, 36*(8), 1427–1450.

Boyle, G., & Harper, P. (1976). Preface. In G. Boyle & P. Harper (Eds.), *Radical technology*. Wildwood House Ltd.

Brus, A. (2012). Behavior of people and policy in a subsiding and flooding area, Semarang Indonesia, Bachelorthesis Geografie, planologie en milieu, June, Nijmegen, Nijmegen School of Management, Radboud University.

Bukvic, A., Zhu, H., Lavoie, R., & Becker, A. (2018). The role of proximity to waterfront in residents' relocation decision-making post Hurricane Sandy. *Ocean and Coastal Management, 154*(March), 8–19.

Build LLC. (2008). *Floating houses*. Retrieved November 15, 2019, from https://blog.buildllc.com/2008/07/floating-houses/.

Byrde, P. (2003). Dress: The industrial revolution and after. In D. Jenkins (Ed.), *The Cambridge history of western textiles* (Vol. II, pp. 882–909). Cambridge University Press.

Chan, F. K. S., Adekola, O. A., Ng, C. N., Mitchell, G., & McDonald, A. T. (2013). Research articles: Coastal flood-risk management practice in Tai O, a Town in Hong Kong. *Environmental Practice, 15*(3), 201–219.

Chatterjee, M. (2009). Slum dwellers response to flooding events in the megacities of India. *Mitigation and Adaptation Strategies for Global Change, 15*, 337–353.

Clement, M. T. (2010). 'Let them build sea walls': Ecological crisis, economic crisis and the political economic opportunity structure. *Critical Sociology, 37*(4), 447–463.

Committee on Climate Change. (2015). *Progress in preparing for climate change: 2015 report to parliament*. Retrieved October 22, 2019, from https://www.theccc.org.uk/wp-content/uploads/2015/06/6.736_CCC_ASC_Adaptation-Progress-Report_2015_FINAL_WEB_250615_RFS.pdf.

Committee on Climate Change. (2019). *Progress in preparing for climate change: 2019 report to parliament*. Retrieved October 22, 2019, from https://www.theccc.org.uk/publication/progress-in-preparing-for-climate-change-2019-progress-report-to-parliament/.

Constable, H. (2019). A floating future? The world's first water-borne dairy farm has been erected on the shores of Rotterdam. What does this say about the future of food, cities and, indeed, cows? *Geographical, 91*(7), 43–48.

The Construction Index. (2019). *Cheesegrater bolt saga finally comes to an end*. Retrieved November 15, 2019, from https://www.theconstructionindex.co.uk/news/view/cheesegrater-bolt-saga-finally-comes-to-an-end.

Cook, H. F., Cutting, R. L., & Valsami-Jones, E. (2015). Flooding with constraints: Water meadow irrigation impacts on temperature, oxygen, phosphorus and sediment in water returned to a river. *Journal of Flood Risk Management, 10*(4), 463–473.

Coomes, O. T., Lapointe, M., Templeton, M., & List, G. (2015). Amazon river flow regime and flood recessional agriculture: Flood stage reversals and risk of annual crop loss. *Journal of Hydrology, 539*, 214–222.

Coover, H. W. (2000). IRI achievement award address: Discovery of superglue shows power of pursuing the unexplained. *Research-Technology Management, 43*(5), 36–39.

Corominas, J. (2003). *Breve diccionario etimologico de la lengua castellana (Concise dictionary of Spanish)* (3rd ed.). Editorial Gredos.

Dawson, A. (2017). *Extreme cities: The peril and promise of urban life in the age of climate change*. Verso.

Delacey, L. (2015). Aboriginal inventions: 10 enduring innovations. *Australian Geographic*. Retrieved December 5, 2019, form https://www.australiangeographic.com.au/topics/history-culture/2015/03/aboriginal-australian-inventions/.

De Lange, W. (2019). *Are 1800 Kapiti homes really threatened by sea level rise?* Retrieved August 29, 2019, from https://www.climateconversation.org.nz/2012/09/are-1800-kapiti-homes-really-threatened-by-sea-level-rise/.

de Silva, S., Vale, B., & Vale, R. (2017). 'Primitive attitudes' and traditional practices: Looking back for sustainable solutions to future flood disasters. In *51st International Conference of the Architectural Science Association (ANZAScA)*, Wellington, 27–30 Nov., 10 pp.

DeSilvey, C. (2012). Making sense of transience: An anticipatory history. *Cultural Geographies, 19*(1), 30–53.

Dhonau, M., Wilson, G., McHugh, A., Burton, R., & Rose, C. (2017). *Homeowners guide to flood resilience*. Retrieved July 18, 2017, from http://www.knowyourfloodrisk.co.uk/sites/default/files/FloodGuide_ForHomeowners.pdf.

Dienel, H.-L., & Schiefelbusch, M. (2000). German commercial air transport until 1945. In *Revue belge de Philologie et d'Histoire* (pp. 945–967). Retrieved July 28, 2020, from https://www.persee.fr/doc/rbph_0035-0818_2000_num_78_3_4472.

Dryland, E., & Syed, J. (2010). Tai O village: Vernacular fisheries management or revitalization? *International Journal of Cultural Studies, 13*(6), 616–636.

Dugan, P., Barlow, C., Agostinho, A., Baran, E., Cada, G., Chen, D., Winemiller, K., et al. (2010). Fish migration, dams, and loss of ecosystem services in the Mekong Basin. *Ambio, 39*(4), 344–348.

EAT-Lancet Commission. (n.d.). *Healthy diets from sustainable food systems*. Retrieved November 7, 2019, from https://eatforum.org/content/uploads/2019/01/EAT-Lancet_Commission_Summary_Report.pdf.

The Economist. (2019). *Graphic detail*. Retrieved January 10, 2020, from https://www.economist.com/graphic-detail/2019/12/14/a-plurality-of-americans-but-not-of-states-want-donald-trump-impeached.

Environment Agency. (2010). *Managed realignment (breach)*. Retrieved September 24, 2019, from http://evidence.environment-agency.gov.uk/FCERM/en/SC060065/MeasuresList/M6/M6T3.aspx?pagenum=1.

References

Eschner, K. (2017). *Before Velcro's patent expired, it was a niche product most people hadn't heard of*. Smithsonian Magazine. Retrieved July 28, 2020, from https://www.smithsonianmag.com/smart-news/velcros-patent-expired-it-was-niche-product-most-people-hadnt-heard-180962701/.

Evans, P., Marschke, M., & Paudyal, K. (2004). *Flood forests, fish and fishing villages*. Asia Forest Network.

European Milk Board. (2016). *What is the cost of producing milk?* Retrieved November 6, 2019, from http://www.europeanmilkboard.org/fileadmin/Dokumente/Milk_Production_Costs/BAL_cost_study_milk_2016_DE_NL_BE_DK_FR.pdf.

Fagan, B. (2013). *The attacking ocean: The past, present, and future of rising sea levels*. Bloomsbury Press.

Florian, U. (2012). Mumbai's suburban mass housing. *Urban History, 39*(1), 128–148.

Galloway, J. A., & Potts, J. S. (2007). Marine flooding in the Thames Estuary and tidal river c.1250–1450: impact and response. *Area, 39*(3), 370–379.

Ganson, B. (2014). *Texas takes wing: A century of flight in the lonestar state*. University of Texas Press.

Gibbard, R. G. (1972). *Beach morphology and sediments of the West Wellington coast: Wanganui to Paekakariki*. Master of Science in Geography Thesis. Massey University.

Gibson, E. (2017). *Officials' long struggle to publish sea level guidance*. Retrieved August 29, 2019, from https://www.newsroom.co.nz/2017/12/20/70263/officials-long-struggle-to-publish-new-sea-level-guidance.

Goemans, T., & Visser, T. (1987). The delta project: The Netherlands experience with a megaproject for flood protection. *Technology in Society, 9*, 97–111.

Goodman, J. (2020). *Spread of coronavirus fuels corruption in Latin America*. Retrieved August 14, 2020, from https://apnews.com/a240ff413fb23220aff30c6d6e6aba4c.

Gordon, K. (2020). *New technologies to solve the earth's oldest problems*. Retrieved September 15, 2020, from https://www.biotechniques.com/plant-climate-science/tech-news_new-technologies-to-solve-the-earths-oldest-problems/.

Gorgolewski, M. T. (2005). The potential for prefabrication in United Kingdom housing to improve sustainability. In J. Yang, P. S. Brandon & A. C. Sidwell (Eds.), *Smart and sustainable built environments* (pp. 121–128). Blackwell.

Grahame, A. (2015). 'This isn't at all like London': Life in Walter Segal's self-build 'anarchist' estate. *The Guardian*. Retrieved January 12, 2020, from https://www.theguardian.com/cities/2015/sep/16/anarchism-community-walter-segal-self-build-south-london-estate.

Gray, G. H. (1946). *Housing and citizenship*. Reinhold.

Greater Wellington Reginal Council. (2019). *Greater Wellington acts on coastal erosion at Queen Elizabeth Park*. Retrieved August 14, 2020, from http://www.gw.govt.nz/greater-wellington-acts-on-coastal-erosion-at-queen-elizabeth-park-2/.

Gropius, W. (1935). *The new architecture and the Bauhaus*. Faber and Faber.

Grundy-Warr, C., & Sithirith, M. (2016). Threats and challenges to the floating lives' of the Tonle Sap. In A. C. Tidwell & B. S. Zellen (Eds.), *Lands, indigenous peoples and conflict* (pp. 127–149). Routledge.

Hadi, S. P. (2017). In search for sustainable coastal management: A case study of Semarang, Indonesia. In *IOP Conference Series: Earth and Environmental Science, Volume 55, 2nd International Conference on Tropical and Coastal Region Eco Development 2016*, 25–27 October 2016, Bali, Indonesia Bristol, IOP Science.

HalFILE. (n.d.) *Paper filing system vs. document imaging system—A cost comparison*. Retrieved January 10, 2020, from http://halfile.com/pdfdocs/Paper_vs_Digital.pdf.

Hall, C. A. S., & Klitgaard, K. A. (2012). *Energy and the wealth of nations*. Springer.

Hallegatte, S., Bangalore, M., Bonzanigo, L., Fay, M., Kane, T., Narloch, U., Rozenberg, J., Treguer, D., & Vogt-Schilb, A. (2016). *Shock waves: Managing the effects of climate change on poverty*. The World Bank.

Harwitasari, D. (2009). *Adaptation responses to tidal flooding in Semarang, Indonesia* Rotterdam. Master's Programme in Urban Management and Development.

Harwitasari, D., & Van Ast, J. A. (2011). Climate change adaptation in practice: People's responses to tidal flooding in Semarang, Indonesia. *Journal of Flood Risk Management, 4*(3), 216–233.

Harper, P. (1976). Autonomy. In G. Boyle & P. Harper (Eds.), *Radical technology*. Wildwood House Ltd.

Hayward, M. (2019). *Dump truck ferry carries up to 100 each day across bridge-less Waiho River*. Retrieved October 28, 2019, from https://www.stuff.co.nz/national/111852077/dump-truck-ferry-carries-up-to-100-each-day-across-bridgeless-waiho-river.

Heidegger, M. (1977). *The question concerning technology*. Harper and Row.

Herrando-Pérez, S., Bradshaw, C. J. A., Lewandowsky, S., & Vieites, D. R. (2019). Statistical language backs conservatism in climate-change assessments. *BioScience, 69*(3), 209–219.

Historic England. (2017). *Conserving historic water meadows*. Historic England.

Hong Kong Tourism Board. (2017). *Tai O Stilt Houses*. Retrieved July 18, 2017, from http://www.discoverhongkong.com/nz/see-do/great-outdoors/outlying-islands/lantau-island/tai-o-stilt-houses.jsp.

Horn, M. J., & Gurel, L. M. (1981). *The second skin: An interdisciplinary study of clothing*. Houghton Mifflin Company.

Inhabitat. (n.d.). *Dutch floating homes By DuraVermeer*. Retrieved October 27, 2019, from https://inhabitat.com/dutch-floating-homes-by-duravermeer/.

IEA (International Energy Agency). (2019). *Renewables 2018*. Retrieved November 15, 2019, from https://www.iea.org/renewables2018/.

IPCC. (2001). *Summary for policymakers climate change 2001: Impacts. Adaptation, and vulnerability*. Retrieved September 6, 2019, from https://www.ipcc.ch/site/assets/uploads/2018/07/wg2TARsummaries.pdf.

Jackson, M. D., Mulcahy, S. R., Chen, H., Li, Y., Li, Q., Cappelletti, P., & Wenk, H.-R. (2017). Phillipsite and Al-tobermorite mineral cements produced through low-temperature water-rock reactions in Roman marine concrete. *American Mineralogist, 102,* 1435–1450.

Kapiti Coast District Council. (2012). *Coastal hazards on the Kapiti Coast*. Retrieved August 25, 2019, from https://www.kapiticoast.govt.nz/media/29809/coastal_hazards_presentation_for_open_days.pdf.

Kapiti Coast District Council. (2019). *Our district: Civil defence/emergency management*. Retrieved August 25, 2019, from https://www.kapiticoast.govt.nz/Our-District/cdem/floods/.

Kapiti Coast District Council. (2020a). *Natural hazards map*. Retrieved July 29, 2020, from https://www.kapiticoast.govt.nz/your-council/planning/district-plan-review/natural-hazards/natural-hazard-maps/.

Kapiti Coast District Council. (2020b). *Coastal hazards: Background information*. Retrieved July 29, 2020, from https://www.kapiticoast.govt.nz/your-council/planning/district-plan-review/coastal-hazards/background-information/.

Kaufman, M. (1963). *The first century of plastics: Celluloid and its sequel*. The Plastics Institute and Illife Books.

Kaźmierczak, A., & Canvan, G. (2011). Surface water flooding risk to urban communities: Analysis of vulnerability, hazard and exposure. *Landscape and Urban Planning, 103*(2), 185–197.

Kelleher, J. (2016). *Dutch approach to dairy comes with big risks attached*. Retrieved November 7, 2019, from https://www.independent.ie/business/farming/dutch-approach-to-dairy-comes-with-big-risks-attached-34334457.html.

Kelly, D. A. (2016). Impact of paved front gardens on current and future urban flooding. *Journal of Flood Risk Management, 11,* 434–443.

Khadiyanta, P., & Dewantari, S. (2016). Settlement adaptation by reshaping dwellings in the degrading area at Genuk Disctrict [sic] of Semarang City, Indonesia. *Procedia - Social and Behavioral Sciences, 227,* 309–316.

Kohler, T. A., Smith, M. E., Bogaard, A., Peterson, C. E., Betzenhauser, A., Feinman, G. M., Oka, R. C., Pailes, M., Prentiss, A. M., Stone, E. C., Dennehy, T. J., & Ellyson, L. J. (2018). Deep inequality: Summary and conclusions. In T. A. Kohler & M. E. Smith (Eds.), *Ten thousand years of inequality: The archaeology of wealth differences* (pp. 289–318). University of Arizona Press.

Kummu, M., & Sarkkula, J. (2008). Impact of the Mekong river flow alteration on the Tonle Sap flood pulse. *Ambio, 37*(3), 185–192.

Kunreuther, H. C., & Michel-Kerjan, E. O. (2009). *At war with the weather: Managing large-scale risks in a new era of catastrophes.* The MIT Press.

Laver, J. (1966). *Dress*. John Murray.

Lee, T. M., Markowitz, E. M., Howe, P. D., Ko, C.-Y., & Leiserowitz, A. A. (2015). Predictors of public climate change awareness and risk perception around the world. *Nature Climate Change, 5*, 1014–1020.

Loh and Civic Exchange. (2002). *Getting heard: A handbook for Hong Kong citizens*. Hong Kong University Press.

Louisiana Resiliency Assistance Program. (2017). *Maasbommel, Netherlands*. Retrieved November 15, 2019, from https://resiliency.lsu.edu/case-studies-blog/2017/11/10/maasbommel-netherlands.

Mcgranahan, G., Balk, D., & Anderson, B. (2007). The rising tide: Assessing the risks of climate change and human settlements in low elevation coastal zones. *Environment and Urbanization, 19*(1), 17–37.

McNamara, D. E., Gopalakrishnan, S., Smith, M. D., & Murrau, A. B. (2015). Climate adaptation and policy-induced inflation of coastal property value. *PLoS ONE, 10*(3). https://doi.org/10.1371/journal.pone.0121278.

Met Office. (2017). *1953 east coast flood - 60 years on*. Retrieved July 18, 2017, from http://www.metoffice.gov.uk/news/in-depth/1953-east-coast-flood.

Milfont, T. L., Evans, L., Sibley, C. G., Ties, J., & Cunningham, A. (2014). Proximity to coast is linked to climate change belief. *PLoS ONE, 9*(7), e103180, 8 pp.

Milfont, T. L., Wilson, M. S., & Sibley, C. G. (2017). The public's belief in climate change and its human cause are increasing over time. *PLoS ONE, 12*(3), 9 pp. https://doi.org/10.1371/journal.pone.0174246.

Ministry for the Environment. (2017a). *Preparing for coastal change fact sheet series*. Retrieved August 29, 2019, from https://www.mfe.govt.nz/publications/climate-change/preparing-coastal-change-fact-sheet-series.

Ministry for the Environment. (2017b). *Sea-level rise*. Retrieved August 29, 2019, from https://www.mfe.govt.nz/sites/default/files/media/MFE_Coastal_Fact%20Sheet%207.pdf.

Ministry for the Environment. (2017c). *Coastal flooding due to storms*. Retrieved August 29, 2019, from https://www.mfe.govt.nz/sites/default/files/media/MFE_Coastal_Fact%20Sheet%202.pdf.

Ministry of Works. (1944). *House construction* (Post-War Building Studies No. 1). HMSO.

Moradibistouni, M., Vale, B., & Isaacs, N. (2018). Evaluating sustainability of prefabrication methods in comparison with traditional methods. In *Proceedings of the 10th International Conference in Sustainability on Energy and Buildings* (SEB'18), 24–26 June 2018, Gold Coast, Australia.

Morris, B. (2014). In defence of oblivion: The case of Dunwich, Suffolk. *International Journal of Heritage Studies, 20*(92), 196–219.

Napier City Council, Hastings District Council, Hawkes Bay Regional Council, He Toa Takitini, Mana Ahuriri, Maungaharuru Tangitu. (2016). *Clifton to Tangoio coastal hazards strategy 2120*. Retrieved August 29, 2019, from https://www.hbcoast.co.nz/assets/Document-Library/Project-Documents/Clifton-to-Tangoio-Coastal-Hazard-Strategy-2120-DRAFT-Aug-2016.pdf.

Netherlands Tourism. (2019). *Is the Netherlands below sea level?* Retrieved October 18, 2019, from http://www.netherlands-tourism.com/netherlands-sea-level/.

Nolan, R. M. (2017). *Late holocene sedimentation on the Southern Kāpiti Coast*. Master of Science in Geology Thesis. Victoria University of Wellington.

Ogburn, W. F., & Thomas, D. (1922). Are inventions inevitable? A note on social evolution. *Political Science Quarterly, 37*(1), 83–98.

Ogilvie, G. (1996). *'Pearse, Richard William', Dictionary of New Zealand Biography. Te Ara–The Encyclopedia of New Zealand*. Retrieved November 16, 2019, from https://teara.govt.nz/en/biographies/3p19/pearse-richard-william.

Olumuyiwa, A., Peyi, S.-A., & Olugbemisola, S. (2019). Gentrification and the challenge of development in Makoko, Lagos State, Nigeria: A rights-based perspective. *Environmental Justice, 12*(2), 41–47.

Osberghaus, D. (2015). The determinants of private flood mitigation measures in Germany—Evidence from a nationwide survey. *Ecological Economics, 110*, 36–50.

Otago Regional Council. (2012). *Natural Hazards on the Taieri Plains, Otago*. Retrieved July 24, 2017, from http://www.orc.govt.nz/Documents/Publications/Natural%20Hazards/Hazards%20on%20the%20Taieri%20Plains/Taieri%20Report%20-%20Introduction.pdf.

Park, S., & Jayaraman, S. (2017). The wearables revolution and Big Data: The textile lineage. *The Journal of the Textile Institute, 108*(4), 605–614.

Pierre, J. (2020). *Why aren't some people taking COVID-19 more seriously?* Retrieved August 14, 2020, from https://www.psychologytoday.com/nz/blog/psych-unseen/202003/why-arent-some-people-taking-covid-19-more-seriously.

Pleijster, E.-J., van der Veeken, C., & Jongerius, R. (2014). *Dutch Dikes*. nai010 publishers.

Polkinghorne, R. K., & Polkinghorne, M. I. R. (1945). *Other people's houses*. George G Harrap and Co Ltd.

Portola Institute, Inc. (1970). *Whole earth catalogue: Access to tools* (Spring). Portola Institute.

Revkin, A. (2019). *Floating cities could ease the world's housing crunch, the UN says*. Retrieved October 28, 2019, from https://www.nationalgeographic.com/environment/2019/04/floating-cities-could-ease-global-housing-crunch-says-un/.

Pollock, E. (2019). *UN brings back controversial floating city concept*. Retrieved October 28, 2019, from https://www.engineering.com/BIM/ArticleID/18941/UN-Brings-Back-Controversial-Floating-City-Concept.aspx.

Rijkswaterstaat. (n.d.). *Room for the river*. Retrieved July 18, 2017, from https://www.rijkswaterstaat.nl/english/water-systems/protection-against-water/room-for-the-river.aspx.

Rose, C. B. (2014). River flood defences. In C. A. Booth & S. M. Charlesworth (Eds.), *Water resources in the built environment* (pp. 223–232). Wiley.

Ruimte voor de rivier. (n.d.). *Room for the river for a safer and more attractive landscape*. Retrieved October 22, 2019, from https://www.ruimtevoorderivier.nl/english/.

Schilderman, T., & Lyons, M. (2011). Resilient dwellings or resilient people? Towards people-centred reconstruction. *Environmental Hazards, 10*(3–4), 218–231.

Schumacher, E. F. (1973). *Small is beautiful: Economics as if people mattered*. Harper and Row.

Semarang City Government. (2016). *Resilient Semarang: Moving together towards a Resilient Semarang*. Semarang City Government.

Smith, A. (2005). The alternative technology movement: An analysis of its framing and negotiation of technology development. *Human Ecology Review, 12*(2), 106–119.

Smith, I., & McAlpine, C. (2014). Estimating future changes in flood risk: Case study of the Brisbane River, Australia. *Climate Risk Management, 6*, 6–17.

Statistics New Zealand. (n.d.). *2013 Census QuickStats about a place: Kapiti Coast District*. Retrieved September 6, 2019, from http://archive.stats.govt.nz/Census/2013-census/profile-and-summary-reports/quickstats-about-a-place.aspx?request_value=14325&parent_id=14322&tabname=.

Steinberg, P. E. (2010). Liquid urbanity: Re-engineering the city in a post-terrestrial world. In S. Brunn (Ed.), *Engineering earth* (pp. 2113–2122). Springer.

Stevens, E. E., Patrick, T. E., & Pickler, R. (2009). A history of infant feeding. *Journal of Perinatal Education, 18*(2), 32–39.

Strickland, M. (1843). *A memoir of the life, writings, and mechanical inventions of Edmund Cartwright...inventor of the power loom, etc. etc*. Saunders and Otley.

Stuff. (2019a). *Floating dairy farm moored in Dutch harbor brings the cows to consumers*. Retrieved October 28, 2019, from https://www.stuff.co.nz/business/farming/113929405/floating-dairy-farm-moored-in-dutch-harbour-brings-the-cows-to-consumers.

References

Stuff. (2019b) *Is another Bailey bridge the best option for Franz Josef's Waiho River?* Retrieved August 17, 2020, from https://www.stuff.co.nz/national/111866240/is-another-bailey-bridge-the-best-option-for-franz-josefs-waiho-river.

Tebboth, M. (2014). Understanding intractable environmental policy conflicts: The case of the village that would not fall quietly into the sea. *The Geographical Journal, 180*(3), 224–235.

The Local de. (2017). *Section of Baltic Sea autobahn to fully close after hole in road threatens to get bigger.* Retrieved November 15, 2015, from https://www.thelocal.de/20171026/section-of-baltic-sea-autobahn-to-fully-close-after-hole-in-road-threatens-to-get-bigger.

United Nations. (2007). *Handbook on housing and property restitution for refugees and displaced persons implementing the 'Pinheiro Principles'.* Retrieved July 30, 2020, from https://www.un.org/ruleoflaw/files/pinheiro_principles.pdf.

UNECE (United Nations Economic Commission for Europe). (2018). *UN Alliance aims to put fashion on path to sustainability.* Retrieved November 22, 2019, from https://www.unece.org/info/media/presscurrent-press-h/forestry-and-timber/2018/un-alliance-aims-to-put-fashion-on-path-to-sustainability/doc.html.

Vale, B. (1995). *Prefabs: A history of the UK temporary housing programme.* E and FN Spon.

Webster, P. J., Holland, G. J., Curry, J. A., & Chang, H.-R. (2005). Changes in tropical cyclone number and intensity in a warming environment. *Science, 309*, 1844–1846.

The World Residence Holdings Ltd. (2019). *The world: Residences at Sea.* Retrieved October 28, 2019, from https://aboardtheworld.com/.

Watts, J. (2018). Brazil's new foreign minister believes climate change is a Marxist plot. *The Guardian*, 15 Nov. Retrieved August 19, 2020, from https://www.theguardian.com/world/2018/nov/15/brazil-foreign-minister-ernesto-araujo-climate-change-marxist-plot.

Wild, J. P. (2003). Introduction. In D. Jenkins (Ed.), *The Cambridge history of western textiles* (Vol. I, pp. 9–25). Cambridge University Press.

Williams-Ellis, C., Eastwick-Field, J., & Eastwick-Field, E. (1947). *Building in Cob, Pisé, and stabilized earth.* Country Life Ltd.

Worldometer. (n.d.). *World population by year.* Retrieved July 29, 2020, from https://www.worldometers.info/world-population/world-population-by-year/.

Wyzga, B., Kundzewicz, Z. W., Konieczny, R., Piniewski, M., Zawiejska, J., & Radecki-Pawlik, A. (2018). Comprehensive approach to the reduction of river flood risk: Case study of the Upper Vistula Basin. *Science of the Total Environment, 631–632*, 1251–1267.

Yarwood, D. (1952). *English costume.* B T Batsford Ltd.

Zhang, Y., Xie, J., & Liu, L. (2011). Investigating sea-level change and its impact on Hong Kong's coastal environment. *Annals of GIS, 17*(2), 105–112.

Chapter 6
Inequality, Collapse and the Built Environment

Part 1: The Problem of Inequality

> *All animals are equal, but some animals are more equal than others*
> George Orwell

Introduction

Previous chapters have discussed how socio-political pressures can lead to societal collapse. As an example of such pressure, this chapter looks into the impacts of inequality on the built environment, with a focus on processes of gentrification in the urban landscape. It discusses the extent to which these might create scenarios that contribute to accelerating the process of collapse.

The problem of inequality is older than human-induced climate change. Scheidel (2017:25–33) proposes that inequality has been always with us, since the time of the first primates but that it really took off after the last ice age, which introduced a time of stability. A group of sociologists (Lobao et al., 2007:1) simplified the definition of inequality into one phrase by saying that is the study of "who gets what and why." The "what" is broad and can cover a wide range of benefits from services and resources, access to health, education and infrastructure, to income or land. The "why" look at fairness, the reason for the inequality, and what transforms different levels of the above into a matter of justice. This chapter looks into inequality in urbanised built environments, which are now the most common habitat for people in the world. Therefore, to the "who gets what and why" definition above, the variable "where" is now added. The specific characteristic of the urban landscape is that it can form a context where inequalities become visible and tangible. This emphasises any problems arising from the way services, opportunities, infrastructure and wealth are distributed. However, a further assumption here is that the built environment is not a passive element that only exposes inequalities, but is also one that could set the conditions for future inequalities, such as accelerating the impact of gentrification that might start in a single street but soon move out to affect a whole neighbourhood. If the built environment has an impact on the inequalities of a community, could modern societies collapse due to high levels of inequality in the built environment? Has any civilisation in the past collapsed due to inequalities in the built environment? Can inequalities in the built environment be linked with the process of collapse?

The Theoretical Link Between Inequality and Collapse

Tainter (1988:4) proposed that collapse is a social and political process manifested as "a rapid, significant loss of an established level of sociopolitical complexity." In this context, the complexity of a society refers to the specialisation of roles and functions as well as its size (see Chap. 3). Compared with modern industrial societies where there are millions of roles, in hunter gatherer societies there were few roles but, more importantly, there was equality between the sexes in terms of choosing with whom to associate. In a modelling study of a hunter-gatherer society based on behaviour in remaining modern such societies, Dyble et al. (2015) showed that where sexes had equal choice groups were mixed, but where one sex had control over choice groups tended to be kinship based. They suggested that gender inequality came with the development of agriculture. Perhaps more importantly, they also noted, "Couples freely moving between camps and sharing interests with kin and affines would be able to maintain cooperation without the need for more complex systems, such as cultural group selection and altruistic punishment."

When a society grows in terms of population and new roles are developed, the complexity achieved is controlled via the generation of hierarchies that become useful for managing and achieving collective goals. This can also be seen in the built environment. Building infrastructure to satisfy basic needs (like food storage or water collection) demands the mobilisation and organisation of a large group of people with different skills, and may even mean bringing in resources from other territories, so other people in other roles have to create the money to be able to do this. Ranking of roles together with managerial positions make the building work of big-scale monuments possible as well as the control and defence of large territories. The bigger the venture the more powerful the ranks and levels as well as the structure and organisation need to be. In an informal settlement, families can self-build a house with the help of relatives and friends. However, hundreds of specialists and workers, usually coming from more than one country, are needed to construct a skyscraper.

The downside of hierarchical structures and positions is that they can create the environment for inequalities derived from privileges. These manifest as unequal access to resources, either material (land, food, houses) or social aspects (status). As noted above, foraging groups were more egalitarian than early farmers or contemporary societies. If hierarchies and inequality are intimately linked to the development of societies from small, homogeneous, loose and equitable communities to large, heterogeneous and unequal societies, then the root of inequality and complexity lies within this development, as noted by Dyble et al. (2015). Through the lens of inequality, the development of societies can be seen in an alternative way, one in which the important thing is not only what is achieved but also how it is achieved and at what cost. Can inequality be considered an unavoidable consequence of development? What role does the built environment play?

Inequality and Collapse of Ancient Societies

The role of inequality as a cause of the collapse of ancient societies is still unclear and is not widely acknowledged. There is no consensus that civil wars and other internal conflicts can be caused by inequality alone. Levitt (2019) declared that the relationship between inequality and collapse deserves more attention from archaeological and anthropological studies and outlined the need for a theoretical framework that could be helpful for digging deeper into the relationship between inequality and collapse. He used the collapse of the Egyptian Old Kingdom in 2184 BCE as an example to explain the role of inequality. The bureaucracy of the Old Kingdom grew alongside its wealth but at the same time corruption led rulers to retain taxes designated for flood control, irrigation, and grain storage for their own purposes. In the face of an important drought, the tax system failed to supply these essential needs, and this led to the collapse of the political system. Even in this example, it is complicated to disentangle and isolate inequality as the only reason for the collapse, since the latter is merged with the political failure of the system, as well as with the natural hazard of a drought. It would seem inequality was an essential unintended consequence of developing a complex society, and it could be very complicated to define its role in the collapse of a society. Are inequality and collapse ever related? Have any ancient societies collapsed due to inequality and what is the role of the built environment in the understanding of inequality and collapse?

Development and Inequality in Ancient Civilisations

In Archaeology, Kohler et al. (2018:293–295) claimed that over a period of 10,000 years, from the Neolithic Age to the sixteenth century, inequality played an important role in the development of early villages, towns and cities in America, Africa, Europe and Asia. The study found that inequalities increased when populations developed from foragers and horticulturalists to agriculturalists. This was because the latter tended to produce more surpluses (Milanović et al., 2007:5) and have lower mobility, becoming more vulnerable to unequal distributions and monopolies. However, Smith et al., (2018:7) note that some hunter-gatherer societies were less egalitarian than others and some states and cities more egalitarian than might be expected. As human groups increased in size this "created opportunities for the emergence of differential wealth and prestige within groups" (Smith et al., 2018:11).

Wealth in hunter-gathering and horticultural communities, the latter being linked to villages and small towns, depended more on physical characteristics such as body weight or relational ties such as food sharing networks. In agricultural societies, which in their turn gave rise to towns and cities, wealth was found to be more materially based, and easier to accumulate and be inherited across generations, and in this way more susceptible to the generation of inequality between the members of such societies (Mulder et al., 2009). Population increases also created more opportunities for political complexity to emerge and with it, power relationships.

Technology also played a role because it permitted ancient societies to increase productivity and generate surpluses, opening the possibility for the concentration of power in fewer hands (see the development of the loom in Chap. 5). From this point of view, it is possible to state "… socioeconomic inequality generally increases with the onset of agricultural production, larger settlements and societal scale, more-hierarchical forms of political complexity, and the development of more-elaborate and more-costly technologies" (Smith et al., 2018:16). Based on this set of assumptions, Kohler et al. (2018:290–291) found that inequality tended to increase in ancient societies under five circumstances. These were resource scarcity produced by changes in the climate, densification of populations, privatisation of resources or enforced policies that limited sharing, production beyond subsistence needs, and the existence of exclusionary institutions. The role of growth in production, creation of surpluses and economic development thus produced the appropriate environment for inequality. However, it is the way growth and surpluses were distributed that created the different scenarios of inequality. For instance, Kohler et al. (2017) found more inequalities in Eurasia after the end of the Neolithic period because of the presence of large mammals that could be domesticated, making agriculture more profitable.

All these findings can be partially linked to the problem-solving mechanism, where societal problems and solutions are tightly bound to escalating investments in resources that produce marginal returns until the system collapses (see Chap. 3). In Kohler et al.'s study (2018), the growth and development from smaller scale foragers and horticultural groups to larger agricultural societies implied an increase in complexity through the creation of more hierarchical societies, with increased political structures, and the use of new technologies. The marginal returns were represented by the inequalities generated through changes in the way of exploiting natural resources. The only point that remains in question is to know if the tipping point of inequality could be the main cause of collapse of a civilisation. Levitt (2019) explains the situation of using the Gini index as a measure of wealth inequality. The Gini index or ratio was devised in 1912 by the Italian statistician Corrado Gini. It usually reveals the distribution of income across a population but sometimes shows the distribution of wealth, with the latter including assets as well as income. A Gini index of zero would mean everyone in the population had the same income. The nearer the index gets to one, the more unequal the distribution of income.

> The role of inequality in collapse does not necessarily imply some teleological tipping points: when a (reliable) Gini exceeds x, collapse ensues. Instead, it involves the build-up over time of grievance, alienation and hostility while elites flourish, creating the conditions in which eventual collapse occurs, triggered by intra-elite conflict, famine, invasion or war. But while

inequality might contribute to collapse, it is not sufficient as a major factor without leadership and organisation of popular uprisings (Levitt, 2019).

Past Inequality and the Role of the Built Environment

The rise of inequality as a result of the way in which societies developed and the surpluses created were distributed can also be traced in the built environment. Houses were, and remain, one of the most important assets of a family. Therefore, researchers have used house sizes as a parameter for studying inequality in ancient societies. Developing Gini indices using house size has almost become a common practice in archaeological research into inequality. Smith et al. (2018:27–28) recognised this approach could be questioned because the role of housing and the size of dwellings varies across cultures and time, but it does provide a measure of wealth in the absence of other data or written records. This is an important factor for researchers wanting to understand the link between the built environment and socio-economic inequality.

Kohler et al. (2018:293–295) produced Gini indices using house size distributions to define the level of inequality in ancient societies using case studies from the town of Catalhoyuk, southern Turkey, in the Neolithic period (9200 BC) to Tenochtitlan, Mexico (1500 BC). When the case studies were compared in terms of the size and function of settlements, they defined the three categories of village, town and city (Kohler et al., 2018:298).

> …a village is defined as a settlement of one thousand or fewer residents that lacks a central feature. A town is defined as either a settlement with more than a thousand residents but lacking a central feature or a settlement of more than two hundred residents that has one or two lower-level (typically small or moderately sized) central features. A city is any settlement with more than one thousand residents and one or more high-level (i.e. large and prominent) central features.

The 'feature' in a settlement in this definition was something whose use extended beyond the boundaries of the settlement, so a temple in a city would draw people from outside the city, or a chief's house in a town might serve villages without such a feature. A feature could also be something the construction of which required skills from outside the settlement (Kohler et al., 2018:297). The frequency and the size of the areas that these different types of settlements served defined the level of complexity of the landscape. In more recent complex landscapes, facilities were bigger and less frequent but they served larger areas than in less complex landscapes where facilities were smaller but appeared with more frequency and served smaller areas. As an example, in medieval Christian societies, every village would have a church but only a significant city would have a cathedral. The comparative analysis by Kohler et al., (2018:309) found that inequality tends to increase when the size and function of a settlement grew and became more complex as a result of agricultural activity. However, lower inequality was found in villages and greater inequality in cities. One reason is that in the agricultural societies of the past, rich people tended to live in the city, creating increasing differences in dwelling size in the built environment. There also tended to be just one rich family in the village—the chief, a few middle people and lots of poorer labourers, so the hierarchy was much simpler.

What is useful in this discussion is the key role that the built environment and the design of facilities played in the materialisation of inequalities in the landscape of ancient societies. Another important factor is the relationship between size and distribution of buildings to create a texture in the landscape that corresponded to a certain level of inequality.

Are Crises and Collapse Levellers of Inequality?

Scheidel (2017:436–442) pointed out that trends towards greater equality are unusual in history and mostly happen in extreme circumstances like mass mobilisation in wars, and during violent revolutions that have led to the collapse of states and civilisations, and alongside lethal pandemics (see also Chap. 9). Scheidel's point is that neither education, development, democracy and economic growth nor economic crises have been an antidote to inequality. Only the painful processes of transformation and collapse at the large scale have succeeded in levelling or partially levelling inequality, such as the transformation of Russia from a Tsarist state to one based on communism, although in this case, it has been argued that it was political inequality and lack of democracy rather than income inequality that led to the uprising (Lindert & Nafziger, 2014).

Revolutions have not always been successful in creating an environment of greater equality. The French revolution promoted the ideals of freedom, fraternity and equality that motivated people around the world to rethink their political environments. The independence gained from the Spaniards in Argentina and other countries of Latin America was influenced by the ideals and actions of the French Revolution. However, did the French revolution achieve the equality sought? According to Piketty (2020:128), the French revolution was not as effective as we think for reducing inequality but rather played an important role in the development and concentration of property ownership during the nineteenth century. Between 1800 and 1810, the wealthiest 1% owned 45% of all types of private property in France. This percentage rose to 65% before World War I. This shows the importance of the relationship between wealth and the built environment as an asset and how both factors impact on the unequal distribution of resources. This particular subject is further developed in Chap. 10.

Considering the evidence that links the development of societies with inequality, the question emerges as to whether it would be possible to have development without inequality. Meaningful reductions of inequality have been rare historically. One of these rare moments is the period from 1914 to 1945, a period that Piketty (2020:418) describes as "the collapse of inequality and private property". The "collapse of the total value of private property in Europe between 1914 and 1945–1950 [...] depended on several factors (destruction, expropriation, inflation) whose combined effect led to an exceptionally large fall in the ratio of private capital to national income" (Piketty,

2020:444). The quote is illustrative in highlighting the link between collapse and inequality and the importance of the built environment. The collapse of perceived value of the built environment was the foundation for the achievement of a more egalitarian society in Europe. During WWII in the UK, the level of literacy reached 90% (in 1939), rationing allowed the poor to eat better, and taxes were imposed on the rich. However, the context in which these processes happened is far from desirable—WWII was very dramatic and cost the lives of many people. However, these examples show that more egalitarian conditions can be achieved in certain circumstances.

Inequality in Contemporary Built Environments: Clustering Processes

The American Professor of urban studies, Richard Florida (2004), gained popularity with a book supporting the idea that talented people or the "creative class" (mainly professionals and artists but more particularly people involved in development of technology), are the real actors responsible for moving the economy. Therefore, cities should make efforts to please the "creative class" to attract it or retain it. However, his later 2017 book *The New Urban Crisis* concerned the idea that the process of concentrating the "creative class" around a selected number of cities in the world, which he had recommended in 2004, has contributed to an increase in the inequality between and within cities, creating "the new urban crisis." While a small group of people benefit from good salaries working in tech companies or similar in big, dense and knowledge-based cities, the rest of the people, mostly blue-collar and service workers, have felt left behind and their affordability has been reduced to the point of struggling to make ends meet after paying for housing costs (Florida, 2017: XXiV). The concentration of the creative class favours ruthless competition between cities and impacts on the accessibility of housing within cities, which in turn increases the inequality gap.

The clustering of elite groups within elite cities has encouraged the creation of a new social class: the super-rich, and this has also had an impact on the built environment. The rise of the rich and super-rich classes has created an inequality gap between a few cities and all the rest, a competition that Florida called "winner-take-all urbanism." Larger cities that get their money out of the financial sector, media, entertainment and technology, which Florida groups as the creative industries, concentrate more people, jobs and generate more growth than the rest in a disproportionate manner. In 2019, London accounted for 23% of the UK economy (Brown, 2019), while London houses about 13% of the UK population. The 2019 average house price in London was £462,000 while that of the North East was only £131,000, meaning you could buy 3.5 average houses in Tyneside for the price of one in London, although this gap in values is predicted to decrease (Lewis, 2019). Similar differences are also found in the USA, where it was reported in 2010 that for the price of one apartment in SoHo, New York City, you could have bought 29 homes in Detroit or 38 in Memphis (Florida, 2015). While clustering of money and people seems to drive economic growth, it increases the competition for land in urban areas that then becomes more and more expensive. In the USA, the average price of 1 acre (0.4 Ha) of urban land in cities with populations over 1 million is between four and five times more expensive than in less populated cities (Florida, 2017:25). The important observation of Florida is that land and housing prices are still important variables that need to be consulted in order to understand the impact the built environment has in the production of inequalities. These inequalities are more difficult to perceive through the macroeconomic lens that does not account for the built environment, which is the setting of everyday life.

The economic growth due to the clustering of jobs, people and companies around a few cities is associated with an increase in real estate values as the competition to be part of this economic growth pushes up the price of land in these locations. This led Rognlie (2015) to state, "the rise in housing's contribution to the capital share can be explained in part as the result of scarcity. The rising real cost of residential investment and the limited quantity of residential land have conspired to make housing more expensive." Capital share is that part of national income which is not allocated to wages. Basically, land scarcity means housing has become more expensive and those who own land in big cities have become much richer. This means the built environment of big cities is a medium and a catalyst that influences the way money is accumulated. The built environment is not only a place for the storage of wealth but also an attractor for investments and the landscape where future investments will be made.

Inequality, Gentrification and Segregation

Florida states that: "gentrification and inequality are the direct outgrowths of the recolonization of the city by the affluent and advantaged" (Florida, 2017:4). The "back-to-the-city movement" is a driving force that is characterised by the attraction of white affluent and educated people to live in and around city centres in order to avoid long commutes and to profit from all their amenities (from libraries to cafés) (Hyra, 2015). This has tended to lead to more compact urban areas in order to put more people close to amenities. Since, in terms of functions, these buildings need to be more diverse in the same plot, old buildings are pulled down and bigger buildings constructed. This increases the overall complexity of the built environment and the urban landscape because more people are attracted to live and work in the same place.

The densification of the urban landscape has an impact on inequality and the segregation of people who cannot be part of the changes and new demands. Florida (2017:120) used an Overall Economic Segregation Index, which included measurements of segregation in terms of economic aspects, education and occupation for metropolitan areas in the USA and found that segregation was closely related to the density of the areas. The most affluent areas, like Los Angeles and New York, have higher economic segregation levels. This trend can also be seen in London and Madrid whose inequality levels (measured with the Gini index) have increased from 2001 to 2011 (Musterd et al., 2015:10). In conclusion, Florida (2017:133–134) found that in the USA:

> ...economic segregation is worse in larger, denser more economically successful, and more diverse metros. This finding reflects the central contradiction that sits at the heart of the New Urban Crisis: the places that are the most productive and offer the highest wages, that have the largest concentrations of high-tech industry and the most talented people, that are the densest and offer the most abundant public transport options, that are the most diverse... nonetheless face the harshest levels of economic inequality and economic segregation.

The paradox is that the cause of urban economic growth and success of cities is also the root of their problems. Cities are very resilient, and regardless of the inequality and segregation they generate they keep on going (Garcia and Vale, 2017:94–113). The competition for land and housing resources is fierce and unfair and creates divided urban landscapes, which, in some places, offer the contrast between wealthy uninhabited "ghost suburbs," bought to make money (NZ Herald, 2016) and poor overcrowded slums. The effect of inequality in cities could be compared with the impact that collapse had in the built environment of ancient civilisations, with the only difference being that ghost villages in the past were the product of long droughts, wars or revolutions which in some combination forced people to abandon a place, rather than the outcome of the accumulation of wealth for the few.

Real Estate and Gentrification

The real estate market plays an important role in the concentration of money in fewer hands and the built environment is the instrument that makes this possible. The ups and downs of real estate prices "have played an important role in the evolution of aggregate capital values during recent decades" (Piketty, 2017:555). However, it is the way institutions favour owners of capital that can help to explain why capital values have been kept so high since the 1980s. Piketty (2017:555) goes on to state "institutional and legal systems have gradually become more favourable to capital owners...and less favourable to tenants and workers in recent decades, in a way that is broadly similar...to the regime that prevailed in the nineteenth century and early twentieth century." This situation contrasts with the legal mechanisms implemented between 1945 and 1980 in Europe as mentioned above that protected and favoured tenants and workers.

The important role that real estate played as a medium for increasing global economic growth has led Stein (2019a) to call it the "real estate state." In his view, the real estate state influences the form and development of the built environment, shapes politics and affects the life of people. The importance of real estate is not new. However, what is new is the scale and impact that real estate has both on politics and the landscape. Real estate shapes politics and decision-making at the municipal scale of government where planning decisions are taken and development occurs with the help of planners. For Stein, who is a geographer and urban planner, planners are wealth managers in cities controlled by the real estate market. The role of planners is to:

> ...raise property values—either because they are low and landowners want them higher, or because they are already high and city budgets will fail if they start to fall...Planners are not just shills to real estate, though; they can and generally want to make spaces more beautiful, sustainable, efficient and sociable. But without control over land the result of their work is often higher land prices, increased rents and ultimately displacement (Stein, 2019b:10).

Leilani (2017:3) in a report for the UN states "The value of global real estate is about US$ 217 trillion, nearly 60 per cent of the value of all global assets, with residential real estate comprising 75 per cent of the total." Most of this capital has been invested in land and buildings even though, for example, in terms of total economic output in the USA, manufacturing is still the most productive sector. The success and growth of the real estate market can be attributed to financial deregulation, support from urbanisation programmes (particularly in the United Arab Emirates, the USA and China, among others) and low interest rates.

A landscape controlled by the rationale of the real estate market makes the built environment more heterogeneous and creates the perfect environment for inequalities to thrive. Living in some places is more expensive than living in others. Some places that belong to the former have big, nice houses and others are totally the opposite in appearance and cost, and all these differences come at a social cost. What transforms these differences into inequalities is their unfairness. Why should living in some places be more expensive than living in others, especially when there can be little physical difference between the two? Such unhappy juxtapositions have become enshrined in the phrase "living on the wrong side of the tracks" suggesting that the railway could be the line that divides somewhere prosperous from somewhere poor.

Inequalities in the built environment create "rent gaps" (Smith, 1987) between the current housing rents in an area and the potential increase in housing rent if design "improvements" are carried out in the same area. Smith goes on to state a rent gap is most likely to occur "...in areas experiencing a sufficiently large gap between actual and potential land values." When this gap is significant it becomes a profitable business to develop new buildings or to update buildings or neighbourhoods because the future rents (or sales) will pay for the cost of the updating in the present and still leave good profits. The built environment only needs a few new buildings that elevate the rent sufficiently to create a significant gap with the surrounding properties and the landscape then becomes susceptible to new investments that will raise the prices again.

When this starts to happen, the problem is that the people living in deprived areas are suddenly not able to cope with the expensive rents and the cost of living in these improving neighbourhoods and they are evicted or displaced. This process of displacement is how the most vulnerable sectors are affected by developments associated with speculations, and this process has become known as gentrification. At the extreme end of gentrification, skyscrapers and high-rise buildings have been significant contributors to increasing housing costs and rents in cities. Moreover, in cities like New York and London, they have contributed to the creation of ghost neighbourhoods where apartments are being sold to super-rich investors who will probably never live in them (see Chap. 9). Andy Yan, an urban planner in Vancouver goes further by saying that high-rise buildings are "machines for money laundering" (Stein, 2019a).

Other Characteristics of Inequality in the Built Environment

The built environment plays an important role in an understanding of inequality and in making inequalities become tangible. The eviction of people, the displacement of minorities and the physical disparities in the quality of housing are events and characteristics that are concrete and tangible. They can easily be seen by simply walking down the street of any neighbourhood. However, there are other inequalities in the built environment that are more difficult to perceive but that are equally important because they affect the everyday life of people. In contemporary built environments, the analysis of inequalities would be very simplistic and incomplete if the focus was only on tangible inequalities. The reality of a wealthy family living in a rich area and a poor family living in a deprived area of the city are probably examples of extreme contrasts, even though both houses could be the same size. A family living in a deprived neighbourhood on minimum wages would probably be renting, be not very satisfied with the poor conditions of the house that would probably come with low environmental comfort, minimum infrastructure and limited accessibility to a good education and job. Such a house could well be situated in a less safe environment. This is a full package of adversities that highlights the great inequalities in the built environment that are not dependent on the size of housing.

There are other inequalities that are non-tangible and more difficult to perceive. As an example, a US study of around 20,000 adolescents and their physical activity found the resources related to this were "…distributed inequitably, with high-minority, low-educated neighborhoods at a strong disadvantage." In turn this led to "…disparities in health-related behaviors and obesity measured at the individual level" (Gordon-Larsen et al., 2006). Functional diversity in the built environment was also found to affect the conduct grades of school children especially males in a study of a Hispanic area of Miami. Children living in mixed-use blocks tended to have better grades than those living in residential blocks (Szapocznik et al., 2006). A relationship of high imbalance between owners and renters in a built environment could also be symptomatic of a hidden set of inequalities in the distribution of wealth and incomes that could have bigger implications for a society.

There are also scales of inequalities: global inequality, and inequalities between countries, within countries, between regions and cities, within cities, between suburbs and within suburbs (with differences between streets and even in the same street). The scales of the built environment could be important in exposing or hiding inequalities. A new housing development with standardised housing spread throughout a suburb could look very homogeneous and egalitarian at a neighbourhood scale but it might hide very unequal relationships between owners and renters that can only be analysed at a plot or block scale. When pictures of the landscape are taken from an aeroplane or drone, sometimes striking differences become visible between a wealthy gated community placed next to an informal settlement (Fig. 6.1).

However, the high inequality existing between neighbourhoods could contrast with the more egalitarian conditions found at the neighbourhood scale. Each scale has a level of heterogeneity that exposes some inequalities while hiding others. When inequalities are found within and across scales in the urban landscape, the built environment increases its complexity and the social and economic context could be the perfect environment for increasing the perception of unfairness, or the perfect disguise for softening its inequalities. These inequalities might never be the sole reason for starting a revolution or collapsing an economy but they have the potential to make the life of the people suffering them without a means of changing this situation miserable.

Fig. 6.1 Gated community of Alphaville meets Carapicuiba, Sao Paulo (adapted from Chicca, 2013:180)

Conclusions

There is a link between inequality and the collapse hypothesis. In the first stage of the development of a society, inequality might play a small role, as it did in pre-agricultural societies. However, when societies become more complex, the relationship between economic growth and inequality produces negative impacts. Societies and communities become more complex and as the current global urban situation reveals, this complexity has led to an increase in inequalities, particularly in wealthy and denser cities. Additionally, these inequalities accumulate and protect the elite at the expense of segregating the more vulnerable sectors of the society through processes like gentrification.

The built environment not only follows the development of societies but also plays a role in increasing their complexity through the real estate market and the speculation that goes with this. These increments in complexity can be seen in the urban landscape of cities, where economic growth and technological advances take shape. Increments in complexity are those related to size, density, economic growth and technology. All these factors encourage economic growth at the cost of increasing speculation and instability. Buildings become assets that play a role in the growth of economic bubbles that in turn increase the risk of economic crises (see Chap. 10 for more discussion of housing and the 2008 global financial crisis). This ends up affecting everybody but especially the people with less money that might end up losing the only house they have. Economic crises might not be a reset button when it comes to inequalities but rather a delay button, since they do not get rid of inequality forever and they do not stop the problem-solving mechanism that will keep on producing inequality.

Even though rich people in contemporary societies can make a lot of money via financial assets such as bonds, funds and stocks, the ownership and size of housing still matter for understanding the urban and spatial inequalities that are all related to the use of a finite resource: land. Since the availability of urban land can be affected by climatic change issues, like flooding, landslides, and sea-level rise, and given the imperative of protecting rural landscapes so they can keep on producing food for a rising urban population, increasing inequality in the distribution of resources can only lead to the fragmentation and ghettoization of cities. Until alternative land controls or taxes are implemented by governments, those who have significant wealth concentrated in land ownership will hold more power, and there will be less chance of tackling inequality in the built environment because changes will not be welcomed. To see the effect of change in the built environment from those wishing to make money from it, Part 2 of this chapter is a case study of gentrification in a suburb of Auckland, New Zealand.

Part 2: Inequality in the Urban Landscape of New Zealand: From the Country to the Plot

> *Gentrification zone—poor people please leave quietly*
> Matt from London/Creative Commons

Introduction

If you do not live in Oceania, it can be easy to have an idealised image of what cities are in this region of the world. Urban designers from other countries would probably have heard that Auckland, the biggest city in New Zealand, has been placed high in the world ranking of cities, while those who live there think it is a city with problems, including traffic congestion (Sankaran et al., 2005) and high house prices (Kendall, 2016). The reality is that New Zealand has been affected by the inequality crisis and this has become visible in Auckland. Part 2 proposes that historical and contextual inequalities can also be traced in the built environment. The hypothesis is that the built environment has played a key role in increasing inequalities and accumulating wealth in fewer hands. The discussion concerns an analysis of inequality in the built environment of New Zealand using multiple scales, from the country scale to the plot scale. The final aim is to discover the impact of inequality in the built environment and the role it might play in the future collapse of a built environment.

Inequalities and New Zealand

Until the mid-1980s, New Zealand has been claimed as an example of income equality in the world and one of the more equal societies across the OECD countries. Concentrations of income in New Zealand in the 1950s and 1960s were said to be one of the lowest among the developed countries (Rashbrooke, 2013:25). Income inequality is often measured using the Gini index. A Gini index of 1 means that all the incomes are concentrated on one hand (extreme inequality) and 0 means that everybody has the same income (total equality). A higher Gini index means greater inequality in income distribution, so with a high index (nearer to 1), high income people have a much larger percentage of the total income of the country. To give an idea of the Gini index, any value below 0.2 represents near perfect income equality, 0.2–0.3 shows relative equality, 0.3–0.4 represents adequate equality, 0.4–0.5 suggests a large income gap, and a value above 0.5 represents a severe income gap. Figure 6.2 shows the pre-taxed income Gini indices for New Zealand from 1935 to 2014.

There are two sets of Gini values in the chart because Creedy et al. (2017) have adjusted the values to allow for the fact that prior to the introduction of pay-as-you-earn (PAYE) income tax, a lot of income was not declared and so did not get

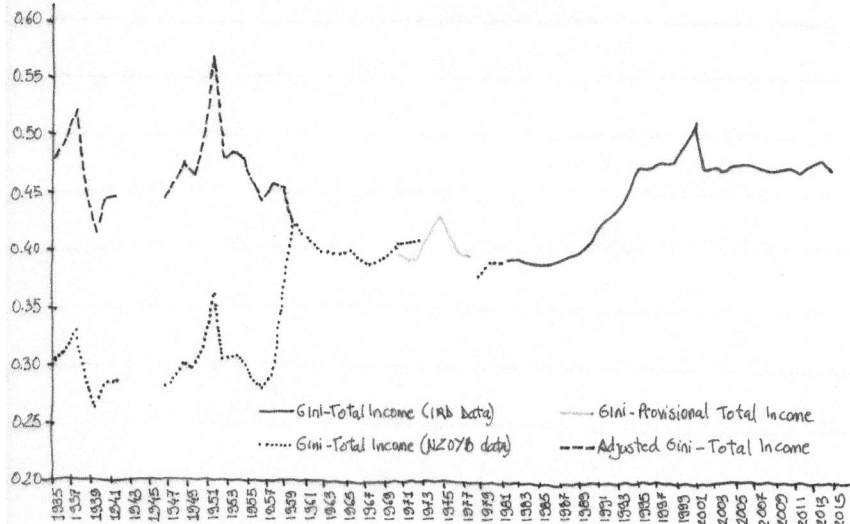

Fig. 6.2 Gini indices for New Zealand—IRD is Inland Revenue Department, NZOYB is New Zealand Official Year Book—an official digest of statistics (adapted from Creedy et al., 2017:14)

counted. The sharp jump in the Gini index calculated according to the NZOYB data in 1958 is when PAYE was introduced and the number of taxpayers increased from between 600,000 and 700,000 to more than 800,000 (Creedy et al., 2017:11). The authors' method to capture non-taxpayers in the period before 1958 gives much higher Gini values in this early period, suggesting that New Zealand has had quite a severe income gap since the 1930s. Considering the picture after 1958, Fig. 6.2 shows a relatively stable situation with less inequality between the 1960s and the end of the 1980s. In 1985, New Zealand entered the globalised world and embraced a "neoliberal" economy and at this point the pre-tax Gini index rises to a new plateau of around 0.465 with a peak in 1999/2000 when tax rates changed. Creedy et al. (2017:9) comment, "It is hard to escape the view that the increase was associated with the reforms which took place during the 1980s." The OECD (OECD.Stat, 2020) also gives a pre-tax Gini value of 0.462 for 2014, which is close to the point where a society is divided by a large income gap.

Another way of looking at inequality is by tracking how wealth has been distributed across different groups in New Zealand society. Figure 6.3 shows the percentage of the population that owned the top 10% share of national income in New Zealand from 1924 to 2014.

Figure 6.3 shows that the share of the top 10% of income has fluctuated from 1924 to 2014. The inequality created by the concentration of money in fewer hands is better in 2014, when it was shared among 31% of the population than in the 1980s when it fell below 27%, but worse than before WWII, in the 1950s or the end of the 1990s. The Gini index by comparison shows a fairly steady level between the late 1950s and the late 1980s. The two charts concur at the start of the twenty-first century, showing a rise in inequality and a sudden decrease in the share of the top

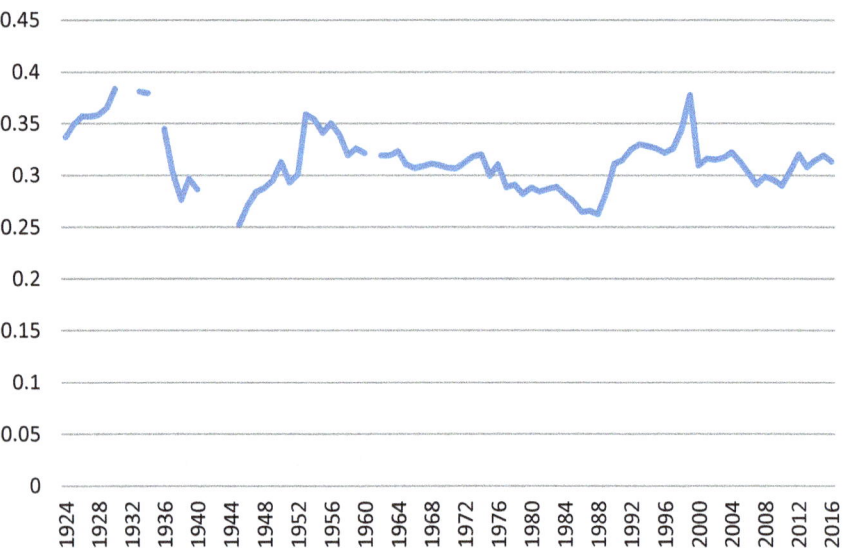

Fig. 6.3 The share of the top 10% of income in New Zealand (Based on data from WIID, 2019)

10% of income both suggesting money was more concentrated in the hands of the few. Figure 6.3 also shows that the concentration of money has been fluctuating.

The Gini index in this case only deals with income and not wealth, which in New Zealand is distributed less equally. Wilkinson and Jeram (2016:27) note for the year ending 2015 "… the top 20% of New Zealand households own about 70% of all household net worth (assets minus liabilities)—the top 10% around 50%, and the top 1% around 18%." However, this distribution was not dissimilar to that of wealth in other similar countries. The situation of wealth is complicated. House price inflation, as has been experienced in New Zealand, tends to equalise wealth distribution to some extent as owner-occupiers have a more valuable asset while at the same time "owner-occupied housing represents a smaller proportion of household assets for the wealthiest quintile" (Wilkinson & Jeram, 2016:30). However, as argued below, increasing house prices can also lead to the perception there are parts of society that 'have' and parts that 'have not.'

Maori

Any brief introduction to inequality in New Zealand would also be incomplete without mentioning a key historical factor: "the alienation of land resources through European colonisation" (Poata-Smith, 2013:148). British migrants arrived in New Zealand in the late 1830s. When the numbers started to grow, land transactions with Maori became more frequent. To secure sovereignty, the British government proposed an agreement between the Crown and the Maori people: The Treaty of Waitangi. However, the development of New Zealand brought pressure from the British people in terms of claims for more land. This process came to a tipping point in 1858, when Waikato was invaded, and land was confiscated from several Maori tribes. Confiscation of land kept happening and, as a result, by the end of the nineteenth century the Maori people had lost most of their land. The situation changed little in the twentieth century with Maori land being appropriated by the government for public works or sold (Orange, 2012). After most of the Maori land assets were taken, the Maori were vulnerable and more dependent on wage labour. This led to a rural–urban migration of Maori to work in industries that were centralised in big cities like Auckland. However, job opportunities for Maori were limited since they were discriminated against and, therefore, they ended up living in poor housing conditions and clustered in specific neighbourhoods that became the most deprived areas of the city (Poata-Smith, 2013). The impacts of colonisation and migration processes have put Maori in a disadvantaged situation. They have not been able to profit from the economic benefits of the development of New Zealand even though their land was an instrumental factor in building that development. All these factors have played an important role in the generation of inequalities in the built environment that have been manifested at multiple scales.

New Zealand Housing

The history of income inequality in New Zealand and the emergences of inequalities in the built environment appear to share a common timeline. Housing accessibility has its own history that makes it important in understanding the context of inequalities in the built environment. In the first decades of the twentieth century, New Zealand experienced significant inequalities in terms of housing conditions and accessibility to housing. Loans for buying houses were only given to the sector of society with stable incomes. This was a time when unemployment was high and many families were living in precarious conditions, in some cases being clustered in a single room, while other families enjoyed a whole house on as much as an acre of land. After the Great Depression weakened the faith in capitalistic monopolies as a mean to benefit everybody, a political change occurred in New Zealand and with it, new ideas about how to make housing more accessible. The Labour Party proposed the state should control the banking system. This led to the State Housing project, which was a programme where the state utilised the banking system to issue credit with the aim of creating new assets for the country (Bolton, 2011). This unprecedented financial move in New Zealand reduced unemployment by 75% in the aftermath of the Great Depression (Bolton, 2011). From 1935, The First Labour Government in partnership with private construction companies built thousands of high quality and affordable state houses for rent (Fig. 6.4) and also facilitated low-interest loans for building new private houses.

By 1939, the government was building 57 houses per week (Schrader, 2012a). In the 5 years from 1938 to 1943, they built 14,600 houses. These houses have become a key part of the urban landscape of New Zealand (an example can be seen in Fig. 6.11). However, by 1941, Maori were still living in poor housing conditions and excluded from having the same benefits as European New Zealanders (Howden-Chapman et al., 2013). Curiously, there is a dip in state house numbers around 1954, which corresponds with a rise in the percentage share of the top 10% of income (Fig. 6.3).

Even though the state began to sell part of this housing stock from the 1950s onwards, the government continued building state houses and providing finance for mortgages. In 1993, for example, there were 69,315 state houses and in 2009 there were 65,583. In this period, 11,504 dwellings were added to the stock, 13,494 were

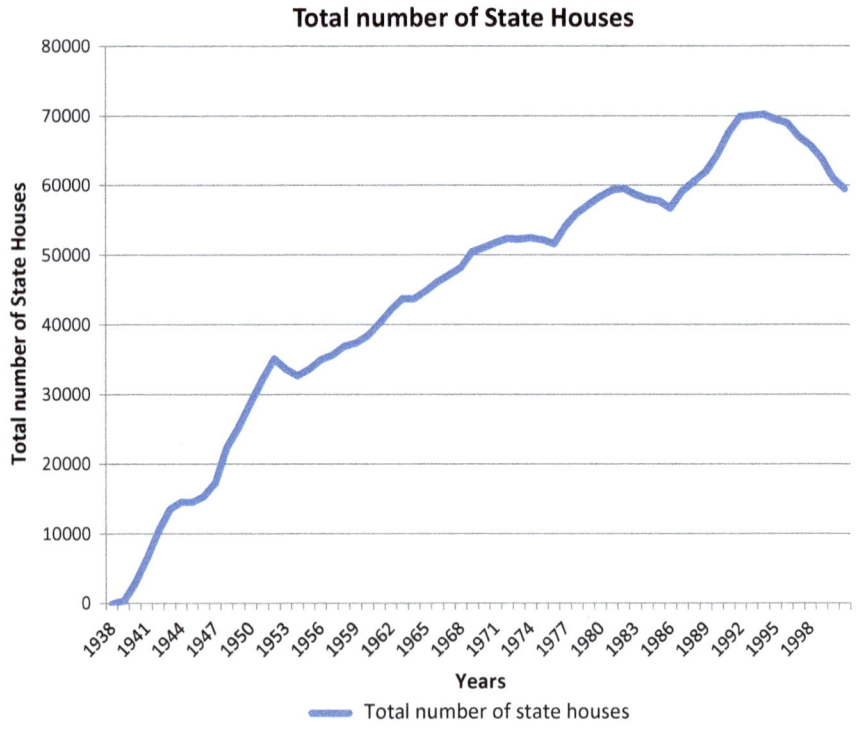

Fig. 6.4 State house numbers (adapted from Schrader, 2012b)

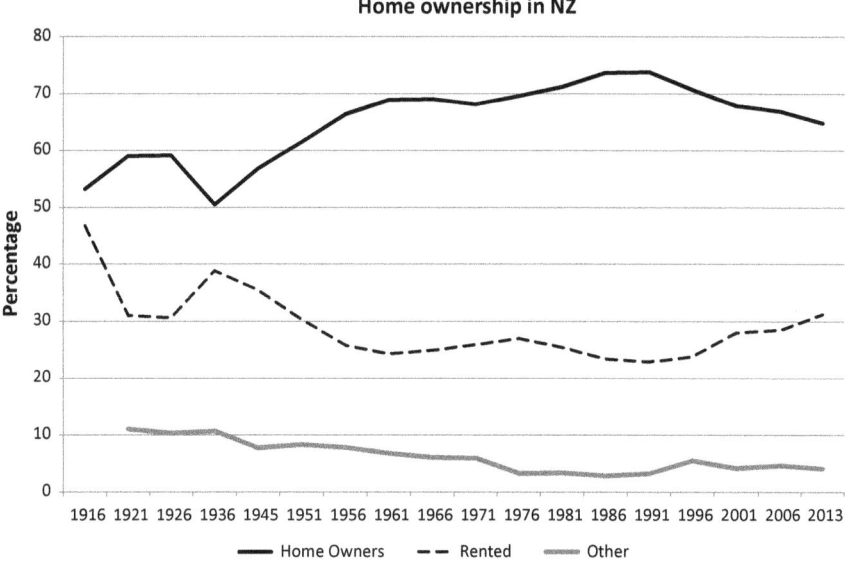

Fig. 6.5 Home ownership in NZ (adapted from Pool and Du Plessis, n.d.)

sold and 1,697 were destroyed (Olssen et al., 2010:26). The efforts to encourage home ownership found continuity in later post-war governments (Howden-Chapman, 2015). This series of housing policies meant that until the 1980s, the state financed and helped many families to get their first house (Howden-Chapman et al., 2013). New Zealand house ownership history is shown in Fig. 6.5.

As a result, from 1936, house ownership rose steadily until the beginning of the 1990s, when this trend was reversed. As noted earlier, the change in the political mindset that happened in the mid-1980s in New Zealand pushed the country to embrace globalisation. In 1991, the National Party (i.e. right of centre) government privatised the financing of housing through the banking sector, a shift that led to less support for households and less building of new affordable housing. Regulated by the market rather than by the state, increases in house prices, mortgages and inflation created the perfect environment for reduced housing affordability.

The Built Environment as a Vehicle for Accumulating Wealth and Increasing Inequalities

Until the 1990s, houses in New Zealand were affordable, namely, they represented less than 3 years of annual household income (Eaqub & Eaqub, 2015:14). From the 1990s to the present, booms in the housing market have pushed values in such a way

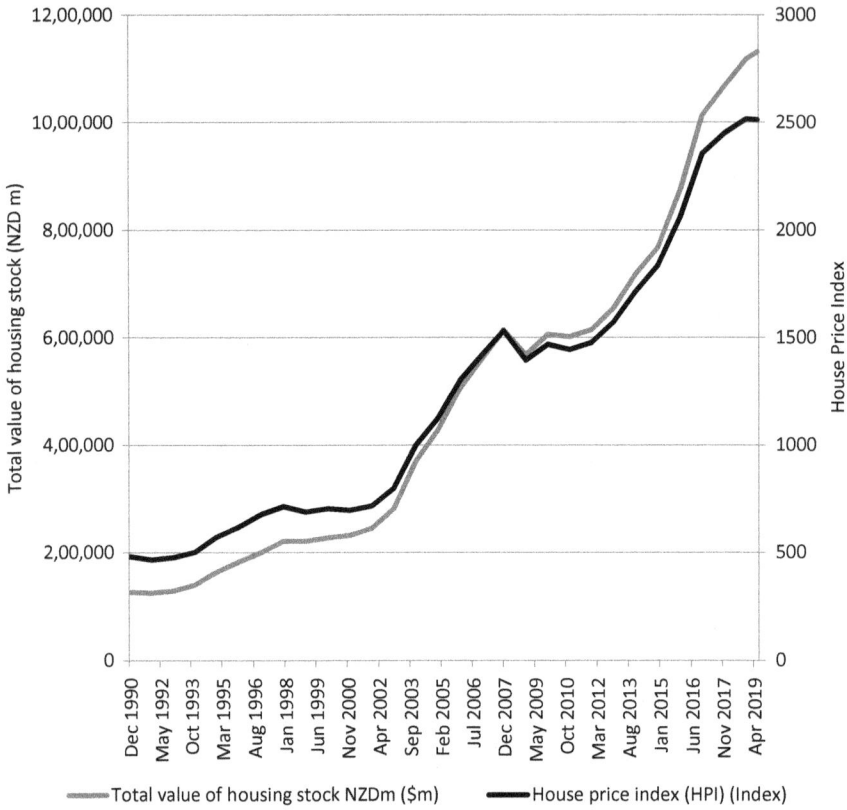

Fig. 6.6 Housing prices in New Zealand from 1990 to 2018 (based on data from RBNZ and CoreLogic, 2020)

that the average price has increased three times. This has played a role in increasing the total value of the housing stock (see Fig. 6.6).

Another factor that could have driven change in housing prices is the change in the number of people per house. Since 1991, New Zealand has experienced a decrease in dwelling occupancy. Until 2009, peaks in the rate of change of the population were buffered by an increase in housing numbers. However, since 2009, the rate of change in the housing stock has been outpaced by the rate of change in the population (Fig. 6.7). These factors and the recent increase in dwelling occupancy could be a reflection of the lack of housing accessibility due to the increase in dwelling prices.

One of the impacts of the rapid increase in housing prices is the lack of affordability of housing. From 1991 to 2018, owner-occupied houses decreased from 73 to 62% of total while rented houses increased from 23 to 34% (see Fig. 6.8). These changes do not seem big in percentage terms but they have massive implications. The 46% change in rented houses from 1991 to 2018 means that the number of rented houses has almost doubled in less than 20 years. But what is the problem of living in a rented

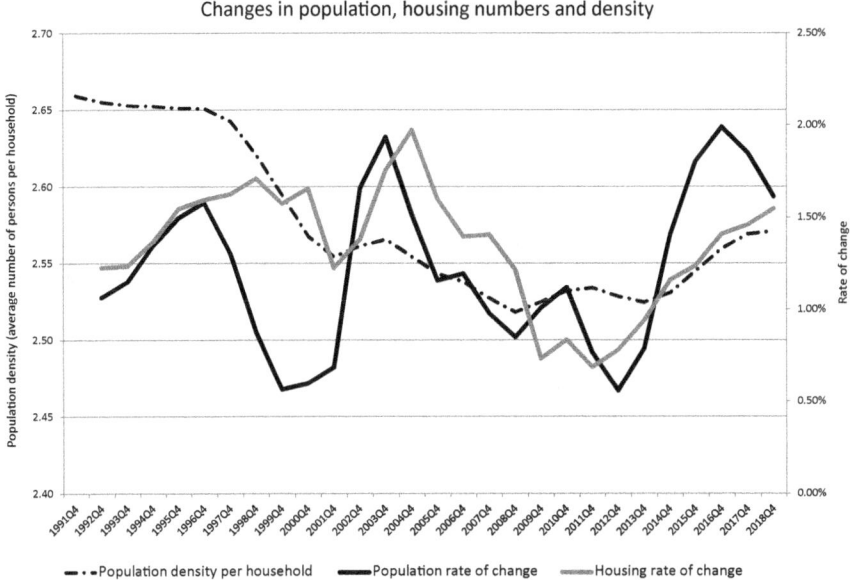

Fig. 6.7 Changes in household density, population and dwelling numbers (based on data from Stats NZ, 2020a)

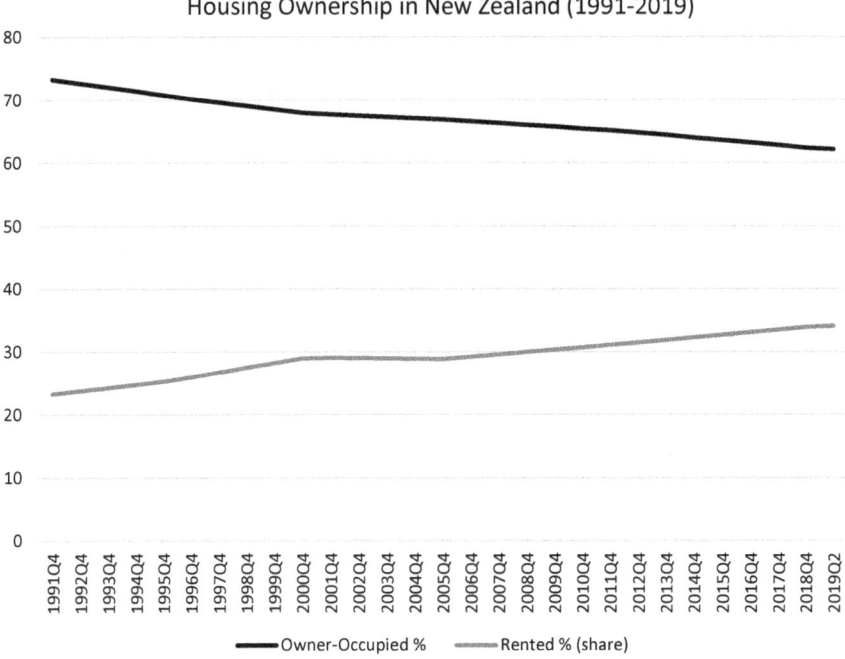

Fig. 6.8 Housing ownership in New Zealand (1991–2019) (based on data from Stats NZ, 2020b)

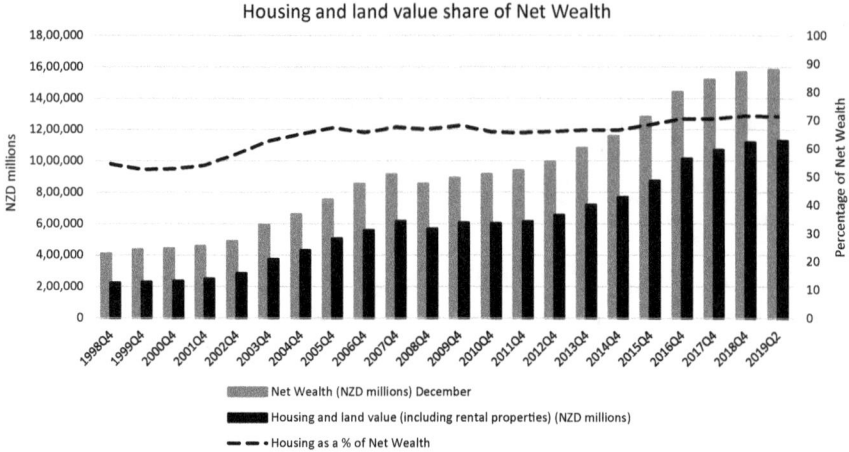

Fig. 6.9 Housing and wealth (based on data from RBNZ et al., 2020)

house? Do rental houses form a problem in the built environment? How does renting link to inequality?

In economic terms, houses are a generator of wealth (see Chap. 10). In New Zealand, the value of housing has increased at least at the pace of the rise in the value of common goods, depreciation is low, and therefore housing tends to produce benefits. Since 2000, the price of housing has produced a marginal return of at least 5% in excess of inflation. Prospective house owning landlords are allowed to borrow money from the bank, up to 90% of the value of the house, which opens new opportunities for investment that are not existing for renters, who find it difficult to save the money for a deposit as rents are also rising (Eaqub & Eaqub, 2015:18–19). Most of the investments of the upper middle class are locked in the built environment, and housing and land values represent 72% of the net wealth of New Zealanders (Fig. 6.9), making real estate one of the biggest industries in New Zealand. This shows the importance of the built environment for the economy of households and the country. It also shows that the built environment can work as an accumulator of wealth.

Since houses tend to last longer than people, the wealth accumulated in the built environment in the form of houses or other infrastructure will be passed on to future generations. In this way, the built environment becomes a vehicle through which wealth is accumulated between generations. At first sight, this seems logical and fair. However, a closer look reveals that the distribution of wealth between generations could be problematic. If younger generations do not have access to the same benefits older generations had when they were younger, then inheritances may contribute to accumulating and concentrating wealth in a disproportionate manner (Edmunds, 2019a).

In New Zealand, people older than 55 own half of the total net wealth of the country (744 billion NZD) (TV NZ, 2020). Having money when you are older is not something bad, however, when much of that wealth is in few hands and invested in housing, changes in the built environment that fail to increase property prices will never see the light. According to data from Statistics New Zealand (2017), in 2016 there were 700,000 people more than 65 years old in New Zealand, 15% of the population, and it is expected that by 2032, 20–22% of New Zealanders will be in this age bracket. However, not all of them are or probably will be wealthy. There is a growing proportion of the elderly who do not own houses, and live alone in cars or campervans, finding it hard to make ends meet every month. The Ministry of Social Development Housing Quarterly report showed that 10% of the applications for social housing are in the 65 plus group. One of the reasons for the poor housing conditions of this group of older people is the lack of housing affordability and the increase in rents. According to the Salvation Army, an agency dedicated to fighting poverty, "the Government would need to build [about] 3000 social housing units a year in order to meet the growing demand of over 65 plus" (Meij, 2018).

Concerns about intergenerational inequality should be taken seriously. From 2001 to 2013, home ownership decreased across all age groups but most for those between 20 and 60 years old (Statistics New Zealand, 2014). Homeowners are predominantly rich, 55 years old or more, New Zealand-European, and most of them are couples living in houses with four or more bedrooms (Statistics New Zealand, n.d.). If inequality keeps on growing and affordability keeps on dropping, in the next decades only the relatives of this half of the population will have easy access to buying a house. This, in turn, will contribute to exacerbating inequalities even more (Eaqub & Eaqub, 2015).

Inequalities in the Built Environment at the Country Scale

The New Zealand economist couple Shamubeel and Selena Eaqub (2015) suggested the fall in house ownership from the 1990s had impacted the identity of a whole generation by imposing on them a single option for access to housing: to be renters. The problem is that renters in New Zealand are prone to less social and financial security. Owning a house increases the bond between households and their community because they can live longer in the same place and become more involved in community activities and interests. The growing division between those who own houses and those who rent, "are a huge part of inequality in New Zealand, in terms of both wealth and health" (Howden-Chapman et al., 2013:105). The quality of rented houses tends to be poorer than owned houses and families are less satisfied, especially families with children. Only a third of renters are satisfied with their houses compared with half of house owners. In the 2018 General Social Survey, a third of renters felt their houses were cold compared with only 15% of house owners. Electric heaters and portable gas heaters were more common for renters, a factor that impacts their electricity and gas bills (Statistics New Zealand, 2019a). Half of the renters stated their houses had mould, a situation that was found in only a third of owner-occupied houses (Statistics New Zealand, 2019b). Renters are also more likely to live in unaffordable accommodation than owners. The less you have the more you pay (comparatively) for having the right to live somewhere, which is an unfair situation.

Households renting are more susceptible to be regarded as people with an inferior status in society. This way of looking at renters is sometimes expressed in the built environment. Renters are not allowed to make transformations in the houses where they live, so maintenance relies on the level of interest that the owners have in keeping the house in good condition. The problem is that this implies a constant investment, which diminishes the profit from the rent. Therefore, rented houses could have no insulation, fences are often in bad condition, facades are not painted, gutters are rotten, which could lead to leakage problems, and the insulation of doors and windows is poor.

The built environment can also be used as a commodity to make even more money, although only a few can play this game, accentuating the privileges that come with more wealth. Buying a house in New Zealand is becoming a business for millionaires or descendants of millionaires since the July 2020 median house price of the country is NZ$660,000 (REINZ, 2020). The situation is worse in Auckland, where the same source gives the median house price as $NZ920,000 NZD for the same date, the next highest median price is $NZ696,800 in Wellington. Therefore, it is no surprise that Auckland is considered one of the least affordable cities in the world in terms of housing, tying for sixth place with Toronto in the third quarter of 2019 (Cox & Pavletich, 2020:3). Speculation in New Zealand has called the attention of the international media. Auckland has been described as a "Flipping hell" after an investigation discovered that two houses in Auckland were sold five times in four days (Aigne Roy, 2017). In real estate jargon, flipping refers to buying a property to renovate it and sell it on or to speculate with the market by buying and selling houses in a short period of time (see also Chap. 10). In 2017, the market was so hot that some people made $NZ150,000 by flipping a house in just four days. At that time, 7% of the property transactions were red flagged as potential flips. However, since the definition of short term is vague, the real percentage of flipping homes could be much higher. The injustice is that only a sector of the population can play this game and make, in 4 days, more money than a whole family working in professional positions for a year. There is no merit, craft, hard work, degree or expertise that can justify the benefits from this type of speculation, which contributes to widening the gap between owners and renters, something that becomes visible in the built environment.

Inequality at City Scale: Auckland

Auckland is probably no more or less deprived than other unaffordable wealthy cities elsewhere in the world but it serves here as an example. From an inequality point of view, Auckland is a divided city where rich and deprived areas are concentrated in specific neighbourhoods. Deprived neighbourhoods are mostly concentrated in South Auckland. This geographic division is accentuated by ethnic contrasts. In the centre and surroundings of Auckland, the population is predominantly European New Zealand. In the south of Auckland, the population is predominantly Maori and Pasifika. This also implies big differences in terms of opportunities, safety, obesity levels and having less accessibility to infrastructure like transportation and high ranked schools (Fig. 6.10).

In the deprivation index, the least deprived 10% of areas would be given a score of 1 and the most deprived a score of 10. In the map (Fig. 6.10) the light areas are the least deprived and dark red the most deprived.

Auckland Topography

Auckland has a very irregular topography with 53 dormant volcanoes scattered throughout the city. Some depressions in the landscape are abrupt and sudden creating big differences between the level of the street and the ground floor of the house. Sometimes the streets are so steep and the landscape so irregular that you can walk along a footpath and see the rooftop of a house at the same level of the street. For this reason, the quality and design of the houses on each side of the street can vary greatly, with one side having big houses set two floors up from the street level with great views and orientation, while on the opposite side houses are submerged in the landscape. The large houses of rich people have traditionally been placed at or near the top of the hills, with great views and orientation while poor people have occupied the lowest areas that tend to be much less sunny and much more damp. In wealthy neighbourhoods, like Ponsonby, the topographic differences mentioned might not create big gaps in terms of the quality of the house or price of the property, however, in middle income and low-income neighbourhoods, these differences are more important.

If the long plots on the depressed side of the street have been subdivided, a series of houses appear connected to a long alley that gets lost deep in the landscape, perpendicular to the street. This is a completely different and hidden façade of the city that is not even perceived from street level. In Auckland, it rains a lot, therefore, the conditions of humidity and exposure to daylight in the last houses of a subdivided plot are less than ideal. These are often the types of house available to those renting. Nonetheless when it comes to renting, all such houses are expensive.

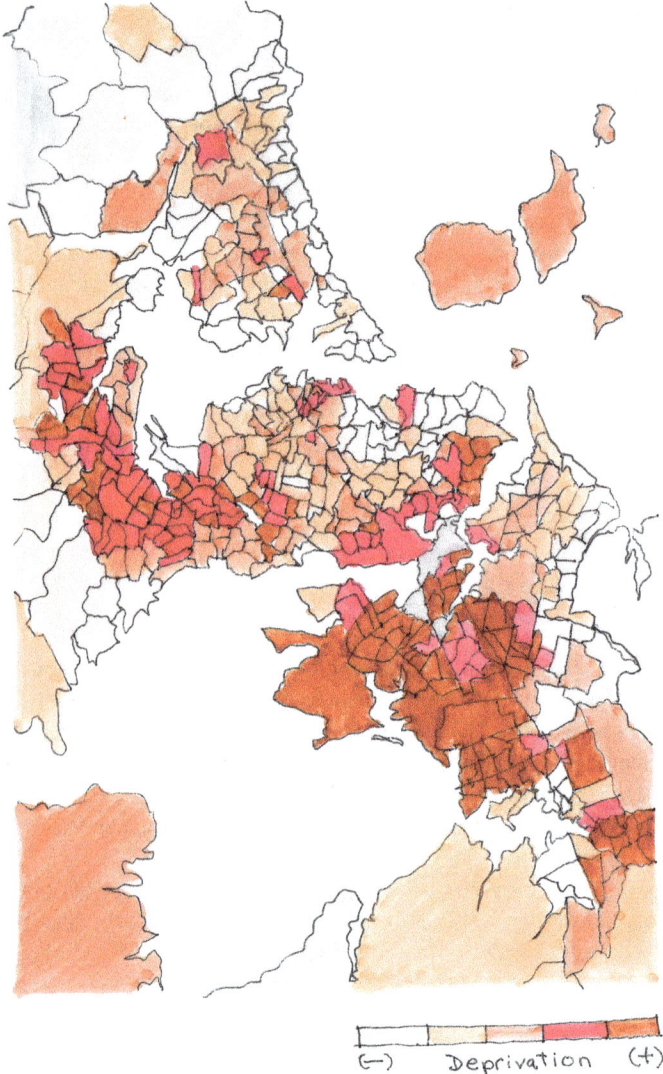

Fig. 6.10 Deprivation map of Auckland. Deprived areas are more frequently found to the South and are particularly clustered in the South East (adapted from https://ehinz.ac.nz/indicators/population-vulnerability/socioeconomic-deprivation-profile/#nzdep-for-2018-nzdep2018)

Renting

The situation of renters in the built environment can range from mild to dramatic depending on the level of deprivation found in each suburb. Inequalities between cities and neighbourhoods can be more pronounced than the inequalities perceived in the same street between two houses. In the traditionally expensive neighbourhoods of Auckland, like Saint Heliers, Parnell or Ponsonby, it is more difficult to differentiate renters from owners by looking at the houses from the street. However, it is still possible to perceive economic differences by just looking at the size of the houses, front yards and high fences. Expensive houses are bigger and have facades with more glazing (particularly facing the back yard) than smaller but well-preserved ex-state houses with fewer and more humble, original windows. In the city centre of Auckland, where more apartment buildings are found, the inequality is disguised behind the walls. Students can easily end up living in apartments that are less than ideal while paying expensive rents (Slack, 2018).

Gentrification

One of the large-scale effects in the built environment of increased rents and housing prices is the displacement of buyers and people from neighbourhoods that became unaffordable, to suburbs, villages or cities that are temporarily more affordable in a process known as gentrification. Presented in this way, it does not look as if this is a problem of fairness or injustice, only a mismatch of opportunities. The problem is that not all people have the same opportunities. When income inequalities grow, what is cheap for a few people become very expensive for the vast majority. The lack of affordability leads to instability and volatility in the urban landscape. The wider the income and wealth gap the bigger the opportunities for rich people to make more money, particularly when they are supported by changes in planning policies, like rezoning and changes in density. These efforts can literally reshape the built environment with or without the need of designers. The action or business we are discussing here is to buy a house or any other type of property at a cheap price, to invest money in its renovation and to sell it at an expensive price.

Changes in the built environment also imply changes in the social structure and behaviours of a place that can lead to friction between residents and newcomers. In 2018, people from Epsom, one of the most historic suburbs in Auckland, well known for having a strong character built on heritage houses, famous residents and good schools, opposed a Housing New Zealand project to build a five-storey building for small families and the elderly. It seems that some neighbours feared the new social housing units would make the neighbourhood unsafe, increase traffic and introduce lifestyles that did not suit the tradition of Epsom (Dunlop, 2018). Phil Twyford, a former Housing Minister, openly said that people who do not want to have affordable housing in their neighbourhoods should live somewhere else (Smith & Ross, 2018). Sadly, these words have not been translated into concrete actions. The statement made in 2018 was probably built on the expectation that the new government intended to build 100,000 new houses in 10 years. At the moment of writing this chapter in late 2019, the government has built less than a thousand houses in a year and the department in charge of executing the plan, Housing New Zealand, has been restructured and renamed. In the meantime, the waiting list for housing keeps on increasing.

The gentrification process shares some similarities with the displacement and abandonment of cities after a collapse. In both cases, there are people abandoning a place with the only difference being that, in the gentrification case, the gentrifiers immediately occupy the place of the people that they displaced. The gentrifiers are the new barbarians.

Another concern related to gentrification is its potential link with climate change in at least two ways: change in lifestyles and the increase in impervious surfaces (see also Chap. 5). The change in lifestyles happens because the newcomers have more money than the previous residents and with it the possibility to consume more things and demand more infrastructures to satisfy their level of consumption. Wealthy people tend to have expensive cars and therefore they demand garages to protect them, which in terms of size is like adding a new living room to a house. This is a draw on materials and land that was not present in state houses that rarely had garages. If the same land was developed for affordable housing, its impermeability would also increase, but the area per person would be smaller and there might be less space provided for private cars.

The change in land use impacts directly on the vegetation of the landscape, which is important for the ecosystem of cities. When two 100-year-old pōhutukawa trees were cut down in the neighbourhood of New Lynn in Auckland, there was growing concern that the building boom and the on-going gentrification in some suburbs were the perfect excuse to bulldoze the landscape. Mels Barton, one of the board members of the Tree Council, a non-profit organisation looking to protect trees, said that Auckland lost a third of its trees between 2012 and 2017 because there was no way of protecting them when private properties were gentrified (Ali & Clent, 2017). The consequences are important since trees are key elements of the ecosystem that absorb water and provide identity and environmental comfort, as well as being a cheap and effective way of buffering the effect of carbon dioxide emissions. The gentrification of greener but poorer areas of the city will increase the quantity of impervious surfaces, a factor that will contribute to the heat island effect. Gentrification is the face of inequality in the built environment. The investment in the expansion of the built environment through gentrification adds complexity since houses need to be rebuilt and new infrastructure will be installed to satisfy more expensive lifestyles. The growth and expansion of the built environment are manipulated to produce wealth that will be accumulated in fewer hands. This is done at the expense of displacing people and consuming non-renewable resources, like land, which are essential for everybody. This is a concrete link between gentrification and the hypothesis of collapse.

Inequality at Neighbourhood Scale: Gentrification in the Suburb of Glen Innes

Glen Innes is a suburb located in the east of the city (Fig. 6.11). In pre-colonisation times, it was an important place because it was flat and situated on a trading route where tribes from the north, south and east came together. The contemporary version of Glen Innes was developed after the Second World War when the Taylor family, who owned 760 acres during the nineteenth century, sold their remaining land to the government for the development of state housing. This along with other land acquisitions from local farmers was the base for the creation of the town centre known as Glen Innes (GIBA, 2016). The suburb was developed by the government with the purpose of housing war veterans and workers from industries in the area, and it became the place with the highest density of state housing in New Zealand (The Spinoff, 2019). The Maori and Pasifika population coming from rural areas and Pacific Islands found in Glen Innes a suitable place because tenancy was granted for life and the housing was affordable. Since the state houses were generous in size, this allowed Maori and Pasifika multi-generational families to become rooted. In Glen Innes, rural migrants found an environment that was a mix of the urban and rural, and they did not feel alienated (Gordon et al., 2017).

However, the situation changed in 1991 when the government introduced new policies that increased state rents to "market levels" and encouraged the sale of state houses to the private sector. This impacted on the potential impoverishment of the area (Murphy & Kearns, 1994) where almost 60% of the housing stock was state housing. Since 2009, the interventions of government in Glen Innes, with the excuse of redeveloping or "regenerating" the suburb, were based on using the state housing

Fig. 6.11 State houses in Glen Innes (adapted from author's photograph)

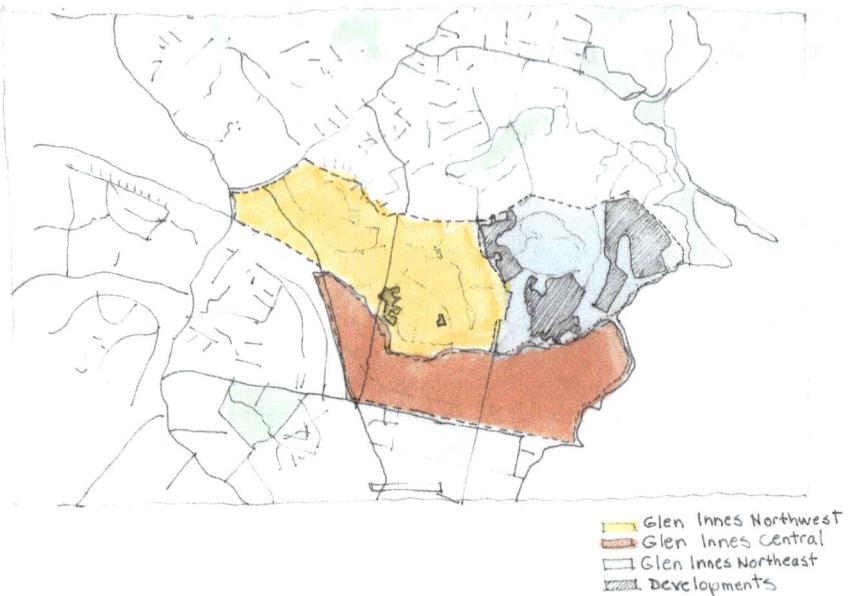

Fig. 6.12 Master plan of the new housing developments in Glen Innes (Tamaki Regeneration) (adapted from https://tamakiakl.co.nz/development/glen-innes)

stock with their big plots as the main asset to give third parties the opportunity to build more houses in the suburb, but at the expense of reducing the percentage of affordable state housing. Figure 6.12 shows the areas designated for new housing.

A firm of architects (Astonmitchell, n.d.) described the situation as one where "156 existing Glen Innes Housing New Zealand properties are being redeveloped into approximately 350 new houses. Of which, 78 will remain social housing alongside 39 affordable houses and a wide mix of new market housing." This means that before there were 156 affordable houses and after the redevelopment, there will be only 39 or 117 (if we assume that social housing should be affordable). In any case, this represents a 30–70% reduction in affordable houses. In the last decade, Glen Innes has moved from being a non-fancy neighbourhood for real estate to one of the hot spots in the city. According to a CoreLogic research study, since 2007 prices in Wai O Taiki Bay in Glen Innes grew 150%, which represents an increase in $793,200 a house, while prices in Glen Innes jumped 134% (Harris, 2018). One way of measuring the affordability is by considering how many median salaries someone will need to buy a house, 3 years being considered an affordable number (Cox & Pavletich, 2020). In 2007, the median income per household was NZ$62,000 and a house in Glen Innes cost NZ$127,000, roughly 2 years of a family income. Currently, the median price in Glen Innes is NZ$920,000 in a population where 50% are renters. The median income per person in New Zealand is NZ$52,000 and 60% of the median households earn more than NZ$100,000. Therefore, if there is one person working in a family, he or she will need 17 years of salary not 3, to buy a house in Glen Innes. Even with two median salaries buying a house in Glen Innes equates to 9 years of income. In both cases, it is clear that the property prices in Glen Innes are no longer affordable (Fig. 6.13).

This regeneration or redevelopment process has been criticised as being one of "state-led gentrification" (Gordon et al., 2017:770). The difference with other forms of gentrification is that state-led gentrification comes with support from the government, and these policies can affect larger areas at a faster pace while depleting state housing and destroying the sense of community that took a long time to develop.

The new plans for Glen Innes faced great opposition from the community but the evictions happened anyway. People received a 90-day warning before being displaced. Some former state house renting households were relocated within the area but these were exceptions. Most of the people were relocated to the periphery of Auckland, or even to other cities like Tauranga and some preferred to move to Australia (Gordon et al., 2017). To have a deeper insight into the fate of the displaced is a challenge. The media and institutions tend to privilege the stories of the positive impact of new residents but these are not measured against the impact suffered by the displaced community.

Apart from the lack of affordability of the new housing units, it is possible to summarise the impact of the gentrification process in Glen Innes in two ways: the first affects people through the economic impact on the cultural diversity of the suburb; the second affects the physical environment through the intensification of impervious surfaces. Both aspects contribute to understanding the complexity of the built environment of Glen Innes and the identity of the suburb and these aspects are developed in the next sections.

Fig. 6.13 New development in Glen Innes East (adapted from author's photograph)

Economic Impact of Gentrification

Statistics New Zealand has a way of collecting information by using a mesh that divides the area into a series of smaller units or blocks, which makes the organisation of data more manageable and provides more insight. In the case of Glen Innes, the suburb is subdivided into the three areas of Glen Innes North, West and East. The present analysis compares data from 2001, 2006 and 2013 censuses. Unfortunately, data from the last census (2018) are not available as there were problems carrying out the census on line. This could be important in understanding the major impact of the last phase of the gentrification process of 2013–2019. The data show that there is an income gap between the people living in Glen Innes North compared with those in the West and East areas (Fig. 6.14).

These inequalities can be seen in the urban landscape since the residential areas have different characteristics. Figures 6.15, 6.16, 6.17 and 6.18 show the contrast between the old state houses in Glen Innes East and West and the new housing developments in the same areas. The old houses are bigger and have generous green areas while the new houses are smaller, two storeys and have very little space for greenery or open areas not designated for cars.

This income gap plays an important role in the changes experienced by those living in Glen Innes since the poorest areas, namely the West and the East, have been targets for greater changes in the urban landscape through the gentrification process. This supports the theory of how the gentrification process happens, namely, by exploiting

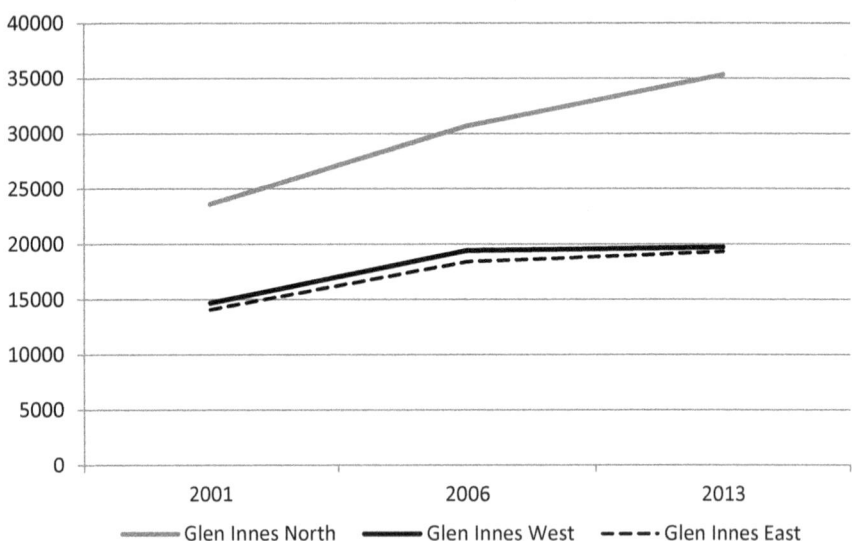

Fig. 6.14 Incomes within Glen Innes. Real values according to the period of each census (based on data from Stats NZ, 2013)

Fig. 6.15 State house in Glen Innes West (adapted from author's photograph)

Fig. 6.16 New housing units in Glen Innes West (adapted from author's photograph)

the weakest points of the system. If prices are defined by the market, there will always be new vulnerable areas that will again be the target of new displacements.

The income inequalities between Glen Innes North and Glen Innes East and West are not the only differences between these areas. The population in Glen Innes North is higher than in the other two areas. Figure 6.19 shows that the number of people in Glen Innes West and East declined between 2006 and 2013 while the national population was increasing. This information matches with the testimonies of school principals who have perceived a decrease in the number of students due to

Fig. 6.17 State houses in Glen Innes East (adapted from author's photograph)

Fig. 6.18 New housing development in Glen Innes East (adapted from author's photograph)

the relocation of families (Gordon et al., 2017). Even though more new houses are in the process of being built, if middle-income families displace Maori and Pasifika multi-generational families, this will have an impact on the population.

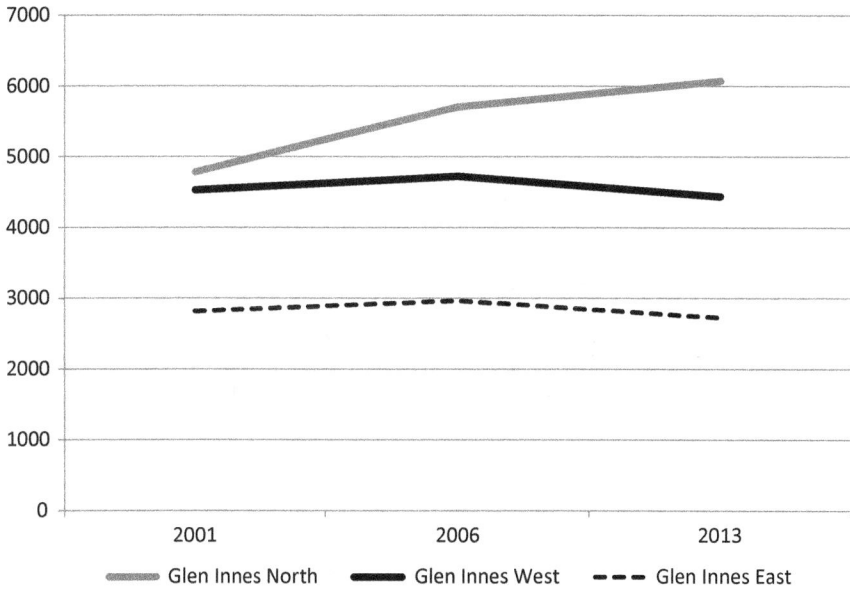

Fig. 6.19 Changes in population (based on data from Stats NZ, 2013)

Another factor is the very different distribution of people by ethnicity. Glen Innes North is predominantly European New Zealand (70%), and in Glen Innes East and West, the Maori and Pasifika population forms more than 50% of total. The data presented in Fig. 6.20 show that the Maori and Pasifika populations in Glen Innes West and East declined from 2006 to 2013, while the European New Zealand population has increased in the same areas. This could be one impact of the changes in the built environment and the displacement of predominantly Maori and Pasifika families from former state housing.

Even though the companies and institutions building the new houses are trying to create designs which blend into the landscape and where the differences between houses are not perceivable, the reality is that differences exist, since the new houses are not affordable and contrast with the remaining state housing and other buildings in the area. One important value of the old state houses is that they were all equally good.

The new plans for the regeneration of Glen Innes along with other private developments in the area are trying to increase the built housing density per plot. Therefore, in areas where there were three state houses, eight or more houses will now be built. This is not a bad intention considering that there is a housing supply crisis in Auckland. What is questionable is the way in which the new developments are designed by prioritising space for cars over space for children to play freely and safely (Fig. 6.21).

Another problem is the quantity of impervious surfaces added with each development. Without counting the impervious surfaces added by new roofs, since the purpose of the project is adding more houses, what is done with the areas between

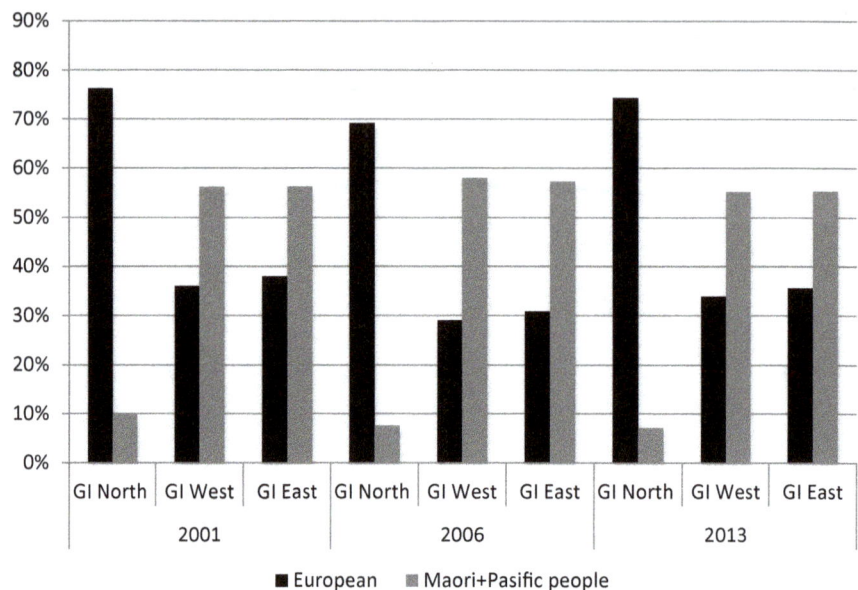

Fig. 6.20 Population changes (based on data from Stats NZ, 2013)

Fig. 6.21 New housing development in Glen Innes. In order to get access to houses located deep in the plot internal streets become wider and longer increasing the impervious surfaces (adapted from author's photograph)

houses is concerning. Unfortunately, current aerial views or satellite maps are not available to compare the previous with present changes. However, the difference with previous development is so evident that a simple picture taken from the street provides concrete evidence, quite literally, of the area of impervious surfaces generated on top of the area for the new houses.

Collapse and Gentrification

The gentrification caused by the interventions in the so-called regeneration of Glen Innes has some similarities and differences with the process of collapse. It is clear that the scales are different: anthropologists' and archaeologists' studies of collapse are more linked to civilisations and states and here we have a case study that works at the scale of a suburb. Nonetheless, it is possible to find similarities if we analyse collapse as a process that can happen at different scales, as in ecological resilience. The on-going changes in the built environment of Glen Innes have happened rapidly, in less than 10 years, and they involve the suffering of people who were displaced in 90 days. The diversity of a built environment is susceptible to playing different roles in inequality and the collapse process depending on the context in which it happens. In a rich suburb, if diversity and inequality are low it means that the suburb is a concentrated cluster of wealth that probably contrasts with the rest of the city. In these suburbs, inequality and diversity might be welcomed in order to integrate people and create more equality at a city level. Perhaps in this context, the social mixing approach could be used to soften the differences, although as noted in the case of Epsom above, this can lead to protests from the existing residents who do not want to see change. In any case, an increase in complexity might be a positive step, even though it might imply the collapse of the identity of the suburb, something that can impact on the exclusivity of a place and its economic value. In a deprived suburb, the situation is completely different. Low inequality and diversity within the suburb could mean that all the people are equally poor and suffering from the same problems. In the first stage, the gentrification process creates diversity by adding a new group of people, new houses and increasing the average income. This is the beginning of the collapse of the suburb. When the social structure is homogenised, income inequality erased, and most of the old houses supplanted by new types, the diversity at social, economic and built environment levels decreases, and it would be possible to say that the complexity of the suburb has also rapidly decreased, thus there is a collapse process at a neighbourhood scale.

Inequality at the Plot Scale: A Divided Garage City

"Home is where the garage is" was the title of an article published in an Auckland Sunday newspaper (Bucknell, 2015) describing the change in the size and function of garages. Garages are bigger than ever before and are now often the same size as an average house in New Zealand in the 1970s. Two-car and three-car garages are becoming the norm, which implies a big change in the footprints of new houses. In New Zealand, the size of new houses has almost doubled between 1974 (108.7 m^2) and 2011 (191.6 m^2) (Khajehzadeh, 2017:6), and a factor that has contributed to this change has been the rise in the number of cars per household. Globally, New Zealand is a country with a high percentage of cars per household (Autofile, 2017). The New Zealand Ministry of Transport state (2015) "In the late 1980s, 40 percent of households had one car and just over 35 percent of households had two. By 2011–2014, this had switched and 40 percent of households had two cars, while 34 percent had one car."

A study done by Vale and Khajehzadeh (2016) shows, not surprisingly, that having more cars implies more garages and parking spaces and that this has an impact on the size of houses. Larger houses tend to have bigger garages and more parking spaces. Garages are an item that might contribute to increasing the price of a house or its rent. House owners, especially middle- and upper-class families with children, have double or triple garages that can have the same build qualities as the rest of the house. The footprint of these garages can easily exceed the parking space per car since the same study found they are normally used for other functions like a laundry, gym and for the storing of gardening equipment and appliances. However, low-income families in deprived areas of Auckland, who own second-hand cars and tend to park in the street or front yards, might also have garages but they can be used in a very different way by being rented out to people with nowhere else to go. In 2016, a New Zealand radio show reported that an Auckland landlord was renting out garages and single rooms, sometimes as housing for a family of five. Garages had been rented out for as much as NZ$400 a week (Ashton, 2016). Unfortunately, this situation is not exclusive to Auckland. A family from Whangarei in the north of New Zealand described how complicated it was living in a garage for 2 years. They paid a rent of NZ$250. As the mother declared "There just aren't enough rentals to go around. And most are through rental companies who want perfect people, and I look pretty alternative" (Edmunds, 2019b). After going through a series of unfortunate events, the family decided to buy a truck and live there, because it offered more space and privacy. It is not surprising to find these cases since rental inflation has gone up. Using data from Statistics New Zealand, Amore (2016:11) showed that the number of people living in severely deprived housing conditions had increased 40% between 2001 and 2013. The number of people living in garages is difficult to estimate because it is hidden within the 70% of deprived people (28,563) "living as temporary resident in severely crowded private dwelling due to a lack of access to minimally adequate housing" (Amore, 2016:11). What is clear is that while some people have decided to live with garages and treat the car like another member of the family, other people are living in garages and treated worse than cars. A part of Auckland looks like one of the most liveable cities of the world for cars and the other part of Auckland is becoming a precarious garage city for people.

Conclusion

The concentration of wealth in New Zealand has not always been the same. It has changed according to the development of contextual factors that include changes in the mind-sets of people as well as governmental policies. Therefore, inequality is not a curse or destiny if there is the will and sense of emergency for reversing the situation. In the spirit of Clifford Hugh Douglas who developed the economic reform social credit movement (Douglas, 1924), "Money is not Wealth but only its token, and tokens cost next to nothing to produce. So what is physically possible and socially desirable can certainly be made financially possible" (Bolton, 2011).

There are four impacts of inequality in the built environment of New Zealand: (1) the lack of accessibility to housing, (2) differences in the housing qualities according to tenancy and location, (3) the displacement and segregation of people and communities and (4) the increase in impervious surfaces and its contribution to climate change-related problems (see Chap. 5 Part 2).

In the built environment, inequality is becoming more visible and this should be a concern. The problem is so important that it is becoming an intergenerational issue. In the future, only old rich people will own their houses with sizeable gardens, and they will have the privilege of allowing their families to build in these. In the worst-case scenario, they might have to sacrifice a pool or a tennis court. However, in South Auckland, more people will end up at the mercy of rapacious landlords or having to ask friends and relatives to allow them to live in their garages. The more divided the tenancy becomes the more explicit the inequalities will be in the built environment. Inequalities are more obvious between neighbourhoods because social and economic factors push the population to cluster in different areas of the city. The moment inequalities in the built environment are found within the same street, it is the moment when things have become ugly because it is the point where the two classes have polarised the urban landscape into the super-rich and the super poor. This is a pressure for collapse that comes from inside the system, rather than outside like climate change. The collapse of Teotihuacan centuries ago has also been linked to internal pressure from unrest in a multi-ethnic society (see Chap. 4), and this should be seen as a warning. Polarisation increases the complexity of the built environment because it multiplies problems and creates parallel universes in the same territory. This complicates governance because the needs of the population are split.

Having a divided city presents a problem since it affects the way inequality is perceived and therefore the way of behaving to reduce the gap between classes. If the wealthy children of one family never play with the poor children of another, how can they understand that there is a reality other than the one in which they live? Moreover, how would they see the need to do something about inequality? The differences between the quality of housing and the satisfaction of people with their own houses are all key variables in this. Renters are more susceptible to living in unhealthy environments, which is costly for them and for the government. Cities that will have to cope with potential collapse need to be big in terms of spaces and opportunities in order to develop some resilience. Inequality only shrinks cities by dividing them into the most liveable urban areas and "surviving-whatever-areas". If inequality worsens, wealthy people will end up living in a portion of the city with the majority of the benefits in terms of infrastructure, quality and services but they will not be able to move far away from that area since they will perceive the disfavoured part of the city as unhealthy, unsafe, ugly and full of unwanted situations and realities. They will live in a small city surrounded by the inequalities that they have contributed to building. Arguably, the rest of the people will be able to move freely in the city but in fact this is far from true. Apart from walking on the sidewalk, the remaining places will become more and more privatised and expensive. Without money, the less wealthy will not even be able to use their cars since parking in the city is costly and public transport in deprived neighbourhoods is increasingly dysfunctional. People who have been displaced or gentrified are still in vulnerable situations and often do not have the money to get by until the end of the month. People with a few more resources might be able to go to work and come home, and that will be all. They will live in a big city with very few options other than staying at home, eating fish and chips in the car park of a supermarket or, if homeless, buying a pie and staying the whole day in a coffee shop.

When speculation, which is led by the greed of one sector of a society (individuals, private businesses and institutions), is the driving force of change, the housing inequality crisis is just an opportunity to profit from someone else's misfortune and misery. The higher the inequality gap between prices, places, income and tenancy, the higher the number of people there will be moving around and displacing other people. It becomes a built environment of predators and prey, or in the real estate jargon, the gentrifiers and the gentrified, a rule that works well in ecosystems but not for people. The critical effect this is having in the built environment is the addition of impervious surfaces, particularly roads and space for cars, within blocks and plots. This is progressively destroying the landscape by bulldozing trees, vegetation and reducing green areas. The consequences are extremely serious because the built environment loses a very efficient and cheap way of absorbing carbon dioxide emissions by replacement with a new and bigger problem. The basic issue is there are too many people chasing a scarce commodity—land.

Such a city will not be a city that can cope with itself, let alone deal with other issues that go beyond the local scale, like the challenges posed by human-induced climate change. Inequality adds complexity to the built environment that is very costly. The more inequality in the built environment the more difficult it becomes to get collective benefits from economic or any kind of growth. The collapse of social housing in New Zealand has turned the city into a costly system whose population is more expensive to keep healthy, where houses are more expensive to keep warm, with a volatile population, unevenly distributed in contrasting environments, with a preference for housing cars before housing people, and with poor people whose best option is to live in a van or a truck.

Inequality is dividing cities by displacing people and clustering resources into fewer hands and places. A global collapse can amplify the impact of these dynamics or be an excuse to reset the economy. The landscape of inequalities presented in the built environment can put people in danger at an exponential rate. At a large scale, inequality is a push to the edge of the cliff, it might not be enough to create a fall but it can leave you hanging on the edge at the mercy of any other disturbance, social, economic or environmental.

References

Aigne Roy, E. (2017). *Flipping hell! New Zealand property frenzy as two houses sold five times in four days*. Retrieved August 12, 2020, from https://www.theguardian.com/world/2017/jan/18/flipping-hell-new-zealand-property-frenzy-as-two-houses-sold-five-times-in-four-days.

Ali, M., & Clent, D. (2017). *Felled 100-year-old pōhutukawa sparks fears of gentrification*. Retrieved August 12, 2020, from https://www.stuff.co.nz/auckland/local-news/western-leader/99673862/felled-100yearold-pohutukawa-spark-fears-of-gentrification.

Amore, K. (2016). *Severe housing deprivation in Aotearoa/New Zealand 2001–2013*. University of Otago, Wellington.

Ashton, A. (2016). *Families housed in garages, cramped rooms, face eviction*. Retrieved August 12, 2020, from https://www.rnz.co.nz/news/national/309571/families-housed-in-garages,-cramped-rooms,-face-eviction.

Astonmitchell. (n.d.). *Northern Glen Innes redevelopment*. Retrieved August 12, 2020, from https://www.ashtonmitchell.com/project/northern-glen-innes-redevelopment/.

Autofile. (2017). *Is NZ's car ownership the highest worldwide?* Retrieved August 19, 2020, from https://autofile.co.nz/nz-car-ownership-tops-global-ranking.

Bolton, K. R. (2011). State credit and reconstruction: The first New Zealand labour government. *International Journal of Social Economics, 38*(1), 39–49.

Brown, J. (2019). *London is still the UK's golden goose – and that needs to change*. Retrieved August 16, 2020, from https://www.theguardian.com/commentisfree/2019/may/20/london-uk-economy-decentralisation.

Bucknell, C. (2015). *Home advice: Home is where the garage is*. Retrieved August 12, 2020, from https://www.nzherald.co.nz/lifestyle/news/article.cfm?c_id=6&objectid=11493674.

Chicca, F. (2013). *Developing a label of excellence in design for urban sustainability*. Ph.D. Thesis in Architecture. Victoria University of Wellington.

Cox, W., & Pavletich, H. (2020). *16th Annual Demographia International Housing Affordability Survey*. Retrieved August 18, 2020 from http://www.demographia.com/dhi.pdf.

Creedy, L., Gemmell, N., & Nguyen, L. (2017). *Income inequality in New Zealand, 1935–2014* (Working Papers in Public Finance 07/2017). Victoria University of Wellington. Retrieved August 16, 2020, from https://www.wgtn.ac.nz/cpf/publications/working-papers/2017-working-papers/WP_07_2017_Income_Inequality_in_New_Zealand.pdf.

Douglas, C. H. (1924). *Social credit*. C. Palmer.

Dunlop, R. (2018). *Housing NZ apartment plans in Epsom draw local ire*. Retrieved August 12, 2020, from https://www.nzherald.co.nz/nz/news/article.cfm?c_id=1&objectid=12053090.

Dyble, M., Salali, G. D., Chaudhary, N., Page, A., Smith, D., Thompson, J., Vinicius, L., Mace, R., & Migliano, A. B. (2015). Sex equality can explain the unique social structure of hunter-gatherer bands. *Science, 348*(6236), 796–798.

Eaqub, S., & Eaqub, S. (2015). *Generation rent: Rethinking New Zealand's priorities*. Bridget Williams Books.

Edmunds, S. (2019a). *Policies 'hard-code' differences between baby boomers and younger generations, researcher says*. Retrieved August 13, 2020, from https://www.stuff.co.nz/business/117339974/policies-hardcoded-differences-between-baby-boomers-and-younger-generations-researcher-says.

Edmunds, S. (2019b). *A year living in a garage: 'I've given up'*. Retrieved August 12, 2020, from https://www.stuff.co.nz/business/109995107/a-year-living-in-a-garage-ive-given-up.

Florida, R. L. (2004). *The rise of the creative class: And how it's transforming work, leisure, community and everyday life*. Basic Books.

Florida, R. L. (2015). *What can you buy for the price of one SoHo Apartment?* Retrieved August 16, 2020, from https://www.bloomberg.com/news/articles/2015-07-14/how-much-real-estate-can-you-buy-for-the-price-of-one-new-york-apartment-in-soho.

Florida, R. L. (2017). *The new urban crisis: How our cities are increasing inequality, deepening segregation, and failing the middle class—and what we can do about it*. Basic Books.

Garcia, E. J., & Vale. B. (2017). *Unravelling Sustainability and Resilience in the Built Environment*. Routledge.

GIBA (Glen Innes Business Association). (2016). *Glen Innes Village: Looking back 60 years 1956–2016*. Retrieved August 12, 2020, from https://www.gleninnesvillage.co.nz/pdf/GI-jubilee-booklet.pdf.

Gordon, R., Collins, F. L., & Kearns, R. (2017). 'It is the people that have made Glen Innes': State-led gentrification and the reconfiguration of urban life in Auckland. *International Journal of Urban and Regional Research, 41*(5), 767–785.

Gordon-Larsen, P., Nelson, M. C., Page, P., & Popkin, B. M. (2006). Inequality in the built environment underlies key health disparities in physical activity and obesity. *Pediatrics, 117*(2), 417–424.

Greater Auckland. (2019). *Deprivation-map*. Retrieved August 18, 2020, from https://www.greaterauckland.org.nz/2019/01/30/right-to-the-city-who-benefits-from-the-crl/deprivation-map/.

Harris, C. (2018). *Where house prices soar—Auckland dominates NZ's top 100 suburbs for property gains*. Retrieved August 12, 2020, from https://www.stuff.co.nz/business/property/106687022/nzs-100-fastestgrowing-suburbs-by-property-price-are-nearly-all-in-auckland.

Howden-Chapman, P., Bierre, S., & Cunningham, C. (2013). Building inequality. In M. Rashbrooke (Ed.), *Inequality. A New Zealand crisis* (pp. 105–117). Bridget Williams Books.

Howden-Chapman, P. (2015). *Home truths: Confronting New Zealand's housing crisis*. Bridget Williams Books Limited.

Hyra, D. (2015). The back-to-the-city movement: Neighbourhood redevelopment and processes of political and cultural displacement. *Urban Studies, 52*(10), 1753–1773.

Kendall, E. (2016). New Zealand house prices: A historical perspective 1. *Reserve Bank of New Zealand Bulletin, 79*(1), 3–14.

Khajehzadeh, I. (2017). *An investigation of the effects of large houses on occupant behaviour and resource use in New Zealand*. Ph.D. Thesis. Victoria University of Wellington.

Kohler, T. A., Smith, M. E., Bogaard, A., Feinman, G. M., Peterson, C. E., Betzenhauser, A., Pailes. M., Stone, E. C., Prentiss, A. M., Bowles, S., et al. (2017). Greater post-Neolithic wealth disparities in Eurasia than in North America and Mesoamerica. *Nature, 519*, 619–622.

Kohler, T. A., Smith, M. E., Bogaard, A., Peterson, C. E., Betzenhauser, A., Feinman, G. M., Oka, R. C., Pailes, M., Prentiss, A. M., Stone, E. C., Dennehy, T. J., & Ellyson, L. J. (2018). Deep inequality: Summary and conclusions. In T. A. Kohler & M. E. Smith (Eds.), *Ten thousand years of inequality: The archaeology of wealth differences* (pp. 289–318). University of Arizona Press.

Lobao, L. M., Hooks, G., & Tickamyer, A. R. (Eds.). (2007). *The sociology of spatial inequality*. SUNY Press.

Leilani, F. (2017). *Report of the special rapporteur on adequate housing as a component of the right to an adequate standard of living, and on the right to non-discrimination in this context*. Retrieved August 1, 2020, from https://digitallibrary.un.org/record/861179?ln=en#record-files-collapse-header.

Levitt, M. S. (2019). The neglected role of inequality in explanations of the collapse of ancient states. *Cliodynamics: The Journal of Theoretical and Mathematical History, 10*(1), 31–53.

Lewis, C. (2019). *North to narrow property price gap with South*. Retrieved August 16, 2020, from https://www.thetimes.co.uk/article/north-to-narrow-property-price-gap-with-South-lwldtvxqr.

Lindert, P. H., & Nafziger, S. (2014). Russian inequality on the eve of revolution. *Journal of Economic History, 74*(3), 767–798.

References

Meij, S. (2018). *Older people forced to sleep in car as housing crisis bites*. Retrieved August 13, 2020, from https://www.stuff.co.nz/national/106063441/rise-in-homeless-older-generation-predicted-as-housing-crisis-bites.

Milanović, B., Lindert, P., & Williamson, J. (2007). Measuring ancient inequality. *Munich personal RePEc archive*. Retrieved August 16, 2020, from https://mpra.ub.uni-muenchen.de/5388/1/MPRA_paper_5388.pdf.

Ministry of Transport. (2015). *25 years of New Zealand travel: New Zealand household travel 1989–2014*. Ministry of Transport.

Mulder, M. B., Bowles, S., Hertz, T., Bell, A., Beise, J., Clark, G., Fazzio, I., Girven, M., Wiessner, P., et al. (2009). Intergenerational wealth transmission and the dynamics of inequality in small-scale societies. *Science, 326*(5953), 682–688.

Murphy, L., & Kearns, R. A. (1994). Housing New Zealand Ltd.: Privatisation by Stealth. *Environment and Planning A, 26*(4), 623–637.

Musterd, S., Marcińczak, S., van Ham, M., & Tammaru, T. (2015). *Socio-economic segregation in European capital cities: increasing separation between poor and rich* (IZA Discussion Paper No. 9603). Retrieved August 1, 2020, from http://ftp.iza.org/dp9603.pdf.

NZ Herald. (2016). *Australia's ghost suburbs a 'national scandal'* 29 Mar. Retrieved August 19, 2020, from https://www.nzherald.co.nz/business/news/article.cfm?c_id=3&objectid=11613193.

OECD.Stat. (2020). *Income distribution database*. Retrieved August 12, 2020, from https://stats.oecd.org/Index.aspx?QueryId=66597.

Olssen, A., McDonald, H., Grimes, A., & Stillman, S. (2010). *A state housing database: 1993–2009* (Motu Working Paper 10–13). Wellington: Motu Economic and Public Policy Research. Retrieved August 17, 2020, from http://motu-www.motu.org.nz/wpapers/10_13.pdf.

Orange, C. (2012). *Story: Treaty of Waitangi*. Retrieved August 12, 2020, from https://teara.govt.nz/en/treaty-of-waitangi.

Piketty, T. (2017). Toward a reconciliation between economics and the social sciences. Lessons from Capital in the Twenty-First Century. In H. Boushey, J. Bradford Delong & M. Steinbaum (Eds.), *After Piketty. The agenda for economics and inequality* (pp. 543–565). Harvard University Press.

Piketty, T. (2020). *Capital and ideology*. Harvard University Press.

Poata-Smith, E. T. A. (2013). Inequality and Māori. *Inequality: A New Zealand Crisis*, 148–158.

Pool, I., & Du Plessis, R. (n.d.). Families: A history. *Te Ara—The Encyclopedia of New Zealand*. Retrieved August 17, 2020, from https://teara.govt.nz/en/graph/30207/rates-of-home-ownership.

Rashbrooke, M. (2013). Inequality and New Zealand. In M. Rashbrooke (Ed.), *Inequality a New Zealand crisis*. Bridget Williams Books.

RBNZ, & CoreLogic. (2020). *House prices and values*. Retrieved August 17, 2020, from https://www.rbnz.govt.nz/statistics/key-graphs/key-graph-house-price-values.

RBNZ, CoreLogic, & Ministry of Education. (2020). *Key household financial and housing statistics*. Retrieved August17, 2020, from https://www.rbnz.govt.nz/-/media/ReserveBank/Files/Statistics/tables/c21/hc21.xlsx?revision=aaa317fc-b278-4e0f-a66b-bbab8d54cf43.

REINZ (Real Estate Institute of New Zealand). (2020). *Residential property data*. Retrieved August 17, 2020, from https://www.reinz.co.nz/residential-property-data-gallery.

Rognlie, M. (2015). Deciphering the fall and rise in the net capital share: Accumulation or scarcity? *Brookings Papers on Economic Activity, 2015*(1), 1–69.

Sankaran, J. K., Gore, A., & Coldwell, B. (2005). The impact of road traffic congestion on supply chains: Insights from Auckland, New Zealand. *International Journal of Logistics: Research and Applications, 8*(2), 159–180.

Scheidel, W. (2017). *The great leveller: Violence and the history of inequality from the stone age to the twenty-first century*. Princeton University Press.

Schrader, B. (2012a). Housing and government. Retrieved August 12, 2020, from https://teara.govt.nz/en/housing-and-government/print.

Schrader, B. (2012b). *Housing and government: State loans and state houses*. Retrieved August 12, 2020, from https://teara.govt.nz/en/graph/32421/total-state-housing-stock.

Slack, M.-M. (2018). *The lows and lows of flat hunting in Auckland*. Retrieved August 18, 2020, from https://www.rnz.co.nz/news/the-wireless/375138/the-lows-and-lows-of-flat-hunting-in-auckland.

Smith, M. E., Kohler, T. A., & Feinman, G. M. (2018). Studying inequality's deep past. In T. A. Kohler & M. E. Smith (Eds.), *Ten thousand years of inequality: The archaeology of wealth differences* (pp. 3–38). University of Arizona Press.

Smith, N. (1987). Gentrification and the rent gap. *Annals of the Association of American Geographers, 77*(3), 462–465.

Smith, S., & Ross, H. (2018). *Phil Twyford rules out 'state-sponsored gentrification' in Auckland*. Retrieved August 12, 2020, from https://www.stuff.co.nz/national/politics/104193297/phil-twyford-rules-out-statesponsored-gentrification-in-auckland.

The Spinoff. (2019). *Lessons on the Auckland housing crisis from Glen Innes*. Retrieved August 18, 2020, from https://thespinoff.co.nz/auckland/inmybackyard/12-11-2019/lessons-on-the-auckland-housing-crisis-from-glen-innes/.

Statistics New Zealand. (n.d.). *Owner-occupied households*. Retrieved August 17, 2020, from http://archive.stats.govt.nz/browse_for_stats/people_and_communities/Households/housing-profiles-owner-occupied/household-income.aspx#gsc.tab=0.

Statistics New Zealand. (2013). *Census map—QuickStats about a place*. Retrieved August 17, 2020, from http://archive.stats.govt.nz/StatsMaps/Home/Maps/2013-census-quickstats-about-a-place-map.aspx.

Statistics New Zealand. (2014). *2013 Census QuickStats about housing*. Retrieved August 17, 2020, from http://archive.stats.govt.nz/Census/2013-census/profile-and-summary-reports/quickstats-about-housing/home-ownership-individuals.aspx#gsc.tab=0.

Statistics New Zealand. (2017). *Population projections overview*. Retrieved August 17, 2020, from http://archive.stats.govt.nz/browse_for_stats/population/estimates_and_projections/projections-overview/nat-pop-proj.aspx#gsc.tab=0.

Statistics New Zealand. (2019a). *Renting vs owning in New Zealand*. Retrieved August 12, 2020, from https://www.stats.govt.nz/infographics/renting-vs-owning-in-nz.

Statistics New Zealand. (2019b). *Wellbeing statistics: 2018*. Retrieved August 12, 2020, from https://www.stats.govt.nz/information-releases/wellbeing-statistics-2018.

Statistics New Zealand. (2020a). *Population summary figures*. Retrieved August 17, 2020, from https://www.stats.govt.nz/assets/Uploads/Population-summary-tables-1991-2019.xlsx.

Statistics New Zealand. (2020b). *Demography dwelling and household estimates—DDE table: Estimated private dwellings, as at quarter ended (qrtly-mar/jun/sep/dec)*. Retrieved August 17, 2020, from http://archive.stats.govt.nz/infoshare/SelectVariables.aspx?pxID=5afd05fa-4145-4435-8295-6457fbfc2a7a.

Stein, S. (2019a). The housing crisis and the rise of the real estate state. *New Labour Forum, 28*(3), 52–60.

Stein, S. M. (2019b). *Capital city: Gentrification and the real estate state*. Verso Books.

Szapocznik, J., Lombard, J., Martinez, F., Mason, C. A., Gorman-Smith, D., Plater-Zyberk, E., Brown, S. C., & Spokane, A. (2006). The impact of the built environment on children's school conduct grades: The role of diversity of use in a hispanic neighborhood. *American Journal of Community Psychology, 38*(3–4), 275–285.

Tainter, J. A. (1988). *The collapse of complex societies*. Cambridge University Press.

TV NZ. (2020). *With NZ's wealth concentrated in Kiwis over 65, the system is broken, researcher says*. Retrieved September 24, 2020, from https://www.tvnz.co.nz/one-news/new-zealand/nzs-wealth-concentrated-in-kiwis-over-65-system-broken-researcher-says#:~:text=Older%20New%20Zealanders%20aged%2055,they%20only%20have%20%24197%20billion.

Vale, B., & Khajehzadeh, I. (2016). The environmental impact of the way residential parking facilities are used in New Zealand. *Paper Presented at the 9th Australasian Housing Researchers Conference*, The University of Auckland. Retrieved August 19, 2020, from https://cdn.auckland.ac.nz/assets/auckland/creative/our-research/doc/urban-research-network/vale-khajehzadeh-residential-parking-facilities.pdf.

References

World Income Inequality Database (WIID). (2019). Retrieved August 17, 2020, from https://www.wider.unu.edu/database/wiid.

Wilkinson, B., & Jeram, J. (2016). *The inequality paradox*. Wellington: The New Zealand Initiative. Retrieved August 17, 2020, from https://nzinitiative.org.nz/assets/Uploads/The-Inequality-Paradox.pdf.

Chapter 7
Growth and Collapse

> *Alice:*
> *Caterpillar:*
> *By the way, I have a few more helpful hints. One side will make you grow taller ...*
> *Alice:*
> *One side of what?*
> *Caterpillar:*
> *...and the other side will make you grow shorter.*
> *Alice:*
> *The other side of what?*
> *Caterpillar:*
> *THE MUSHROOM, OF COURSE*
> *Lewis Carroll*

Introduction

The previous chapters have linked growth in complexity with collapse. This chapter sets out to look at the issue of growth itself, although it is obvious that as a society grows it does become more complex just to be able to function. A link between physical growth and the collapse of a past society was made by Taagepera (1978), who noted the following: "Few empires or states in human history have lasted for a thousand years. Empires have tended to collapse or shrink after only a few centuries or even years at near-maximum size, to be replaced by new entities with a different name, ethno-cultural identity and geographical focus." Taagepera's interest in size was because it could be measured although he noted that area did not necessarily correspond with political power (Taagepera, 1979). By mapping the area covered by empires in Asia and Europe against time, including the Roman (Fig. 7.1), Taagepera (1979) showed a steady growth in the area of the empire, which was followed by a stable period and then a rapid decline in the area. This growth followed by a period of decline is mirrored in the direction of the curves in the adaptive cycle (see Fig. 4.4).

Tainter (1988:125–126) reproduced Taagepera's curves and described the situation further: "Growth begins slowly, accelerates as the energy subsidy is partially invested in further expansion, and falls off when the marginal cost of further growth

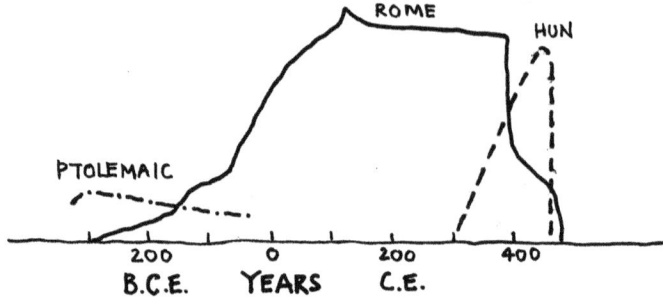

Fig. 7.1 Areas covered by three empires over time (adapted from Taagepera, 1979)

becomes too high." He further comments that these early civilisations lacked the technical knowledge to solve some problems such as exploiting fossil fuels for energy, so acquiring more land was a way of obtaining an energy supply, for example through exploiting forests, but this came at the cost of having to expend energy in order to collect energy, much like the case of energy return over investment discussed in Chap. 5. Once the marginal returns are exceeded, there is nowhere else to go but retrenchment, which if severe appears as a collapse. More recently, Taagepera (2014) made the point that "…individual civilisations have collapsed before, and having now a single world-wide civilisation makes humankind singularly vulnerable." The close connectivity could lead to a domino effect whereby the failure of one country or region could trigger another, something that was seen with the 2008 global financial crisis. This raises the question of how much growth is possible in the modern world to avoid collapse and what might be the effect of this growth on the built environment? Taagepera originally mapped the land area of empires but Tainter would argue it is not physical area but the growth in complexity leading to negative marginal returns on the investments made to solve problems that precedes the possibility of collapse. However, neither of these views, rather like Alice trying to find out which side of the circular mushroom would make her grow, really addresses what is growing.

As populations grow then the built environment has to change to accommodate more people, either by growing in area or by being used more intensively. However, economies also grow, not only to enable the people involved to have what they need for life (with luck it might be a healthy and fulfilled life) but also so that more wealth can be generated, however this might eventually be distributed (see Chap. 6). The built environment plays a role in economic growth as it can be changed to facilitate wealth generation. Money invested in building urban roads can help speed goods and services around an economy but roads also have an effect on urban pollution, which is normally not costed as part of wealth generation. In turn, the generation of wealth can affect the environment, an example being the growth in house areas in some developed countries such as New Zealand and Australia (Fuller & Crawford, 2011) (see also Chap. 6). Bigger houses use more resources and require more land to contain them. This chapter looks at these various issues in an effort to understand how and why the built environment grows and, more importantly, what form this

Introduction

growth might need to take to best survive any future substantial change as might come with the issues facing modern global society, such as climate change.

Growth

People grow in height until the point where they stop, usually around the age of 19 at the end of puberty (Taylor-Miller & Simm, 2017). Toward the end of life people also lose stature (Dahl et al., 2010). Cessation of growth is also observed in other flora and fauna. As a tree grows the path to conduct water from the soil to the topmost branches increases and these "water transport dynamics" ultimately determine the final height of the tree (Domec et al., 2008). There is thus for a tree a trade-off between the benefits of growth—better access to sunlight for photosynthesis—and the problems of supplying water and nutrients from the roots to the rest of the tree. This balance leads to an ultimate height and cessation of growth. It thus appears that unlimited growth is not part of how the natural world, and humanity as part of this, works. Nevertheless, in a planet abundant with natural resources, growth can happen and has happened in a way that makes it appear to be infinitely possible to keep growing. Kahn and Simon (1984:3) state: "We are confident that the nature of the physical world permits continued improvement in humankind's economic lot in the long run, indefinitely." This optimism occurs despite examples of the disappearance of former abundance. Bardi (2017:98) gives a current example of this by noting the lack of fish at Stintino in Sardinia, the Italian island that gave the sardine its name (BBC, 2020).

> What makes the town a truly unique place in the world is the fleet of old sailboats kept and maintained in perfect shape by the descendants of the ancient fishermen. These boats were once used for fishing, but, today, nobody in Stintino lives on fishing anymore and the boats are kept for pleasure trips only.

A similar story of the growth in fishing leading to loss of fishing grounds is found in the history of the American whaling industry, which produced oil leading to the need to "…supply the demand that had become rapacious" (Starbuck, 1878:96). This led to opening up new whaling grounds, specifically the mostly uncharted Pacific Ocean, whereas previously in 1600, when the industry first started in the New England colonies "…the whales were very numerous, both along the coast and in deep water" (Starbuck, 1878:5). Despite this history of overfishing leading to the need to search out new fishing grounds, or new ways of deploying the investment in sea-going equipment, as in the example from Sardinia (possibly the only island to have a fish named after it), the oceans are still overfished (Barkin & Sombre, 2013:1). The same authors also point out that the fish only belong to someone once they are caught, so investing in the biggest and fastest boats increases the chances of being in possession of the fish. This investment cannot lie unused so more investment is made in technology to make sure these vessels can catch the remaining fish, depleting the stocks even faster (Barkin & Sombre, 2013:2–3). This unchecked growth in the ability to catch fish is predicted to lead to the collapse of some fish stocks as "…the percentage of stocks fished at biologically unsustainable levels increased from 10 percent in 1974 to 33.1 percent in 2015, with the largest increases in the late 1970s and 1980s" (FAO, 2018:12). This means fishing needs to be controlled to avoid loss of these endangered fish stocks and as an industry fishing cannot continue to grow forever.

Another view of growth is given by Ulanowicz (1986). He defined growth from a network perspective, which is a way of analysing the flows in a system by using their elements and the interactions between elements. From this point of view, the "size" of a system is not only linked to its spatial extension (the number of nodes or parts) but also with the total output of flows running through the system. Therefore, a system will be growing when both its parts and its total flows increase. In this sense, development is an increase in the organisation of a system, not just an increase in its size. The hypothesis is that in an organised system, the flows between nodes will be more predictable than in a less organised system. Imagine one street with 20 neighbours who all usually drive to work in the same place and there is only one route to take to get to work. If, alternatively, the same neighbours have 20 possible routes, which of the two cases will make it easier to predict when all 20 reach the workplace? This will be the first case. The first case would be considered "less connected" but "better articulated" since the information about what flows are leaving a node and what are arriving at another one is easier to predict than in the second case. Therefore, the first case will be described as the more developed. In order to describe growth and development combined, Ulanowicz (1986:102) uses the term "ascendency," which refers to "an increase in network size and organisation." Earlier in the book Ulanowicz (1986:5) states, "The observable drives of living systems towards coherency, efficiency, specialisation and self-containment are argued to be implicit in the "principle" of optimal ascendency." Conversely (Ulanowicz, 1986:7) the things that get in the way of growth and development are "…diversity, variability, senescence, and dissipation." It is useful to compare Ulanowicz's theory with the later work of Holling and his followers on the adaptive cycle (see Fig. 4.4). In the latter, the increase in capital, which could be read as an increase in complexity, is limited since it leads to the release phase. The theory of ascendency, perhaps as a challenge to ecological resilience theory, only deals with the upward curve of the adaptive cycle (exploitation and conservation) without recognising there could be release and reorganisation. Development in an ecological sense means growing to maturity, which is followed by death and recycling of the remains so the process and can start again (Garcia & Vale, 2017:22). Looking at Ulanowicz's idea that diversity impedes growth and development, diversity increases the complexity of a system but that, in turn, may push it to the release phase of the adaptive cycle. What the adaptive cycle shows is that growth and development have cycles, something that observance of the seasons in temperate climates makes obvious.

Economic Growth

Given that for a resource like fish in the world's oceans, which seems endless, growth in the fishing industry has to be curtailed, why do countries insist on continued growth in their economies? Victor (2010) noted that the prominence given to economic growth is a relatively new phenomenon. "Regular estimates of GDP by governments date back only to the 1940s, and the measure was initially used in support of specific objectives, such as stimulating employment. Only in the 1950s did economic growth become a policy priority in its own right." Whatever its history, those who propose there is no need to worry about economic growth, such as the quote from Kahn and Simon (1984) given above, put forward the idea that through human ingenuity, there will be less reliance on the natural resources that are recognised as finite. Referring to those who have predicted world doom, such as the Club of Rome in their report *Limits to Growth* (see Chap. 1), Ramez (2013:104) states: "It's not that they were all wrong. It's that they were all wrong in the same way, and for the same reason. They all ignored or underestimated the most critical human faculty that exists, and the most important source of our prosperity. Innovation." Blair (2015:18–19), in writing about the need to rethink economic growth to take account of biophysical limits, summarises this position.

> This model presents two possible methods for decoupling output (Y) from resource input (R): either labor (L) and capital (K) inputs may be substituted for natural resources, or technological progress (A) may shift the entire production function, allowing more output per unit of input. Since there are no inherent limits (in neoclassical theory) to either substitution or technological progress, there are no inherent limits to resource decoupling.

However, as long ago as the nineteenth century, Jevons noted that technological progress that allowed for more output per unit of input did not necessarily lead to less use of resources. In his book on coal published in 1865, Jevons noted that the more efficient use of coal led to an increase rather than a decrease in its consumption (Jevons, 1865:VII.3). This has become known as the rebound effect where improved efficiencies do not necessarily lead to reductions in resource use. Examples given by the UK Energy Research Centre (2020) describe the person who buys a more efficient car and then spends the money saved on fuel driving further and the household that insulates their loft and goes overseas for a holiday on the money saved. Rees and Moore (2013:14) take this further by stating "However, historically, in the absence of policies to negate the so-called 'rebound effect', efficiency gains have actually stimulated consumption by lowering costs and enabling wage increases," but without explaining the detail of this. However, in a report on the effects of energy efficiency measures to meet climate change commitments on the Canadian economy, Dunsky Energy Consulting (2018:1) state this will "…add 118,000 jobs (average annual full-time equivalent) to the Canadian economy, and increase GDP by 1% over the baseline forecast, over the study period (2017–2030)," suggesting that investment in energy efficiency will stimulate the economy.

To combat the rebound effect Canadian economist Peter Victor (2018:306) argues that emphasis on growth needs to be replaced with an emphasis on human development that takes account of natural limits. This idea of limits will be developed further in the section on growth in the built environment. In his simulation model of the Canadian economy based on these principles, Victor (2018:306) found that:

> ... it is possible to develop scenarios over a 50-year time horizon for Canada in which full employment prevails, incomes are more evenly distributed, people enjoy more leisure, greenhouse gas emissions are drastically reduced as are other environmental burdens, and household and government debt is contained, all in the context of a cessation of economic growth.

More leisure occurs because work is distributed across the population so that everyone has some but work is not necessarily full time. In this way, it could be said there would be more equity in such a no-growth society. Whether this would lead to a more equitable distribution of income is questionable. Piketty (2015:66–99) argues that the reasons behind income inequality are complex, briefly being a compounded mixture of access to education and willingness to be educated and prejudice against certain groups in society. Taxing the higher paid to transfer money to the lower paid is a tested method of dealing with income inequality in an attempt to create a more equal standard of living. However, rather than probing further how a no-growth society might operate the first consideration should be the effect of continuing economic growth on the built environment. To do this, it is first necessary to consider how urban areas grow.

Growth and Buildings

Growth in the built environment can mean a number of different things. Jane Jacobs (2000) distinguished the terms growing, expanding, extending and developing in her discussion of economies, and these terms are often linked to the built environment, so exploring them further using the simple example of a single house could be useful in order to understand what is growing. A household may start with just two people and they might begin their life together in a small terraced house with two rooms and a kitchen downstairs and two bedrooms and a bathroom upstairs. Their indoor space is limited to this. If they then have a child the household has grown as where there were two people there are now three. In Jacobs' terms, the family is developing, as something new—a child—has come out of an existing situation—a couple. Jacobs (2000:17) also states development is an "open-ended process, which creates complexity and diversity," a situation with which most new parents would agree. At this stage in this imaginary story, the assumption is the same level of economic activity in the family can sustain this development. However, there may well be a change in the level of available income if one parent opts to stay at home as sole carer while the other works. At this stage there is a free bedroom for the child. If and when the parents have a second child, they can grow the number of beds so two children now sleep in the one room. Again, the assumption is the level of economic activity has not changed. However, a third child could lead to the need to expand the house up into the loft space and this will require funding. This might happen with a new job for the sole carer that pays more. For the house, the built form is still the same shape but the interior space has expanded within the same envelope to cope with a household that has grown from two to five people. According to Jacobs (2000:43), development and expansion "...are tightly locked. They make each other possible," although, in the example of the family, it was the development led by the increase in family members that led to an expansion of floor area. People are the agents of change and it is their decisions that made the house, in this example, and continue to make cities more complex. Without people, the built environment remains a tool not in use. There is also a parallel with resilience theory in this example. The house has become more diverse as it now has more bedrooms but this has added to its complexity as these same bedrooms imply more time spent cleaning them and finding the resources to decorate and furnish them.

To continue the example, as the children reach their teenage years, they keep asking for a bedroom each so at this stage the house is extended into the rear garden to create an additional ground floor bedroom. Fortunately, the sole carer is now free to work while the children are at school, so the family income has also developed making the expansion of the house possible. There is now a larger built footprint and less open space. The parents also decide to upgrade the kitchen, which is now looking shabby and as their eldest child has now grown up and left home, they extend the old kitchen into what was the downstairs bedroom to create a fashionable entertainers' open plan kitchen. In doing this, they could be described as developing the potential of the house and making it into something more desirable than previously. In this example, first, it is the population that grows but keeps occupying the same space and only later does the built space grow. This is important as in any discussion of the growth of a built environment it is essential to know what is growing.

Would this house ever collapse? Hopefully, unless there is some type of neighbourhood re-zoning that requires demolition, for instance to make way for a new road, the house will remain. However, it may not prove suitable for the family once the children have left home, leading to the parents selling up and moving on. In this case, what has happened is the collapse of the family/house system, leading to a reorganisation with the children leaving home and the parents moving elsewhere, which again links to the idea of the adaptive cycle. The house remains and will now become the base for a new family/house system. This is the real issue for the built environment, it is the systems within it that change and sometimes the systems collapse rather than the built fabric.

Urban Growth Patterns

Just as the house above changed in typical ways—extension upwards into an existing loft space followed by building more house on the back—there are also patterns in the growth of urban areas. The important thing to note about urban growth is that most parts of most cities are old rather than new, the latter being visible as the limited number of new building works. As Whitehand (1994) stated, "We are living in built environments many of which have been bequeathed to us by our ancestors and which we shall in turn pass on to our descendants." Whitehand goes on to describe a pattern of urban growth that starts with a boom in house building, which is followed by a drop in land prices that leads to a different form of development on the urban fringe, as land is now cheaper. This might include institutions with landscaped grounds, parks and cemeteries. Whitehand's argument is that the successive rings of development "leapfrog" (i.e. jump over) the fringe area to start a new area of development further out from the city centre. However, across this cycle is also redevelopment, whether wholesale or piecemeal, which again changes the appearance of the urban fabric. This is what Conzen observed in his study of Alnwick leading to the "burgage cycle" (see Chap. 4). This has produced the so-called doughnut effect where the city centre, which would be expected to have the highest land values loses out to the economic activity in the new suburban areas with their shopping malls and other attractions, at least until the city centre area is redeveloped and again becomes an attractive place to live, thus starting the cycle again.

Whitehand (1994) also makes the point "…that the landscapes that we are altering today are not just our landscapes; they are also the landscapes of our ancestors and our descendants." We have managed to squeeze higher speed travel into this pattern of growth, and continue to do so with investment in roads, but in a time of climate change when there is a need to drastically curtail the use of fossil fuels should we not be building cities that are appropriate for handing on to generations who may have to travel differently? Creutzig et al. (2015) suggest that taxes on fossil fuels could help produce the urban compaction that would aid a return to walking or cycling to work and the switch to electric vehicles, and that together these actions could see the transport sector meet its CO_2 reduction targets, but that without a rapid effort to do this "transport may remain a roadblock to the world's efforts to mitigate climate change." This suggests we have reached the point rather like that of the growing family with the house described above, who moved once they had exhausted all the possibilities offered by their existing house. Perhaps urban growth should now be focussed inwards by lowering the land area per citizen but in such a way as to avoid creating places where people do not want to live. One way of doing this is to house more people on the same area of land although this does not necessarily mean high-rise dwellings (see Chap. 8). This, however, may be culturally dependent as people currently living in older European cities and in Chinese cities where space per person is less may be more flexible to change than those living in cities characterised by urban sprawl in the USA or Australasia (Dong et al., 2019). However, once walking and cycling become the norm, the perception of the city will change from that of car-using citizens and much land currently given over to roads, carparks and garages could be freed for other purposes.

How Its Citizens Perceive the City

As noted above, if cities grow through an increase in urban population then the built environment must also change to accommodate these extra bodies. This happens either through the city growing outward or the city growing upward but such change is also fluid in terms of the location of the residential, commercial and industrial buildings that go to make up a city. People need to live somewhere and they need to work somewhere so not only is the distance between work and home critical but also where that work is located. If all work is centralised to take advantage of proximity and the potential innovation this encourages as espoused by those who believe in Smart Cities (see Chap. 2), then a system has to be set up to ensure all workers can reach the central district in reasonable time and at a reasonable cost, something that few modern cities achieve.

In 1994, Marchetti formulated the principle named after him that people are willing to have a daily one-way commute of 30 min. This was based on the idea that human beings are territorial animals that wish to expand their territory but have to balance this with "…the physical exertion of moving over larger distances, and because moving means to be in the open, under the possible threat from enemies and predators" (Marchetti, 1994). For those only moving on foot, this produced a territory, which if it were a circle would have a radius of 2.5 km and an area of 20 km^2. This he showed was a typical pattern for the distribution of villages in Greece. An earlier study had suggested that irrespective of culture people were commuting for an hour each day, and using the fastest means of transport available to them given cost and location (Zahavi, 1979). Based on this, Marchetti explored different types of settlement from villages to walled cities, showing that each was based on walking for an hour a day. The introduction of motorised transport changed the size of cities, but based on the commute of an hour a day, Marchetti could "…muse about a city of 1 billion people, which would require an efficient transportation system with a mean speed of only I50 km/h." This is the speed of the original high-speed train—the Japanese Bullet Train. Construction of a specific line for this train began in 1959 for an opening in 1964 for the Tokyo Olympics. Within a year, it had an average operational speed of over 160 km/h including two intermediate stops (Amos et al., 2010:3–4).

Obviously, it is not possible to run such a high-speed rail service as a means of dealing with commuting from all residential areas to places of work, but how people get about in a city is part of how they perceive themselves as its citizens. If you are wealthy and can afford high-speed travel the whole city is yours, whereas if you are poor and are reliant on walking or public transport, the "city" you inhabit will be a lot smaller, again as people found in the COVID-19 lockdown (see Chap. 9). It also depends on how near to the station you live and how far your workplace is from the station at the other end. In the 1950s, Shell built their new London headquarters right next to Waterloo Station so that their office staff could commute by train.

This issue of access means as a city grows its growth will be perceived differently by its citizens, also noting that it is the wealthy with their ability to move who will be generating far more of the transport-associated greenhouse gas emissions than the poor. In their survey of all modes of travel including air travel and associated GHG emissions for Oxford in the UK, Brand and Boardman (2008) found a "...strong link between emissions and personal income". Chancel and Piketty (2015:32) found the same connection and further observed that between 1998 and 2013 "...the group representing the 2% lowest CO_2e emitters in the world, saw its per capita CO_2e emissions level decrease by 12% between 1998 and 2013", showing that global warming is not caused by the poorest in the world's societies.

Design of the Built Environment

It might seem obvious that congestion resulting from the growth of cities could be controlled by design. All that is needed is to make streets wider so that traffic can flow more freely. However, it seems that building more or wider roads just leads to more traffic and eventually congestion. The introduction of highways into a city to cope with increased car numbers has also been linked to rise in crime, by cutting off areas of the city and also destroying their socio-economic systems through the demolition of key buildings, such as a market, in order to create road space (Gómez et al., 2020). In a study of US highways and highway vehicle kilometres travelled (vkt), Duranton and Turner (2009:28) found the "...data show that building roads elicits a large increase in vkt on those roads." This suggests that the redesign of the urban fabric is not an answer to traffic congestion, but neither is the design of more efficient vehicles. In a meta-analysis of the effect of improved energy efficiency in transport, Dimitropoulos et al. (2016:7) found that the rebound effect meant that people travelled more by private transport leading to fewer energy savings and also "... contributing to mileage-related externalities, such as non-exhaust air pollution, noise, congestion and traffic accidents." Congestion is led by human behaviour not design. Design is involved in the growth of urban areas but it cannot necessarily solve the problems that come with growth such as crime and congestion. Design also cannot deal with the non-built environment problems that come with city growth. "The type and scope of these impacts vary but include air and water pollution, land change, loss of natural habitats, strain on water resources, higher demand for energy, and rising greenhouse gas emissions" (Brelsford et al., 2017). This re-echoes the view that solving one problem leads to bigger problems to solve, and hence an increase in overall complexity.

At first sight, it might seem that design could deal with greenhouse gas emissions in the built environment since it is possible to make a building that is zero energy in its operation. Even here, there are issues as more materials may be required to make this possible, so a life cycle energy analysis is needed alongside a life cycle greenhouse gas analysis in order to see the real impact of a building designed to be zero fossil fuel energy over its lifetime (see discussion of SIPs in Chap. 2). This adds not just cost but complexity to the project as a specialist has to be called in to do the analysis, and even here no such analysis is possible without using the work of others who have calculated the energy and CO_2 coefficients of materials. These coefficients also change according to the national energy mix and types of processes used to make building materials and so need updating, leading to further complexity. Obviously, if humanity is to deal with climate change and fossil fuel reduction then all new buildings need to be zero fossil fuel energy but this, as shown above, adds complexity and hence cost to the design process. Once ideal ways of achieving a zero fossil fuel building with minimal impact on the environment have been worked out for each particular location and climate, it can become the normal way of building but this will take time, and as buildings last a long time, to turn a whole city into zero fossil fuel buildings may be a long process. Even if it happens, there will still be emissions unless all the travel between the zero energy buildings is on foot or bicycles, or in renewably powered vehicles.

Growth and Collapse

The question is, can cities grow indefinitely or will growth always lead to some form of system collapse? In this case, this would be the collapse of the socio-economic system within a built environment. This means it is hard to talk about collapse in the built environment without further discussion of economies.

In Jacob's discussion of economies, the relationship between growth and collapse is linked to the capacity of an economic system to act continuously in a way she called "self-refuelling" so as to avoid running out of energy, just as a cow needs to eat to avoid dying. The economy needs energy in the form of fuel and resources, whether the latter are human or non-human. This "fuel" is transformed within the economy in order to do things. More growth is possible if waste from these transformations is minimised. However, as discussed earlier (Chaps. 1 and 5), it takes energy to make energy so there needs to be a reasonable return on the energy investment if growth is to continue. Once the energy that fuels the economy starts to dry up then the economic system becomes inert or disintegrates (Jacobs, 2000:85). Effectively it collapses. The aviation industry in the 2019–20 coronavirus pandemic is a good example of such collapse, where collapse is defined as a drastic change to the established situation. Without the passenger traffic to "fuel" aviation, especially overseas journeys, there has been a near 95% drop in air traffic compared with the same time the previous year (Associated Press, 2020). The world tourism industry, responsible for 1 in 10 jobs globally, has also been described as being brought to the brink of collapse by the pandemic with the drastic change in travel patterns (Lam, 2020).

For Jacobs, the way of avoiding collapse is by using "self-corrections" to keep a system stable. It is the ability of a system to cope with self-corrections that is a measure of its resilience. This is something we do all the time. If we overspend the household budget one week, the next week we have to retrench spending so as to keep the household system going if we do not want to go into debt. This is an example of negative feedback. Jacobs proposes another strategy used by systems to keep going is that a system can "bifurcate" whereby the system forks and a new direction appears with the old path being left behind. An example would be the industrial revolution whereby machines replaced craft-based practices, noting that such a revolution happened at different times in different socio-economic systems. The key issue here is that after an industrial revolution, the nature or identity of a society changes, moving from one that was largely rurally housed and rurally focussed to one that increasingly became housed in the expanding urban areas and was thus city focussed. This idea of moving in a new direction links to the theory of resilience and how adaptation works through cumulative change over a number of adaptive cycles. This offers another link to the collapse of societies as, "...the investigation of collapse in ancient states and civilizations really entails identifying the various kinds of social reorganization in these types of societies and so viewing collapse as part of the continuous process of boundary reconstruction" (Eisenstadt, 1988:242). The link between the process of change and identity is made clear in the following (Kintigh et al., 2014):

> From technologies and houses to landscapes and cyberspace, the processes of making, doing, sensing, inhabiting, and relating to things and beings are intimately connected to human neurological development, cultural values, identity formation, social structure, and political change.

This suggests the built environment plays a role in the formation of identity but at the same time because it is generally around a long time it also plays a role in memory and the use of memory as part of recreation after a crisis. Middleton (2017:47) further argues:

> ...in every collapse, there is also continuity and innovation. Implicit in this view is that collapse is rarely the apocalyptic catastrophe it is often presented as: there are still people and social memories and bonds, these people remember the past and recreate or discard elements of it as it suits them in their negotiation of new and meaningful identities and relationships.

After the 1931 earthquake in Napier, New Zealand, which destroyed the town, the existing street pattern was retained with the exception of one change that was already planned to happen. The built environment thus plays an important role in what might happen after a collapse as memories are associated with it, it being the place where people live and create such memories. This has an echo in resilience theory where collapse is part of the adaptive cycle, and is thus closely related to the process of growing, which rather than being something linear is more a process of growth and change, as Whitehand described in the discussion above. The implication is that at certain scales, things need to change radically to keep the stability of the whole urban system. The point here is that growth is not a straight line, it has bumps with ups and downs. It also includes moments of not growing to be able to keep on growing later. The point of Jacobs is that natural growth is a continuous fight to evade collapse whereas in resilience theory it seems that collapse and the regeneration that comes afterwards is a part or phase of the process of growing. If a system is not dynamic it will deteriorate and collapse. But even the resilience of a system has limits defined by its level of complexity. Cities also have limits since they rely on their economies and these, in turn, rely on natural resources that are finite. Even when cities appear diverse and prosperous, their capacities to generate exports from imported products are limited by their capacity to "self-refuel" in this way.

How Should We Grow the Modern Built Environment?

The answer to this question comes back to the idea of limits and which urban form requires the fewest resources related to sustaining the socio-economic system it contains. If the big issue is climate change and the need to move from an economy based on fossil fuels to one based on renewable energy then there will be changes in moving around the urban environment and even in how buildings are placed to make use of solar energy. This was something the Ancient Greeks used in ordering cities like Olynthus and Priene (Butti & Perlin, 1981:5–10). Will tall buildings still be viable given that they depend on electricity for lifts, or should there be a return to walk-up blocks of six or seven storeys, such as found in the older quarters of many European cities? Will all-glass towers still be viable in both hot and cold climates without the fossil fuel energy to feed their HVAC systems? Transport energy is another big issue as to whether there are sufficient resources for all existing transport to be replaced with versions running on electricity from batteries. Predictions suggest in 2030 that the world will contain 1.4 billion passenger vehicles, with only 8% of these being electric. This figure should rise to 31% by 2040 (Bloomberg NEF, 2020). Whether it would be a good idea simply to change from fossil fuel cars to electric cars is another issue. Intensification of urban environments means that more use can be made of the resources that go into the built environment but this means there would be no room for everyone to have a car and somewhere to park it as well and people would have to get around on foot or by cycling. If the vehicle fleet is simply electrified then cities have to have lower densities and give over space to roads that only some people can use at any one time. This is where the less intensely used and looser built environment can offer more resilience because it offers more ways in which it could be used or more diversity (Fig. 7.2).

In Fig. 7.2, the same area of land has been divided into 4 and 16 squares. The latter allows for more diversity in the relationships between the different functions accommodated. However, there is less room for each function, which makes them more vulnerable. The smaller scale not only makes for more diversity but also makes for increased vulnerability. To explain diversity further, imagine you could dig up your current garden (assuming you have one) and grow enough potatoes to last you the year. This leaves no room for growing all the other vegetables you might want to eat. Alternatively, you could divide the garden into 16 patches in order to grow a wide variety of vegetables, including potatoes. A small patch might grow all the peas you need but you cannot grow sufficient potatoes, so diversity gives choice but at the expense of not having enough of some things. However, should potato blight strike, in the less diverse system you could lose the whole crop, whereas in the more diverse system you only lose 1/16th. What resilience really teaches about growth and

Fig. 7.2 Built environment divisions

development is that there is no win–win situation and you cannot win on all fronts. If society insists on growth, this will be at a cost somewhere.

Diversity has advantages but there is still a limit to how much you can grow, limited by the size of the garden, even with diversity. All growth has limits unless there is a sudden technological advancement that would allow for greater production of crops per unit area, such as by using genetically modified seed. This could be the answer but it also introduces risk. Vegetables have been grown for thousands of years and how to do this is well established at the scale of the garden. The same is not true of repeated use of genetically modified seed, it could be good but it might also have unintended effects. This was true of the use of DDT in the 1950s, which was seen as a useful product against insect-borne diseases such as malaria but which was later banned as a probable human carcinogen (EPA, 2017).

Diversity in a landscape can be measured (Garcia & Vale, 2017:192–197). A more diverse built environment might convey benefits that a less diverse one, such as the city of Hong Kong with its proliferation of high-rise buildings, might not have. The NZ lockdown in 2020 in the face of COVID-19 allowed people to go for walks provided they maintained a social distance. This was only possible because the roads existed (and traffic was vastly reduced) so people could pass each other safely (for further discussion see Chap. 9). As another example, a passive solar city, along the lines of the Ancient Greeks, might be a way of building to mitigate climate change, but would it be sufficiently diverse? That would perhaps depend on the size of the solar 'city'. One thing we can learn from the study of collapse is that diverting resources into building large monuments that can only be enjoyed by the few can precede a collapse. A more modest built environment that can be enjoyed by the many might be a better way to start.

References

Amos, P., Bullock, D., & Sondhi, J. (2010). *High-speed Rail: The fast track to economic development?* The World Bank.

Associated Press. (2020). *Air travel drops nearly 95% from 1 year ago amid pandemic*. Retrieved April 20, 2020, from https://ktla.com/news/nationworld/air-travel-drops-nearly-95-from-1-year-ago-amid-pandemic/.

Bardi, U. (2017). *The Seneca effect: Why growth is slow but collapse is rapid*. Springer.

Barkin, J. S., & De Sombre, E. R. (2013). *Saving global fisheries: Reducing fishing capacity to promote sustainability*. The MIT Press.

BBC. (2020). *Good food: Sardine*. Retrieved March 27, 2020, from https://www.bbcgoodfood.com/glossary/sardine-0.

Blair, F. (2015). *Rethinking Economic Growth from a Biophysical Perspective*. Springer.

Bloomberg NEF. (2020). *Electric vehicle outlook 2020*. Retrieved August 2, 2020, from https://about.bnef.com/electric-vehicle-outlook/.

Brand, C., & Boardman, B. (2008). Taming of the few—The unequal distribution of greenhouse gas emissions from personal travel in the UK. *Energy Policy, 36*(1), 224–238.

Brelsford, C., Lobob, J., Handa, J., & Bettencourt, L. M. A. (2017). Heterogeneity and scale of sustainable development in cities. *Proceedings of the National Academy of Sciences of the United States of America, 114*(34), 8963–8968.

Butti, K., & Perlin, J. (1981). *A golden thread*. Marion Boyars.

Chancel, L., & Piketty, T. (2015). *Carbon and inequality: From Kyoto to Paris*. Retrieved August 2, 2020, from http://www.ledevoir.com/documents/pdf/chancelpiketty2015.pdf.

Creutzig, F., Jochem, P., Edelenbosch, O. Y., Mattauch, L., van Vuuren, D. P., Mccollum, D., & Minx, J. (2015). Energy and environment. Transport: A roadblock to climate change mitigation? *Science, 350*(6263), 911–912.

Dahl, A. K., Hassing, L. B., Fransson, E. I., & Pedersen, N. L. (2010). Agreement between self-reported and measured height, weight and body mass index in old age—A longitudinal study with 20 years of follow-up. *Age and Ageing, 39*(4), 445–451.

Dimitropoulos, A., Oueslati, W., & Sintek, C. (2016). *The rebound effect in road transport: A meta-analysis of empirical studies* (OECD Environment Working Paper No. 113). OECD.

Domec, J.-C., Lachenbruch, B., Meinzer, F. C., Woodruff, D. R., Warren, J. M., & McCulloch, K. A. (2008). Maximum height in a conifer is associated with conflicting requirements for xylem design. *Proceedings of the National Academy of Sciences of the United States of America, 105*(33), 12069–12074.

Dong, T., Jiao, L., Xu, G., Yang, L., & Liu, J. (2019). Towards sustainability? Analyzing changing urban form patterns in the United States, Europe, and China. *The Science of the Total Environment, 671*(June), 632–643.

Dunsky Energy Consulting. (2018). *The economic impact of improved energy efficiency in Canada*. Retrieved March 27, 2020, from http://cleanenergycanada.org/wp-content/uploads/2018/04/TechnicalReport_EnergyEfficiency_20180403_FINAL.pdf.

Duranton, G., & Turner, M. A. (2009). *The fundamental law of road congestion: Evidence from US cities* (NEBR Working Paper 15376). National Bureau of Economic Research.

Environmental Protection Agency of the United States (EPA). (2017). *DDT—A brief history and status*. Retrieved April 20, 2020, from https://www.epa.gov/ingredients-used-pesticide-products/ddt-brief-history-and-status.

Eisenstadt, S. N. (1988). Beyond collapse. In N. Yoffee, & G. L. Cowgill (Eds.), *The collapse of ancient states and civilizations* (pp. 236–243). The University of Arizona Press.

FAO. (2018). *The state of world fisheries and aquaculture (in brief)*. Retrieved March 24, 2020, from https://reliefweb.int/sites/reliefweb.int/files/resources/ca0191en.pdf.

Fuller, R. J., & Crawford, R. H. (2011). Impact of past and future residential development patterns on energy demand and related emissions. *Journal of Housing and the Built Environment, 26*, 165–183.

Gracia, E., & Vale, B. (2017). *Unravelling sustainability and resilience in the built environment*. Routledge.

Gómez, J. A., Haarhoff, E., & García, E., et al. (2020). Highway expansion and crime: challenges on urban development for sustainability. In H. Bougdah (Ed.), *Urban and transit planning* (pp. 133–143). Springer.

Jacobs, J. (2000). *The nature of economies*. Vintage Books.

Jevons, W. S. (1865). *The coal question: An inquiry concerning the progress of the nation, and the probable exhaustion of our coal-mines*. Macmillan and Co.

Kahn, H., & Simon, J. L. (1984). *The resourceful earth: A response to global 2000*. Basil Blackwell.

Kintigh, K. W., Altschul, J. H., Beaudry, M. C., et al. (2014). Grand challenges for archaeology. *American Antiquity, 79*(1), 5–24.

Lam, S. (2020). *Coronavirus: The travel industry is in danger of collapse: How will it weather the storm of Covid-19?* Retrieved April 20, 2020, from https://inews.co.uk/inews-lifestyle/travel-industry-aviation-cruises-hotels-staycations-covid-19-coronavirus-crisis-2521239.

Marchetti, C. (1994). Anthropological invariants in travel behaviour. *Technological Forecasting and Social Change, 47*, 75–88.

Middleton, G. D. (2017). *Understanding collapse: Ancient history and modern myths*. Cambridge University Press.

Ramez, N. (2013). *The infinite resource: The power of ideas on a finite planet*. University Press of New England.

Rees, W. E., & Moore, J. (2013). Ecological footprints and urbanization. In R. Vale & B. Vale (Eds.), *Living within a fair share ecological footprint* (pp. 3–32). Routledge.

Piketty, T. (2015). *The economics of inequality*. Harvard University Press.

Starbuck, A. (1878). *History of the American whale fishery from its earliest inception to 1876*. Washington. Retrieved March 24, 2020, from https://archive.org/details/historyofamerica00star/page/n3/mode/2up.

Taagepera, R. (1978). Size and duration of empires: Systematics of size. *Social Science Research, 7*, 108–127.

Taagepera, R. (1979). Size and duration of empires: Growth-decline curves, 600 B.C. to 600 A.D. *Social Science History, 3*(3/4), 115–138.

Taagepera, R. (2014). A world population growth model: Interaction with Earth's carrying capacity and technology in limited space. *Technological Forecasting and Social Change, 82*, 34–41.

Tainter, J. A. (1988). *The Collapse of Complex Societies*. Cambridge University Press.

Taylor-Miller, T., & Simm, J. P. (2017). Growth disorders in adolescents. *Australian Family Physician, 46*(12), 913–917.

UK Energy Research Centre. (2020). *The rebound effect report*. Retrieved March 27, 2020, from http://www.ukerc.ac.uk/programmes/technology-and-policy-assessment/the-rebound-effect-report.html.

Ulanowicz, R. E. (1986). *Growth and development: Ecosystems phenomenology*. Springer-Verlag.

Victor, P. A. (2010). Questioning economic growth. *Nature, 438*(7322), 370–371.

Victor, P. A. (2018). *Managing without growth: Slower by design, not disaster*. Elgaronline. https://doi-org.helicon.vuw.ac.nz/10.4337/9781785367380

Whitehand, J. W. R. (1994). Development cycles and urban landscapes. *Geography, 79*(1), 3–17.
Zahavi, Y. (1979). *The 'UMOT' project*. Washington: US Department of Transport; Bonn: Ministry of Transport Fed. Rep. of Germany. Retrieved March 31, 2020, from http://www.surveyarchive.org/Zahavi/UMOT_79.pdf.

Chapter 8
Growth and Resources

This land is your land, this land is my land
Woodie Guthrie

Introduction

The aim of this chapter is to think further about the type of built environment we should be constructing, particularly given the predicted increase in people living in cities, and then to consider the effect this might have on resources and the natural environment. This will be achieved through the ecological footprint (EF) as one possible measure of environmental impact. The usefulness of the ecological footprint comes from the fact that it can be a way of measuring the impact on the environment of aspects of modern living. The EF works at all scales, it can be used to measure the impact of a whole nation (GFN, 2020a) or of an individual action, such as playing golf (Vale & Vale, 2009:255–556).

The EF measures the resources needed to produce a product or provide a service in a sustainable way, not relying on finite resources, but on land. Because land is the ultimate resource, the EF is measured in a unit known as global hectares (gha). A global hectare is defined as "a biologically productive hectare with world average biological productivity for a given year" (GFN, 2020b). As productivity can change from year to year, this means the EF is at its best when comparisons are made between footprints calculated on the same basis and at the same time. This is somewhat easier to do for a nation where national input and output statistics can be combined with annual crops grown to give a sense of how much biologically productive land is needed to support a national population. It is much harder to define the boundaries of calculation for a city, the majority of which do not include any, or very little, agriculturally productive land.

Using input and output statistics is generally referred to as a "top-down" calculation, whereas the EF of a city generally has to be calculated in a "bottom-up" way, often through surveys of what people do, by working out how much energy and materials a city consumes and then converting this to the equivalent global average carrying capacity of land (Moore et al., 2013). Because it takes time to collect and collate the local statistics required for this method, EFs are generally a measure of what happened in the immediate past. The other problem with comparing the EFs of cities is that the boundaries of what is included in the calculation may not always be the same. However, where a series of calculations have been made on the same basis, then a fairer comparison is possible (Chicca et al., 2018:10). The same is true of the carbon footprint, usually measured in greenhouse gas (GHG) emissions, which forms more than 50% of the EF. These issues will be borne in mind in the following discussion.

City Living

In 2018, globally 55% of humanity lived in urban areas and this is predicted to rise to 68% by 2025, with most of this increase happening in Asia and Africa (UN, 2018). Globally, rural population has fallen from just over 66% of the total in 1960 to just under 45% in 2018 (The World Bank Group, 2019a). However, urban living also tends to be associated with developed countries. As an example, Fig. 8.1 shows the percentage change from urban to rural population in the USA (the change in 1970 signals a difference in the definition of an urban population but the trend remains) (United States Census Bureau, n.d.).

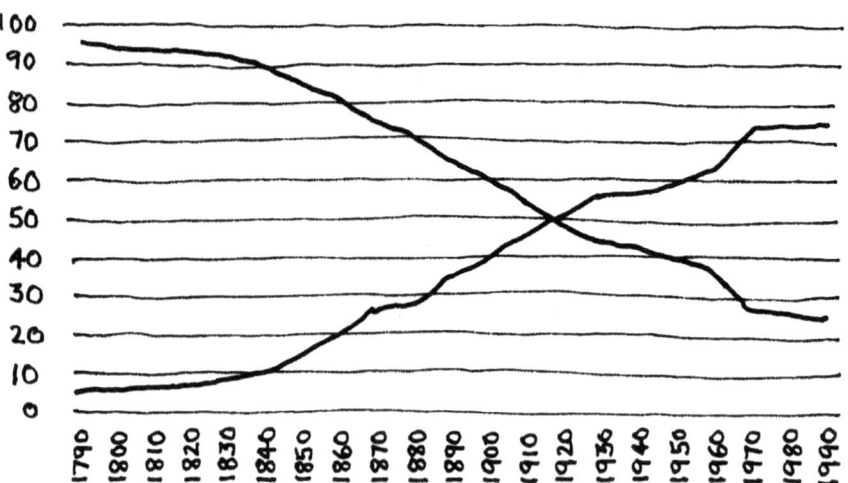

Fig. 8.1 Percentage change in US population with rural population declining and urban increasing (based on United States Census Bureau, n.d.)

According to The World Bank Group (2019a) in 2018, the US urban population was 82% of the total. The same source also states that in 2018 the average urban population for high-income countries was 81%, falling to 33% for low-income countries. Urban living is, and always has been, associated with wealth generation. Owens and Green (2016:44–45) in discussing the accumulation of wealth point out that from 1800 "for many European countries, industrialization, urban growth and the development of commercial and territorial interests overseas created new opportunities for accumulation." Coming forward to the present, Zenghelis (2017:1) states:

> Over the next fifty years, most new wealth will be accumulated in cities; this includes physical infrastructure (such as road, rail, electricity, telecommunications, and sanitation), productive capital (such as houses, offices, factories, and markets), and skills, know-how, and ideas (so-called 'knowledge capital').

Thus, the wealthier countries in the world tend to be those that urbanised first, such as the USA (Fig. 8.1). Globally, the point when urban and rural populations, both of which have grown as the population has grown, were equal occurred sometime between 1990 and 2018 (UN, 2019:23). Dr. Joan Clos, a former executive director of UN-Habitat gives the year as 2008 (Pictet Group, 2020), six decades after the USA reached this point of urban-rural equality.

The wealth that comes with urbanisation leads to the consumption of more resources making the environmental impact of cities normally higher than the average for the nations in which they stand. In a 2001 study, Hubacek et al. (2009) using a top-down approach found that the EF of China was 1.78 gha/person while that of a citizen of Beijing was 4.99 gha/person. Whereas the impact of agriculture accounted for just over 50% of the total EF of China, the main contributions to the impact of the lifestyle in Beijing were manufacturing and energy, accounting for 60% of the total, and transportation and the services sector for a further 20%. They also predicted the Beijing EF would almost double by 2020. Another study that used a different method of EF calculation showed the EF of Beijing in 2020 would be twice that of 1996 while at the same time, the GDP of Beijing would be 19 times that of 1996 (Lui & Lei, 2018), again demonstrating the importance of cities in economic performance. More people are predicted to be living in cities. This tends to lead to larger EFs that are far higher than the average global carrying capacity of the earth, which in 2013 stood at 1.7 gha per person (Chicca et al., 2018:7). Should this be a cause for concern? Will the compaction of people into cities lead to any type of collapse because of exceeding the earth's carrying capacity? Before attempting to answer these questions, it is necessary to look at how people live at the higher densities found in cities and how this might affect EF.

High-Rise Buildings

High-density living in cities conjures up images of tall buildings in close proximity, as might be found in Hong Kong or on Manhattan Island. However, high-density living is equally true of many slum communities, which are anything but high rise, being made by hand from materials to hand. As an example, in 2013, Hong Kong had 352 persons/hectare (Atlas of Urban Expansion, 2016), while in 1995, the most densely populated slum in Abidjan on the Ivory Coast had a population of 340 persons/hectare (United Nations Human Settlements Programme, 2003:200). At approximately the same date, the average population in slum areas in the main cities of Bangladesh was 336 persons/hectare, with the comment "This density figure is extraordinary, given that almost all residential structures in slum areas were single storey" (Islam et al., 2009). Allowing 10% of a slum area for circulation would mean there is approximately 27 m^2 per person for all residential and other activities in the average slum in Bangladesh. Against this, in 2019, the average living space of someone in public rental housing in Hong Kong was 13.3 m^2 per person (Statista, n.d.). Obviously, this is not a true comparison as it excludes the tenant's share of all the other buildings related to the economic activity of Hong Kong, but it certainly demonstrates that high-density living does not necessarily mean high rise living.

Putting more people into the same area of land, hopefully in a way that does not lead to slum conditions, inevitably for the designer increases the difficulty of ensuring they have access to their daily needs such as food and water, work and schools (see also Fig. 7.2). Delivering water to the top of a high-rise apartment block, using pumps and zones with intermediate tanks, is a much more complex process than delivering water to a single house. You can walk upstairs in a house but high-rise apartments need lifts. If the electricity fails, a house will be easier to live in, as seen when Hurricane Sandy struck the US east coast on 29 October 2012 (Manuel, 2013).

> Loss of power presented a distinct threat to people living in the region's many high-rise apartments. In normal times, those living on upper floors consider themselves lucky to enjoy the views. But when the electricity went out in these buildings, the elevators stopped working, and many of those same people—physically unable to descend the stairways—were trapped for days and even weeks on end.

The materials used for high-rise buildings also lead to more energy being embodied in the building per unit of area compared with low rise. In a study of five Melbourne office buildings of 3, 7, 15, 42 and 52 storeys, Treloar et al. (2001) found the two highest buildings embodied approximately 60% more energy per unit of floor area than the two low-rise buildings. The embodied energy figures in GJ/m^2 were 10.7, 11.9, 16.1, 18.0 and 18.4, respectively. Both more energy-intensive materials and more materials overall were required to make the high-rise structures, which were subject to greater wind loads than the low-rise buildings. In a discussion of skyscrapers, Boyle et al. (2016) note that when it comes to construction "…building higher structures often requires new types of cranes, cement pumping systems…building higher often requires newer and faster elevators, lighter cables, new efficiencies in moving water and sewage, space-saving temperature control systems etc." Tall buildings are also less flexible when it comes to changes in lifestyle. Houses are adaptable as walls can be moved and the simple structure changed, something that is more difficult in apartments unless they have deliberately been designed, to be changed by the users (Habraken, 1972). Extensions are also something that is almost impossible in apartments, although creatively converted and

City Living 211

Fig. 8.2 Russian balcony extension (adapted from https://weirdrussia.com/2015/08/27/balconies-in-russia/)

extended balconies are features of some Russian apartment blocks (Weird Russia, 2015) (Fig. 8.2).

Taken altogether, tall buildings cost more per unit of floor area than low-rise (up to seven storeys) buildings, noting that the seven-storey block is the historic urban form of many European cities, such as Paris. A study of building costs in the three main cities of New Zealand (Auckland, Wellington and Christchurch) showed that low-rise three-storey 75 m^2 apartments were cheaper per m^2 than 10-storey concrete frame apartments of the same size. Averaging costs across the three cities meant moving from low-rise apartments to concrete frame high-rise increased the cost by over 20% and moving to timber high-rise by 7% (Meade et al., 2018:60). Much of this cost comes from having to incorporate basement parking, which shows solving one problem (more people on the same area of land) leads to another (where to put their cars). This again suggests that densification has to be accompanied by a change in how people move around urban areas. The six- and seven-storey walk-up blocks worked in Europe, just as the similar *insulae* worked in Ancient Rome, because people did not have cars.

The added expense of building higher is not a new problem. Talking about the problems of living in flats in the 1960s in London, Downing and Calway (1963) state: "The irony is that the flat is more expensive to build than the house and with the modern tendency to go higher and higher so that the street is in fact up-ended into the vertical plane, more expensive still to produce." Currently in New Zealand, buying an apartment (based on TradeMe figures) can cost nearly twice as much per square metre as a house though it could come with better fittings (Edmunds, 2017). However, no one can avoid seeing a tall building in the urban landscape and this means a high-rise building can have a "value" beyond the provision of office or living space (Barr et al., 2015). Many have become landmarks and tourist attractions, such as the 83 storeys of the 1988 Petronas Towers in Kuala Lumpur and the 95 storeys of the 2013 Shard in London.

Not everyone necessarily wants to live in a monument and much has also been written about the problems of high-rise living in societies where people have not been used to it, such as in the UK (Anon, 1967; Downing & Calway, 1963; Stewart, 1970), although often UK problems have arisen because of rehousing in flats and the loss of former community ties (Young, 1962). High-rise living is also not just a UK problem. Writing of the 1968–75 estate of Bijlmermeer in Amsterdam, based on the modernist ideals of CIAM, Helleman and Wassenberg (2004) state "High-rise estates are associated with problematic living conditions, deprived areas, isolated locations, a poor population, a negative image, social isolation, pollution and crime." However, where living in apartments is much more common, as in parts of Asia, attitudes are different. In a study of Singapore where the majority live in apartments, Yuen (2011:129) stated "Singapore has carefully and comprehensively planned its public high-rises to provide quality living environment" (Yuen, 2011:144). This was achieved by ensuring apartments had good facilities nearby and were maintained and upgraded as necessary with resident input. In a country that is short of land and with a rising population like Singapore, ensuring high-rise living is successful is essential, even if this means careful planning of housing as "Creating a bond between resident and the high-rise is critical to enhancing liveability" (Yuen, 2011:145).

Another issue that has emerged with higher density housing in cities like London and New York is the buying up of properties by very wealthy absentee owners. "In Manhattan, the *New Yorker* had a look at Census Bureau numbers, which revealed that in midtown nearly 1 in 3 residences are unoccupied at least 10 months a year" (Booth & Laadam, 2017). This increases local complexity as the available land is not being used for housing as many people as it should, with a consequent effect on the local economy. Such property investments are, in resource terms, a very expensive way of banking money.

Tall buildings may seem like an answer to housing people in cities, but they come with added complexities in terms of how they are made and operated in an effort to accommodate more people on less land in a way that still provides people with a satisfactory living experience. This is useful as long as the additional complexities do not outweigh the gains from using less land. Is it, therefore, reasonable to be concerned about the increase in high-rise living? As an example of the latter and based on 2016 census data, apartments now form 30% of all dwellings in Sydney, a city of 4.4 million people, which was a 17% increase in 5 years. Of the remaining dwellings, 17% are townhouses and 55% are detached houses (McCrindle, 2017:5). In Sydney, "Apartment dwellers are not only more likely to be mobile and untethered to a mortgage but are also younger and more formally educated than those in detached houses" (McCrindle, 2017:6). At present, it would seem, at least in Sydney, that the gains from building high and putting more people on the same area of land outweigh any disadvantage such as cost. However, whether the land as measured by the ecological footprint in gha is less in such a high-living situation remains an area worth further investigation.

Ecological Footprint and GHG Emissions

The usual reason given for high-density and high-rise living is that it will save on emissions from transport as people will no longer be commuting from the suburbs and more use makes public transport more effective (Newman & Kenworthy, 1989). However, if high-density buildings require more materials, then there will also be more emissions from these materials, offsetting the possible lower use of cars. It is vital to know which urban pattern is associated with the lowest emissions.

There is a relationship between greenhouse gas emissions, the carbon footprint and the ecological footprint. All the energy that goes into what we do and the materials, goods and services that we use is the carbon footprint and currently this forms 60% of the EF. Energy in all forms, whether used directly for heat and electricity or indirectly in the goods and services people consume, accounts for 72% of all greenhouse gas emissions (Center for Climate and Energy Solutions, n.d.). Energy in all forms is also the most rapidly growing component of the EF, and globally there has been an 11-fold increase in the human carbon footprint since 1961 (GFN, 2020c), although part of this increase is because of the rise in the human population. However, although the human population has risen 152% since 1961, based on Worldometers (n.d.a), the carbon footprint has risen 1,100%. The carbon footprint is converted to land through assessing how much land of global average productivity would be required to sequester the emissions.

> Energy land...means the area of carbon sink land required to absorb the carbon dioxide released by per capita fossil fuel consumption (coal, oil and natural gas) assuming atmospheric stability as a goal. Alternatively, this entry could be calculated according to the area of cropland necessary to produce a contemporary biological fuel such as ethanol to substitute for fossil fuel. This alternative produces even higher energy land requirements. (Wackernagel & Rees, 1996:11).

The value given by Wackernagel and Rees (1996:73) is that annually it takes 1 hectare of forest of average productivity to sequester the CO_2 emissions from 100 GJ of consumed fossil fuel energy. This reveals the difficulty of calculating a carbon footprint as the first thing to discover is whether the energy consumed comes from fossil fuels or from renewable resources. Apart from sitting directly in the sun to keep warm, when your footprint is the land you occupy, all energy sources have an environmental impact, whether this is the energy that goes into making and maintaining a 3 MW wind generator, or the land taken up and the materials used to make the dam and energy-generating components of a hydro system, as well as the effect on rivers and the methane generated from decomposing organic matter when the land is flooded to make the reservoir behind the dam (Hertwich, 2013). A life-cycle study by Hertwich et al. (2015) that examined the resource requirements and environmental impacts of the generation of low carbon electricity, including fossil fuel production with carbon capture and storage (CCS) (see Chap. 2), found that "…an electricity supply with a high share of wind energy, solar energy and hydropower would lead to lower environmental impacts [including GHG production] than a system with a high share of CCS." Although the renewable technologies of wind and solar used more materials, "The pollution caused by higher material requirements of these technologies is small compared with the direct emissions of fossil fuel-fired power plants." In a further life-cycle study of fuels and greenhouse gas emissions but omitting other pollution effects, Pehl et al. (2017) found fossil fuel plants with carbon capture would emit 78–109 gCO_2eq/kWh, compared with 504 gCO_2eq/kWh for the current average global electricity from fossil fuels, while the emissions from nuclear, wind, photovoltaics and concentrating solar power were only 3.5–11.5 gCO_2eq/kWh. They also found the life-cycle emissions for hydropower and bioenergy were much larger $\sim$100 gCO_2eq/kWh but that this figure was "highly uncertain."

More important than all these figures are the current world dependences on fossil fuels, with their high environmental impacts. In 2017, in descending order of importance, world primary energy consumption was approximately 31% crude oil, 26% coal, 22% natural gas, and then a huge drop to 7% traditional biofuels, 6% hydropower, 4% nuclear, 2% wind, 2% other renewables and 1% solar (Ritchie & Roser, 2018). Since most transport relies on oil in some forms, finding settlement patterns that would reduce the need for transport would seem to be a high priority. To reduce the current carbon footprint in the UK, one study (Harrabin, 2020) found the change that would make the most impact was to live car free, which suggests a radical change to the way the built environment is used. Much less use of the private car suggests the idea of the compact city (Jenks et al., 1996), which means that more people will be living on less land to bring them closer to work and other facilities, in other words having people live at higher densities.

At first sight, high-density living would seem to have advantages in terms of environmental impact, as many more people can be accommodated on a given area of land. An early study found that high-rise living had the potential to have a lower environmental and cost impact than accommodating people in low-density energy-efficient housing (RERC, 1974). Not only are there fewer people per hectare in urban sprawl but the houses need infrastructure and the nature of sprawl means those who live there have to travel more, often much more, than those closer to work and facilities near the city centre. However, this is urban sprawl in a developed country, whereas the issues in a developing country might be the rapid but haphazard development of higher density settlement (Wilson & Chakraborty, 2013).

The argument as to whether high-rise living is less sustainable than suburban living as claimed in 2011 by Tony Arnel, former chairman of the World Green Building Council (Australian Design Review, 2018) is a much-debated point, but without any statistics to back up what is being said. One exception is the science looking at the effect of increased density and the urban heat island effect (UHI), caused by the hard surfaces of buildings and streets absorbing solar energy. In warm climates, this induces the use of air conditioners, which raises GHG emissions, something that can be avoided by living at lower densities where natural airflow is far less impeded. On the other hand, in colder climates, the UHI could reduce the energy for heating in winter. However, Santamouris (2014) states, regarding the UHI effect, "Energy increase is very significant in cooling dominated zones where the rise of the summer energy needs is much higher than the possible decrease of the heating needs in winter." This is a situation that is compounded by global warming. Additionally, the efficiency of air conditioners, measured by their coefficient of performance (COP), falls with an increase in air temperature (Santamouris et al., 2001). In a review of studies of UHI, global warming and the effect on buildings, Santamouris (2014) found, for the period 1970–2010, that the average total energy load for heating and cooling rose by 11%.

A comprehensive comparative study was made of higher density housing with suburban housing in Adelaide, Australia (Perkins et al., 2009). They compared the life cycle energy consumption and GHG emissions of households in inner city apartments and those in the inner and outer suburbs in terms of buildings and transport. The energy embodied in the buildings and vehicles was also included in the analysis. This is important as high-rise buildings have a higher embodied energy per square metre than low-rise buildings, partly because of the energy embedded in the structural materials and partly the need to provide lifts and circulation spaces (Rickwood et al., 2008). As might be expected, when it came to the results, those living in city centre apartments in the Adelaide study had far fewer transport emissions. "The substantial reduction in travel energy use by city centre residents, even when compared with inner suburban residents, reflects a combination of a high proportion of shorter trips and a preference for walking" (Perkins et al., 2009). In terms of housing type and based on households, the outer suburban households had the highest total emissions (23.0 tonnes CO_2e/annum), mainly coming from transport operational and embodied emissions, while apartment households had slightly higher emissions (16.2 tonnes CO_2e/annum) than inner suburban households (14.0 tonnes CO_2e/annum). However, on a per capita annual basis, the situation changed, with the highest emitters being those living in city centre apartments (10 tonnes CO_2e/annum), compared with inner suburbs (7.0 tonnes CO_2e/annum) and outer suburbs (7.4 tonnes CO_2e/annum). The reason for the change was the low occupancy rate of the inner-city apartments. Another factor might be the type of person choosing to live in inner-city apartments. A study of transport emissions in Montreal found the "...number of workers and retirees at the household level play an important role in the contribution to GHG emissions (102% *increase* by adding one worker and 51% *decrease* by adding a retiree to the household)" (Zahabi et al., 2012). Retirees living in the outer suburbs could affect the figures, and in New Zealand, it has been the young people who have been attracted to apartment living. "Almost half of all apartment dwellers were aged 20–29 years at the time of the 2006 Census" (Statistics New Zealand, 2010:8). This shows how complicated the issue of urban form and reduction in environmental impact is, and it is not just a question of inner city living good—low-density suburbs bad.

GHG Emissions and Density

Perkins et al. (2009) argued that it might be relatively simple to reduce energy use in apartment blocks through design while reducing transport emissions for the outer suburbs would require better public transport (and people using it) or more efficient vehicles. Another option would be less commuting into the city centre for work, either through more working at home or through decentralisation of work opportunities, with local centres within the fabric of the city. What this study points out is it is not a simplistic case of densification being good. It depends on what is built, how it is built and importantly, how it is occupied. The fact the inner suburbs in the Adelaide investigation had the lowest emissions in both household and per capita terms argues for the decentralised work hub model with such centres being scattered throughout the city surrounded by their "inner" suburbs. Working from home would also be a good solution. A looser arrangement like this might also help to combat increased emissions through UHI, as in suburbs there is more opportunity for tree planting to shade hard surfaces.

A bottom-up study in Halifax in Canada (Wilson et al., 2013a) found that household emissions were similar in both suburbs (5–10 km to the centre; 20.5 kg CO_2e/person/per day) and inner city living (5 km or less to the centre; 20.2 kg CO_2e/person/per day), although living in outer suburban/rural areas within a 50 km commute to the centre, termed the exurbs, had the highest emissions. The Halifax study also stated "…the notion of a single flow from suburbs to inner city for work and shopping…is incorrect." They noted that in Halifax not only had the most recent office space been built in the suburbs but also that those living in the inner city would travel to the suburbs for shopping, work and other facilities.

The fact the suburbs with their lower density allow for more tree planting also has an effect on GHG emissions. The flow of CO_2 was studied in a woodland area, an urban area and a suburban area situated in Southern England, and all with similar climate characteristics and latitude. The urban site was 81% covered with roads and buildings, this reduced to 49% for the suburban site and zero for the woodland. The net exchange of CO_2 between the surface and the atmosphere was measured for each site. Estimates were made of the human-induced emissions for the urban and suburban sites using available data. The woodland area produced a small amount of CO_2 at night and in the winter with the uptake of CO_2 when the trees were in leaf. The urban area produced CO_2 all year round with most occurring in winter, while in the suburban area, the greatest emissions still occurred in winter but the vegetation showed the potential of suburban areas to take up CO_2, especially in the growing season. However, Ward et al. (2015) concluded that "The huge anthropomorphic emissions associated with cities means that variation in annual CO_2 exchange among urbanised study sites is about ten times that observed among vegetated ecosystems." The authors concluded that the major effect cities have on global CO_2 emissions should be noted, though perhaps this is no more than saying where there are people there will be emissions. Others have noted that higher emissions are associated with wealthier neighbourhoods and communities (Wilson et al., 2013a; Heinonen & Junnila, 2011; Chancel & Picketty, 2015:21).

In a study of Finnish cities, Heinonen and Junnila (2011:6) found little correlation between density and emissions. Their method looked at consumption and GHGs so as to avoid the problem in many studies of only looking at buildings and transport. Their top-down approach took in how people live as well as where they live. The authors state "Interestingly, the results show that the type of the urban structure, whether a dense metropolitan core with apartment buildings or a less dense suburban area with primarily detached housing, has quite a small effect on the carbon emissions." They also found a connection between wealth and emissions. "One interesting notion about the relation of income and carbon consumption is that emissions seem to grow as income grows, but with decelerating speed." When it came to transport, which formed a small part of total emissions using this method, there was more use of private transport in the lower density areas but increased use of public transport in the higher density areas. However, transport, buildings and green space to absorb CO_2 are not the only issues when it comes to urban form and environmental impact. Food is another essential to consider.

Food and Urban Settlement

Food forms a significant component of the personal EF, and the poorer you are the greater the contribution of food to your overall EF (Table 8.1).

Food has always been such an important issue for human society that it has even become bound up in how land is measured. In England, before the Norman Conquest of 1066, the Hide was the area of land required to support a household and the unit varied from 60 to 180 old acres, depending on the locality and hence the soil and climates. A quarter of a Hide was a Virgate, or sufficient land to support one person (Direct Line Software, 2010). This is measuring land in terms of what lifestyle it can support, as with the EF, and understanding this might vary with climate and local terrain, as with the global hectare. We often forget that the same problems have faced humanity in the past.

Table 8.1 Food as a proportion of bottom-up EFs (food, energy, transport, housing and consumer goods) calculated on the same basis (adapted from Chicca et al., 2018:149)

Country	Lifestyle	EF gha	Food EF gha	Food as % of EF
Mozambique	Rural	0.13	0.11	85%
India	Rural self-sufficient	0.33	0.12	36%
Cuba	Urban	0.83	0.71	86%
Argentina	City-urban	3.43	1.75	51%
Malaysia	City-suburban	4.38	2.39	55%
UK	Town-suburban	7.61	1.62	21%

The figures in Table 8.1 are affected by the food that is eaten. The Indian food EF is low because the families are vegetarian and grow their own food, and it is the consumption of meat and dairy that raises the food EF. In addition, the poorer you are in terms of disposable income to buy goods, the more food forms the major part of your EF. Everyone has to eat and everyone's stomach is about the same size, but goods and services are consumed in parallel with available income, so as income rises so does consumption of goods and services. What Table 8.1 also makes clear is the significance of food in the EFs of a wealthy society such as the UK. An earlier and more comprehensive EF analysis of the city of Cardiff in the UK was conducted on a different basis and including the impact of being a citizen of Wales (each person's share of infrastructure, educational buildings etc.). This study produced an EF of 5.6 gha/person in 2001 (Collins et al., 2005), of which 1.33 gha (24%) was food in the form of a reasonable mixed diet for a developed country (Vale & Vale, 2009:39).

The history of settlement is also the history of the provision of food for those not involved in agriculture. Chicca (2013:79–80) describes how the Ancient Greek city was dependent on the surrounding villages so that city growth was determined by the water and food supplies available, meaning there were limits to growth. Cities are still dependent on land to grow the food for their citizens, even if this land is in a different country. In 2008, the city of Wellington, the capital of New Zealand, occupied an area of approximately 290 km^2 and would have needed a circular area of land, 56 km in diameter (approximately 2,460 km^2 or 8.5 times the area of the city), to provide a diet like that of Cardiff (Vale & Vale, 2009:55). When it comes to a densely occupied city, like Hong Kong, on the same basis, the area required to provide its food would be 91,000 km^2, which is 300 times the built area of the city or 85 times its total area. "Thus the area of land needed to feed a city is likely to be between 10 and 100 times the area of the city itself, depending on both its density and on the diet of its residents" (Vale & Vale, 2009:56).

The problem of Hong Kong is that its high density means it is virtually impossible to grow local food, something that can be achieved with lower density settlements. In a study of Auckland, New Zealand, Ghosh et al. (2007) found the typical detached houses on plots of 500 m^2 offered the lowest environmental impact when solar energy collection, carbon sequestration, vegetable and fruit growing, waste absorption through composting and commuting to work by car were all quantified. As a result, the authors stated, "…lower-density residential development may have more potential to be sustainable because of the ability of residents to grow food and to make use of on-site renewable energy technologies." This highlights the importance of future thinking and deciding not what urban environment is needed currently but what might be required if change is deemed to be important.

EF and Urban Settlement

Given the size of the food component of an EF, it is obvious that an urban built environment requires very much more land than is occupied by its buildings and infrastructure. Rees and Moore (2013:3) claimed a typical modern city occupied land of up to a thousand times the area of the city itself, in terms of the resources that needed to be supplied to its citizens and the wastes that had to be absorbed. The same authors calculated the EF of the city of Vancouver, finding it to be "...200–300 times larger than the city's physical footprint" (Rees & Moore, 2013:13).

Another EF study of the town of Oakville, Ontario, found a wide variation in EFs across one community, ranging from 5.4 to 15.2 gha, with an average of 9.0 gha (Wilson et al., 2013b). Dividing the EF into six components, two (*shelter, energy* and *shelter, non-energy*) were calculated bottom-up and the remaining four (*consumer goods and services, mobility, food,* and *government*) were calculated top-down using available statistics. The resultant EF was more than 100 times the area of the town, showing that, in environmental terms, it is not so much high or low density but much more profound changes to lifestyle that are needed. As the authors state, "...municipal governments cannot directly influence vehicle fuel efficiency standards, industrial greenhouse gas emissions, the embodied energy associated with food supply chains, the number of televisions a household has or the temperature at which they set their thermostat."

If you are going to share the world's available renewable resources equitably, every citizen, whether urban or rural, needs to live within a fair share EF of 1.7 gha/person, called one planet living. In 2015, only eight countries met the two required criteria for sustainable development of having high human development and an ecological footprint lower than 1.7 gha/person (GFN, 2020d) (Table 8.2).

Although the approximate population density figures for the largest cities are not necessarily all on the same basis, what Table 8.2 shows is that urban density is not the key to having a fair share EF of around 1.7 gha/person, as the population density varies by a factor of nearly 20. With the exception of Sri Lanka, all the other countries in Table 8.2 have urbanised populations around or significantly higher than the world

Table 8.2 Countries with near fair share EFs

Country	Population (million)[b]	% urbanised[a]	Largest city	People/km² in the city (approximately)[c]
Algeria	43.9	73.0	Algiers	10,790
Colombia	50.8	80.4	Bogota	4,310
Cuba	11.3	78.3	Havana	2,890
Ecuador	17.6	63.0	Guayaquil	7,830
Georgia	4.0	58.1	Tbilisi	2,320
Jamaica	3.0	55.4	Kingston	1,380
Jordan	10.2	91.5	Amman	2,380
Sri Lanka	21.4	18.4	Colombo	20,070

[a] All figures for 2020 (Worldometers, n.d.c)
[b] World Population Review, 2020)
[c] These figures were calculated using data from Wikipedia

average of 56.2% (Worldometers, n.d.c). It is not where we live but how we live that is the issue.

Table 8.3 was created to investigate what these low EFs might mean for the quality of life. It compares the most urbanised countries of Table 8.2 (Algeria, Colombia, Cuba and Jordan) with similarly urbanised countries with higher EFs (UK, Japan, Canada, United Arab Emirates).

The Human Development Index has also been added to Table 8.3. This is calculated from life expectancy at birth, expected years of schooling, mean years of schooling and GNI (Gross National Income per capita). Any value of 0.8 or above is considered to be Very High Human Development, while 0.7–0.8 is High Human Development. Of the four low EF countries in Table 8.3, Cuba has the highest level of human development. Everyone wants to live as long as possible and as one factor in the HDI is life expectancy, this has also been added to Table 8.3. Apart from the obvious fact that women tend to live longer than men, the female life expectancy in Cuba, the country closest to a fair share EF, is 80.7. It rises to 83.0 in the UK for a 58% increase in EF, to 87.5 in Japan for a 60% increase in EF, and to 84.3 in Canada for a 131% rise in EF. Despite the large difference in EF, female life expectancy in the UAE with an EF of 8.9 is just less than in Cuba. Having a high environmental impact does not guarantee a long life, at least on average.

Table 8.3 again makes it clear that it is wealth that has a very big influence on environmental impact, not the way the urban environment is configured, as a

Table 8.3 EFs, wealth and human development index

Country	% urbanised[a]	Buildings (150m+)[b]	EF (2016)[c]	GDP per capita US$[d]	Wealth/ adult US$ (2019)[e]	HDI (2018)[f]	Life expectancy	
							F	M
Algeria	73.0	1	2.4	4,114.7	9,348	0.759	79.9	75.5
Colombia	80.4	23	2.0	6,667.8	16,411	0.761	79.9	74.3
Cuba	78.3	–[g]	1.8	8,821.8	n/a	0.778	80.7	76.8
Jordan	91.5	1	2.1	4,241.8	7,465	0.723	76.2	72.7
UK	83.2	22	4.4	42,962.4	280,049	0.920	83.0	79.5
Japan	91.8	257	4.5	39,290.0	238,104	0.915	87.5	81.3
Canada	81.3	111	7.7	46,234.4	294,255	0.922	84.3	80.3
United Arab Emirates	86.0	253	8.9	43,005.0	117,060	0.866	79.2	77.1

[a](Worldometers, n.d.c)
[b](Council on Tall Buildings and Urban Habitat (2020)
[c](GFN, 2019)
[d](World Bank, 2019b)
[e](Credit Suisse, 2019:19–22)
[f](UNDP, 2019:300–310)
[g]The Tallest Building in Havana is Edificio FOCSA, Built in 1956 at 130 m (Emporis, 2020)

significant correlation is between personal wealth and EF. There are some correlations between buildings over 150 m and high EF, but then tall buildings are a reflection of wealth in a society. Maybe all that we need to measure is wealth, rather than going through the intricacies of calculating EF and carbon footprint? This will ignore those who choose to live, even in a wealthy country, in a way that has less impact on the environment. The difference the numbers of those living a fair share EF lifestyle in a developed country, whether through choice or poverty, would make to the national average EF is a question yet to be investigated. As a starting point in the UK, 10% of the population holds a 51.9% share of its wealth (WidWorld, n.d.) so a study of the EFs of the top 10% might be a useful starting point. To look into those who choose to try and live within a fair share EF, a bottom-up study was made of the Hockerton Housing Project (HHP) in the UK. In the HHP, the residents live in zero fossil fuel houses, are autonomous for water and sewerage, are only allowed one fossil-fuelled vehicle, and grow food on site. The study showed the average EF across four houses was 50% less than the UK average of 5.5 gha (Vale & Vale, 2018:265). Even within the already low Hockerton household EFs, there was a 20% variation (highest 3 gha/person, lowest 2.4 gha/person).

Perhaps the biggest signal from considering wealth as a measure of environmental impact is the simple fact that to reduce human impact to one planet living, there is the need to reduce wealth. As suggested in Chap. 11, the wealthy could afford to pay for climate change mitigation. If this could be achieved, their wealth would be used for reducing environmental impact rather than further consumption of luxury, but then, of course, they could lose interest in being wealthy.

Another thing to note from Table 8.3 is that the EFs of countries that were claimed as living within a fair share EF in Table 8.2 have all risen above the 1.7 gha/person threshold, although that of Cuba comes closest. Cuba also has the highest GDP and HDI of the four countries and the smallest EF. One planet living is simply humanity having the lifestyle of Cuba, although Cuba does have the advantage of a tropical climate. The other observation from Table 8.3 is that a very large increase in GDP is needed to raise the HDI from high to very high and this could perhaps be viewed as another example of marginal returns. Obviously, no country sets out to raise its HDI by increasing GDP as HDI is just a reflection of a few national statistics, but to reduce the current global use of the resources from 1.75 planets to the 1 planet available will need some serious thinking about wealth, and hence how economic activity is organised.

However, since this chapter sets out to look at how cities might need to change, it might be useful to look at the urban environment of Cuba, while retaining the thought that it is the lifestyle that is the key to low EF living rather than bricks and mortar.

Urban Cuba

The tourist image of Cuba is probably defined by its largest city Havana, founded by the Spanish in the sixteenth century. In its Spanish colonial-style buildings, the wealthy lived on the upper floors above street level. "This privileged gaze was constructed by second-storey living spaces positioned directly above the social heterogeneity and commercial activity in the streets and public spaces below" (Neill, 2015:34). However, Havana is also home to "...a range of architectural styles, from Renaissance, Moorish, and Baroque to Neoclassical and Mid-Century modern" (McKnight et al. (2012). This heady mixture led Rattenbury (2006) to claim that Havana was to an extent the "...ideal of the dense mixed city: organised, organic, and chaotic." Athough Havana is home to over 2 million people it is not the totality of urban Cuba. Other smaller cities set up by the Spanish serve as administrative centres in the various provinces. However, there was a change of focus after Fidel Castro came to power in 1959.

Under the communist regime, Cuba adopted the soviet ideal of planned regional development rather than allowing cities to grow unchecked. The "primacy of the city" was viewed as a capitalist ideal (Edge et al., 2006). This also entailed bringing unused land into production and diversifying agriculture. In turn, this led to the establishment of 83 new rural villages.

> These settlements fulfilled several purposes: to improve public services to rural areas, to provide housing to isolated peasants by replacing their homes with more permanent structures, and to create the foundations of a gradual restructuring of the rural economy that better suited it to the needs of a long-term development program. (Barkin, 1979).

The aim was to strengthen rural society and reduce migration to the city. Barkin goes on to state "...cities should not be conceived as instruments of concentration and centralization of these resources, but as tools that facilitate their integrated use throughout the island." Is it this different view of the city that has led to a society of low EF and high HDI? Perhaps the key to the low environmental impact of urban Cuba is the elephant in the room, at least the elephant for most of the western developed world—the fact Cuba was a communist society with an interest in seeing as many people as possible had access to the resources of the country.

Fidel Castro is quoted as saying the following (Barkin, 1979):

> I raise this with Cubans as follows: I firmly believe that we should not think about increased consumption. It is important for us to strengthen our economic structure so that we depend less on imports for growth of production and above all, that we have less dependence on the capitalist areas...we should maintain our (consumption) levels that we had in times of crisis...we should not talk about rising living standards.

Under the leadership of Raul Castro, who has already initiated measures such as allowing private enterprise, whether there will be a change to what has been described as "state-sponsored capitalism," like that of China (McKnight et al., 2012), and the effect this might have on Cuba's EF remains to be seen.

Density and Collapse

The perceived wisdom that increasing urban density will be the best sustainability measure has to be questioned because it is too simplistic. Ottelin et al. (2015) make the point that inner city areas often contain small studio apartments, each with its own full set of services, which raises the carbon footprint. They also suggest:

> For example, low-rise areas are less dependent on existing infrastructure, and because of this, new energy efficiency measures and other technological solutions can spread quickly in such areas, as our findings suggest. Low energy housing, heat pumps, HEMS [home energy management systems], photovoltaic panels, and hybrid and electric vehicles are good examples of such technologies.

Ottelin et al. (2015) also propose there is a need to target carbon footprint reduction in different ways for different types of development. For example, small apartment inner city living could benefit from more shared facilities. In contrast to the problem of small apartments, the luxury apartment market has also experienced a boom in the USA (Borodovsky, 2017), with those with money seeking the convenience of living in such apartments in the inner-city areas. There will be no sharing of facilities in such luxury apartments, however desirable this might be. Those with most disposable income tend to have higher carbon footprints, even if they do live in the inner city.

The EF analysis of different types of community also shows that it is lifestyle, which in turn is driven by available income, and not urban form that drives the EF. All types of urban form need to be the target of trying to reduce environmental impact.

When it comes to collapse, perhaps the easiest way of looking at the problem of high density/low density is to consider what happens in a major earthquake. Is it better to be living in the suburbs or in a high rise in the inner city when the earthquake strikes? Where buildings are close together they can bump into each other, known as pounding, which can lead to building damage in an earthquake and even collapse (Cole et al., 2011). Such pounding damage is unlikely to occur to small houses in gardens. A looser urban form can also aid recovery after an earthquake. Allan and Bryant (2014) analysed recovery after the 1906 earthquake and fire in San Francisco, noting that the existing urban landscape had a resilience to natural hazards as the "...regularity of the plan, laid over hilly terrain, resulted in a 'city of villages' where each neighbourhood was supported by a diverse network of connected open space." The 200,000 made homeless were able to set up temporary accommodation in "streets, parks, private gardens or vacant lots." Recovery was also aided as the "...city's wide streets also encouraged the rapid resumption of commercial activities, allowing makeshift shopfronts to be erected directly in front of damaged ones." The same study contrasted the urban form of Conceptión and the recovery of the city after the 2010 earthquake and tsunami in Chile. Conceptión was described as a modular city but one where the modules were not autonomous but relied on connections between the parts. When various bridges were destroyed, these connections were broken, which hindered recovery. In areas such as the wealthier leafy suburbs where there was space, people had somewhere to set up home, just as in San Francisco in 1906. Local communities also set up roadblocks so that the resources of a much smaller neighbourhood could be pooled and recovery of the community aided. This again suggests that small, diverse, autonomous for everyday needs, local communities with access to some open space would seem to be a resilient form of urban development. Perhaps the most striking comment made by Allan and Bryant (2014) with regard to Conceptión and its resilience was the following. "The roots of this problem are embedded in the disparity of income equality; a malaise that will always affect the resilience of a city, irrespective of its urban structure."

Marginal Returns, Urban Complexity and Collapse

Daly (2005) states that in everyday life people are very good at judging when the marginal returns are so small it is better to stop doing something, noting "People stop at the point when the marginal cost equals the marginal benefit. That is, it is not worth spending another dollar on ice cream when it gives us less satisfaction than a dollar's worth of something else." Sadly, this common-sense approach does not seem to apply to wealth-generating economies, which are expected to grow every year, and to cities which are seen as the means through which growth can be achieved.

Tainter (1988:124) makes the point that modern economic growth has been predicated on tapping new sources of energy, first fossil fuels and then atomic power. However, as discussed in Chaps. 1 and 5, EROI (energy return on investment) means that once easy supplies of the new fuels are consumed, more energy has to be put into extracting more difficult supplies so the marginal returns are reduced. However, adding a new fuel to the system can also create unexpected problems, such as climate change in the example of fossil fuels. This may lead to increased costs through the need to capture and store carbon, so the marginal returns are again reduced. Ultimately all expansion, including economic growth, comes with the danger of diminishing marginal returns. "A complex society pursuing the expansion option, if it is successful, ultimately reaches a point where further expansion requires too high a marginal cost" (Tainter, 1988:125). Ignoring this and pursuing expansion can lead to collapse. Could this idea also apply to the growth of cities?

The real issue here is that given the desirability of wealth, growth in wealth and marginal returns go together. This is made visible in the EF where high EFs tend to go with wealthy societies and vice versa. To achieve a very high HDI also requires considerable investment in resources and hence goes with a high EF. Wealth is not directly related to urban form, apart perhaps from the investment in high-rise and very high-rise buildings, which tends to go with wealthier societies, and the diminishing marginal returns that result from such investments. The issue for urban form is that wealth creation is linked to the concentration of people into cities, and with the latter comes added complexity.

Urban complexity comes down to the set of relationships between everything that goes to make up the city, so must include people and their systems as well as the physical form (López Baeze et al., 2017). The implication is that as cities grow larger this complexity also increases making management much harder. This has a parallel with the theory of diminishing returns as the basis of collapse. One way forward might be to manage growth so as to ensure that the facilities that are needed every day—work, education, market—are available in the centre of multiple clusters through the city so that they can be accessed in the easiest way possible—on foot. Facilities that are to do with "free" time could be positioned at the edges of clusters so they can be shared. In this way, the complexity is broken down into a series of many smaller communities within the overall city. The current problem of centralisation of shopping in suburban malls, and work in the CBD, means the linkages between people and facilities are much longer and more intertwined. However, this is still a simplistic view and fails to deal with the issue of the resources needed to support those living in cities or to deal with the issue of wealth accumulation. It might be a way of providing a built environment that can support city living in a more localised rather than globalised manner but only if other substantial changes also occur.

Earlier we have put forward the evidence that building high is a more expensive option. The resources for such buildings have to be found from outside the city boundary, just as the food and other consumer goods for the people living and working in such buildings. All this raises the average EF of the citizen and eventually, the marginal returns of putting people in more costly buildings in terms of both resources and money will outweigh the returns. There is no sense yet that the height of buildings is seen as having reached a desirable limit and cities are still vying to be the place with the highest building in the world. This is currently held by Dubai, with its Burj Khalifa (Fig. 8.3), a structure of 828 m with 163 habitable floors and a spire on top with no use, being there just to make the building taller (WorldAtlas, 2020).

These huge buildings are made possible by the technological advancements in structural materials, items such as lift cables, and simulations of wind effects, together with "…easier access to capital for developers in a low interest-rate environment, plus multimillionaire clients with the means to pay for a view" (Van Praet, 2016). Thus wealth, through the ability to make these very tall buildings for both residential and commercial uses, is at the same time creating denser cities with the view that densification is supposed to increase wealth. No one is yet asking the question as to whether increasing wealth is something that should be encouraged in this way. Unfortunately, increasing wealth goes with increasing environmental impact, which can be viewed as the diminishing return on the investment in cities. It should come as no surprise that countries with a shorter working week have a smaller EF (Bregman, 2017:142). This suggests that one way forward might be for everyone to work less so that less wealth is generated. Kallis et al. (2013) state the following. "In economies that progress technologically and in which capital becomes increasingly productive, workers should work less and less. This has not happened in the past because the surplus has been reinvested in new goods and more consumption, rather than more leisure and increases in wages." However, they also caution that reducing work hours needs to be accompanied by ensuring that everyone has sufficient income for a decent standard of living, though given humanity's inability to deal with inequality that may be difficult to achieve.

Fig. 8.3 Burj Khalifa, 2010, in the Dubai skyline (adapted from https://en.wikipedia.org/wiki/Burj_Khalifa)

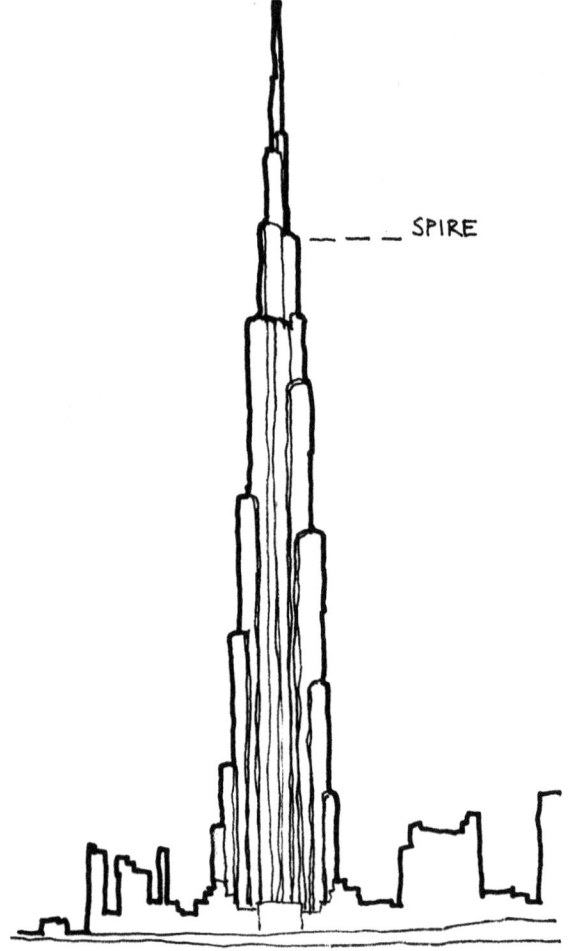

When it comes to the built environment, we can say that building high and increasing population density is not going to be the simple answer to a sustainable built environment unless the complex way in which high buildings are associated with wealth generation is also unravelled. The relationship between wealth and the built environment is explored further in Chap. 10.

References

Allan, P., & Bryant, M. (2014). The attributes of resilience a tool in the evaluation and design of earthquake-prone cities. *International Journal of Disaster Resilience in the Built Environment*, 5(2), 109–129.

References

Anon. (1967). Living in flats. *British Medical Journal, 4*(5576), 376.

Atlas of Urban Expansion. (2016). *Hong Kong, Hong Kong*. Retrieved May 18, 2020, from http://www.atlasofurbanexpansion.org/cities/view/Hong_Kong_Hong_Kong.

Australian Design Review (2018). *GBCA chair claims vertical living is a sustainability myth*. Retrieved 5 July, 2021, from https://www.australiandesignreview.com/news/gbca-chair-claims-vertical-living-is-a-sustainability-myth/.

Barkin, D. (1979). La transformación del espacio en Cuba post-revolucionaria. *Boletín de Estudios Latinoamericanos y del Caribe, 27*, 77–95.

Barr, J., Mizrach, B., & Mundra, K. (2015). Skyscraper height and the business cycle: Separating myth from reality. *Applied Economics, 47*(2), 148–160.

Borodovsky, L. (2017). *Daily shot: Hard times hit high-end housing: Rising demand for luxury apartments has spurred an epic building boom and a new threat to the U.S. housing market*, Wall Street Journal (online), May 14. Retrieved May 12, 2020, from https://search-proquest-com.helicon.vuw.ac.nz/docview/1898555773?rfr_id=info%3Axri%2Fsid%3Aprimo.

Booth, W., & Laadam, K. (2017). London struggles with 'ghost mansions'; Empty properties hot button amid housing shortage. *National Post (Index-Only)*. Retrieved June 2, 2020, from https://search-proquest-com.helicon.vuw.ac.nz/docview/1942461925?accountid=14782.

Boyle, E., Engelhardt, L., & Thornton, M. (2016). Is There such a Thing as a skyscraper curse? *The Quarterly Journal of Australian Economics, 19*(2), 149–168.

Bregman, R. (2017). *Utopia for realists*. Bloomsbury Publishing.

Center for Climate and Energy Solutions. (n.d.). *Global emissions*. Retrieved May 24, 2020, from https://www.c2es.org/content/international-emissions/.

Chancel, L., & Picketty, T. (2015). *Carbon and inequality: From Kyoto to Paris*. Paris School of Economics.

Chicca, F. (2013). *Developing a label for excellence in design for urban sustainability*. Ph.D. Thesis, Victoria University of Wellington.

Chicca, F., Vale, B., & Vale, R. (2018). Introduction. In F. Chicca, B. Vale, & R. Vale (Eds.), *Everyday Lifestyles and Sustainability: The environmental impact of doing the same thing differently* (pp. 1–13). Routledge.

Cole, G. L., Dhakal, R. P., & Turner, F. M. (2011). Building pounding damage observed in the 2011 Christchurch earthquake. *Earthquake Engineering and Structural Dynamics, 41*(5), 893–913.

Collins, A., Flynn, A., & Netherwood, A. (2005). *Reducing Cardiff's ecological footprint*. Cardiff: WWF Cymru, Sustainable Development Unit Cardiff Council and The Centre for Business, Relationships, Accountability, Sustainability and Society.

Council on Tall Buildings and Urban Habitat. (2020). *Cities ranked by number of 150m+ completed buildings*. Retrieved August 3, 2020, from https://www.skyscrapercenter.com/cities.

Credit Suisse. (2019). *Global wealth databook 2019*. Retrieved August 3, 2020, from https://www.credit-suisse.com/about-us/en/reports-research/global-wealth-report.html.

Daly, H. E. (2005). Economics in a full world. *Scientific American, 293*(3), 100–107.

Direct Line Software. (2010). *Surveying units and terms*. Retrieved June 2, 2020, from http://www.directlinesoftware.com/survey.htm.

Downing, G. L. A., & Calway, J. P. T. (1963). Living in high flats—Problems of tenants and management. *Royal Society of Health Journal, 83*(5), 237–243.

Edge, K., Scarpaci, J., & Woofter, H. (2006). Mapping and designing Havana: Republican, socialist and global spaces. *Cities, 23*(2), 85–98.

Edmunds, S. (2017). *The high life: Apartments cost more per metre*. Retrieved May 18, 2020, from https://www.stuff.co.nz/business/93697402/the-high-life-apartments-cost-more-per-metre.

Emporis. (2020). *Tallest buildings in Havana*. Retrieved June 5, 2020, from https://www.emporis.com/statistics/tallest-buildings/city/101285/havana-cuba.

GFN (Global Footprint Network). (2019). *Compare countries*. Retrieved May 27, 2020, from http://data.footprintnetwork.org/#/compareCountries?cn=all&type=EFCpc&yr=2016.

GFN. (2020a). *Country work*. Retrieved May 1, 2020, from https://www.footprintnetwork.org/our-work/countries/.

GFN. (2020b). *Glossary*. Retrieved May 1, 2020, from https://www.footprintnetwork.org/resources/glossary/.

GFN. (2020c). *Climate change*. Retrieved May 21, 2020, from https://www.footprintnetwork.org/our-work/climate-change/.

GFN. (2020d). *Only eight countries meet two key conditions for sustainable development as United Nations adopts sustainable development goals*. Retrieved may 26, 2020, from https://www.footprintnetwork.org/2015/09/23/eight-countries-meet-two-key-conditions-sustainable-development-united-nations-adopts-sustainable-development-goals/.

Ghosh, S., Vale, R., & Vale, B. (2007). Metrics of local environmental sustainability: A case study in Auckland, New Zealand. *Local Environment, 12*(4), 355–378.

Habraken, N. J. (1972). *Supports, an alternative to mass housing*. Architectural Press.

Harrabin, R. (2020). *Climate change: Top 10 tips to reduce carbon footprint revealed*. Retrieved May 26, 2020, from https://www.bbc.com/news/science-environment-52719662.

Heinonen, J., & Junnila, S. (2011). Implications of urban structure on carbon consumption in metropolitan areas. *Environmental Research Letters, 6*. 9pp.

Helleman, G., & Wassenberg, F. (2004). The renewal of what was tomorrow's idealistic city. *Cities, 21*(1), 1–25.

Hertwich, E. G. (2013). Addressing biogenic greenhouse gas emissions from hydropower in LCA. *Environmental Science and Technology, 47*(14), 9604–9611.

Hertwich, E. G., Gibon, T., Bouman, E. A., Arvesen, A., Suh, S., Heath, G. A., Bergesen, J. D., Ramirez, A., Vega, M. I., & Shi, L. (2015). Integrated life-cycle assessment of electricity-supply scenarios confirms global environmental benefit of low-carbon technologies. *Proceedings of the National Academy of Sciences of the United States of America, 112*(20), 6277–6282.

Hubacek, K., Guan, D., Barrett, J., & Wiedmann, T. (2009). Environmental implications of urbanization and lifestyle change in China: Ecological and water footprints. *Journal of Cleaner Production, 17*(14), 1241–1248.

Islam, N., Mahbub, A. Q. M., & Nazem, N. I. (2009). *The urban slums of Bangladesh*. Retrieved May 18, 2020, from https://www.thedailystar.net/news-detail-93293.

Jenks, M., Burton, E., & Williams, K. (1996). *The compact city: A sustainable urban form?* E and FN Spon.

Kallis, G., Kalush, M., O'Flynn, H., Rossiter, J., & Ashford, N. (2013). "Friday off": Reducing working hours in Europe. *Sustainability, 5*, 1545–1567.

López Baeze, J., Cerrone, D., & Männigo, K. (2017). Comparing two methods for urban complexity calculation using the Shannon-Wiener index. *WIT Transactions on Ecology and the Environment, 226*, 369–378.

Lui, L., & Lei, Y. (2018). An accurate ecological footprint analysis and prediction for Beijing based on SVM model. *Ecological Informatics, 44*, 33–42.

Manuel, J. (2013). The long road to recovery: Environmental health impacts of Hurricane Sandy. *Environmental Health Perspectives, 121*(5), 153–159.

McCrindle. (2017). *Sydney lifestyle study*. McCrindle Research Pty Ltd.

McKnight, J. M., Wilner, T., & Raskin, L. (2012). Havana: Bracing for a boom. *Architectural Record, 200*(2), 21.

Meade, L., Hutley, N., Van der Merwe, L., Parker, H., & Phai, C. (2018). *Cost of residential housing development: A focus on building materials*. Retrieved August 26, 2020, from https://www2.deloitte.com/content/dam/Deloitte/nz/Documents/Economics/nz-en-DAE-Fletcher-cost-of-residential-housing-development.pdf.

Moore, J., Kissinger, M., & Rees, W. E. (2013). An urban metabolism and ecological footprint assessment of Metro Vancouver. *Journal of Environmental Management, 124*, 51–61.

Neill, P. B. (2015). *Urban space as heritage in late colonial Cuba: Classicism and dissonance on the Plaza de Armas of Havana, 1754–1828*. University of Texas Press.

Newman, P., & Kenworthy, J. (1989). *Cities and automobile dependence: An international source book*. Gower.

References

Owens, A., & Green, D. R. (2016). Historical geographies of wealth: Opportunities, institutions and accumulation, c. 1800–1930. In I. Hay & J. V. Haverstock (Eds.), *Handbook on wealth and the super-rich* (pp. 43–67). Edward Elgar Publishing.

Ottelin, J., Heinonen, J., & Junnila, S. (2015). New energy efficient housing has reduced carbon footprints in outer but not in inner urban areas. *Environmental Science and Technology, 49*(16), 9574–9583.

Pehl, M., Arvesen, A., Humpenöder, F., Popp, A., Hertwich, E. G., & Luderer, G. (2017). Understanding future emissions from low-carbon power systems by integration of life-cycle assessment and integrated energy modelling. *Nature Energy, 2*, 939–945.

Perkins, A., Hamnett, S., Pullen, S., Zito, R., & Trebilcock, D. (2009). Transport, housing and urban form: The life cycle energy consumption and emissions of city centre apartments compared with suburban dwellings. *Urban Policy and Research, 27*(4), 377–396.

Pictet Group. (2020). *The role of cities in the global economy*. Retrieved May 4, 2020, from https://www.group.pictet/wealth-management/role-cities-global-economy.

Rattenbury, K. (2006). Viva Havana. *Building Design April, 21*(1718), 8–9.

Real Estate Research Corporation (RERC). (1974). *The cost of sprawl: Detailed cost analysis*. US Government Printing Office.

Rees, W. E., & Moore, J. (2013). Ecological footprints, fair earth-shares and urbanization. In R. Vale & B. Vale (Eds.), *Living within a fair share ecological footprint* (pp. 3–32). Routledge.

Rickwood, P., Glazebrook, G., & Searle, G. (2008). Urban structure and energy—A review. *Urban Policy and Research, 26*(1), 57–81.

Ritchie, H., & Roser, M. (2018). *Energy*. Retrieved May 24, 2020, from https://ourworldindata.org/energy#all-charts-preview.

Santamouris, M. (2014). On the energy impact of urban heat island and global warming on buildings. *Energy and Building, 82*, 100–113.

Santamouris, M., Papanikolaou, N., Livada, I., Koronakis, I., Georgakis, C., Argiriou, A., & Assimakopoulos, D. N. (2001). On the impact of urban climate on the energy consumption of buildings. *Solar Energy, 70*(3), 201–216.

Statista. (n.d.). *Average living space of public rental housing tenants in Hong Kong from 2007 to 2019*. Retrieved May 18, 2020, from https://www.statista.com/statistics/630746/hong-kong-public-rental-housing-average-living-space-per-person/.

Statistics New Zealand. (2010). *Apartment dwellers: 2006 census*. Statistics New Zealand.

Stewart, W. F. R. (1970). *Children in flats—A family study*. National Society for the Prevention of Cruelty to Children.

Tainter, J. A. (1988). *The collapse of complex societies*. Cambridge University Press.

Treloar, G. J., Fay, R., Ilozor, B., & Love, P. E. D. (2001). An analysis of the embodied energy of office buildings by height. *Facilities, 19*(5/6), 204–214.

United Nations (UN) Department of Economic and Social Affairs. (2018). *News*. Retrieved May 1, 2020, from https://www.un.org/development/desa/en/news/population/2018-revision-of-world-urbanization-prospects.html.

United Nations (UN) Department of Economic and Social Affairs. (2019). *World urbanization prospects: The 2018 revision*. United Nations.

United Nations Development Programme (UNDP). (2019). *Human development report 2019*. Retrieved May 27, 2020, from http://hdr.undp.org/sites/default/files/hdr2019.pdf.

United Nations Human Settlements Programme (2003). *The Challenge of Slums: Global Report on Human Settlement 2003*. Earthscan Publications Ltd.

United States Census Bureau (n.d.). *Urban and rural*. Retrieved May 2, 2020, from https://www.census.gov/programs-surveys/geography/guidance/geo-areas/urban-rural.html.

Vale, R., & Vale, B. (2009). *Time to eat the dog? The real guide to sustainable living*. Thames and Hudson.

Vale, B., & Vale, R. (2018). The Hockerton housing project, England. In R. Vale & B. Vale (Eds.), *Living within a fair share ecological footprint* (pp. 262–274). Routledge.

Van Praet, N. (2016). Thin dizzy: Growing urban population plus less open space equals the rise of the super-skinny skyscraper. *Report on Business Magazine Sept. 16*.

Wackernagel, M., & Rees, W. E. (1996). *Our ecological footprint*. New Society Publishers.

Ward, H. C., Kotthaus, S., Grimmond, C. S. B., Bjorkegren, A., Wilkinson, M., Morrison, W. T. J., Evans, J. G., Morison, J. I. L., & Iamarino, M. (2015). Effects of urban density on carbon dioxide exchanges: Observations of dense urban, suburban and woodland areas of southern England. *Environment Pollution, 198*, 186–200.

Weird Russia. (2015). *Balconies in Russia*. Retrieved June 2, 2020, from https://weirdrussia.com/2015/08/27/balconies-in-russia/.

WidWorld (n.d.). *World inequality database*. Retrieved June 5, 2020, from https://wid.world/world/#shweal_p90p100_z/GB;WO/last/eu/k/p/yearly/s/false/43.218/100/curve/false/country.

Wilson, B., & Chakraborty, A. (2013). The environmental impacts of sprawl: Emergent themes from the past decade of planning research. *Sustainability, 5*, 3302–3327.

Wilson, J., Spinney, J., Millward, H., Scott, D., Hayden, A., & Tyedmers, P. (2013a). Blame the exurbs, not the suburbs: Exploring the distribution of greenhouse gas emissions within a city region. *Energy Policy, 62*, 1329–1335.

Wilson, J., Tyedmers, P., & Grant, J. (2013b). Measuring environmental impact at the neighbourhood level. *Journal of Environmental Planning and Management, 56*(1), 42–60.

WorldAtlas. (2020). *10 tallest buildings in the world*. Retrieved June 8, 2020, from https://www.worldatlas.com/articles/10-tallest-buildings-in-the-world.html.

The World Bank Group. (2019a). *Rural population (% of total population)*. Retrieved May 2, 2020, from https://data.worldbank.org/indicator/SP.RUR.TOTL.ZS.

The World Bank Group. (2019b). *GDP per capita (current US$)*. Retrieved May 27, 2020, from https://data.worldbank.org/indicator/NY.GDP.PCAP.CD.

World Population Review. (2020). *World population review*. Retrieved May 27, 2020, from https://worldpopulationreview.com/.

Worldometer (n.d.a). *World population by year*. Retrieved May 21, 2020, from https://www.worldometers.info/world-population/world-population-by-year/.

Worldometer. (n.d.b). *Cuba population*. Retrieved May 26, 2020, from https://www.worldometers.info/world-population/cuba-population/.

Worldometer. (n.d.c). *Population: World*. Retrieved May 26, 2020, from https://www.worldometers.info/population/world/.

Young, M. D. (1962 rev. ed.). *Family and kinship in East London*. Penguin Books.

Yuen, B. (2011). Liveability of tall residential buildings. In B. Yuen & A. G. O. Yeh (Eds.), *High-rise living in Asian cities* (pp. 129–147). Springer.

Zahabi, S. A. H., Miranda-Moreno, L., Patterson, Z., Barla, P., & Harding, C. (2012). Transportation greenhouse gas emissions and its relationship with urban form, transit accessibility and emerging green technologies: A Montreal case study. *Procedia - Social and Behavioural Sciences, 54*, 966–978.

Zenghelis, D. (2017). Cities, wealth, and the era of urbanization. In K. Hamilton & C. Hepburn (Eds.), *National wealth: What is missing, why it matters* (27 pp.). Oxford Scholarship Online.

Chapter 9
Epidemics, Pandemics and Collapse

Atishoo, atishoo, we all fall down
Nursery Rhyme

Introduction

The pandemic caused by the rapid spread of the severe acute respiratory syndrome coronavirus 2 (SARS-CoV-2 or COVID-19) has disturbed everyday life across the globe. At the time of writing this chapter (3 August 2020) and since the end of December 2019, there have been 18,056,310 reported cases and 689,219 deaths (ECDC, 2020). Even though the virus has had an impact on the global scale, some countries like Italy, the USA, Spain, France and China have been hit harder than others countries or regions such as Iceland and Oceania. Regardless of social, economic and political disparities, the lack of an available vaccine up to this time has pushed non-pharmaceutical strategies based on social distancing and quarantine as the only ways of dealing with the virus. This has caused the lockdown of billions of people and with it the sudden disruption of urban life as we knew it.

This is not the first pandemic in modern times. The Hong Kong flu of 1968 led to the deaths of 1 million people (CDC, 2019), and humankind has found ways to come out from epidemics and pandemics before, but only at the cost of many lives. Can epidemics and pandemics thus contribute to the collapse of civilisations? What lessons can be learned from these experiences?

At the moment of writing this chapter, there is still no available vaccine against COVID-19. The changes experienced by most of the countries that have moved quickly into lockdown bring into question the current faith in economic growth and technology to solve global issues, like the present pandemic or climate change. The current situation deserves attention because it sets a precedent for things that can and cannot be done in a short time and what are essential and non-essential matters in a critical time. The exploration of the current situation could shine a light on what qualities are necessary for collapsing gracefully.

Epidemics and the Collapse of a Civilisation: The Case of Tenochtitlan

When the Spaniards arrived in America, they brought with them a series of novel diseases for the indigenous people. Among these, smallpox became one of the deadliest weapons of the conquest and colonisation of a vast territory from Mexico to Paraguay (Livi-Bacci, 2006).

Tenochtitlan in what is now part of Mexico City was founded in 1325 on an island in a lake and became the centre of the Aztec empire. It was the economic, social and political headquarters for a series of settlements that co-existed under a coalition in the central area of Mexico. Tenochtitlan was a very populated city when the Spaniards arrived. Its then population was estimated to be 400,000 in 13km^2 (Encylopaedia Britannica, 2020), giving it a population density of almost 30,800/km^2, which is somewhat greater than the population density of Manhattan, the densest urban area in the USA (Márquez Morfín & Storey, 2016:192). From an urban design point of view, it was a very strange city placed in a muddy landscape almost like an island in the middle of Lake Texcoco and connected through bridges to the hinterland. The situation provided direct access to the resources in the lake and the city was easy to defend as destroying the bridges compromised accessibility to the city. This was a problem for Cortes, the leader of the Spanish troops, when he decided to try to take Tenochtitlan. The first battle ended up with the "Sad night" where the Aztecs defeated the Spaniards, so not a sad night for the Aztecs. After that defeat Cortes decided to change strategy and start creating alliances with local communities oppressed by the Aztecs. It is during this process that the smallpox epidemic outbreaks occurred. Smallpox could have played a key role in debilitating the defences of Tenochtitlan, the capital of the empire that finally fell to the Spaniards in 1521 (Márquez Morfín & Storey, 2016:195). This event defined the beginning of the collapse of the Mesoamerican civilisation.

Estimations of the indigenous Aztec population before and after the arrival of the Spaniards are still contested because the historical documents have incomplete information and much of the anthropological evidence of Tenochtitlan lies underneath the present Mexico City, (Márquez Morfín & Storey, 2016:5). However, a conservative estimate shows that in 1519 the total population could have been between 1.5 and 3 million inhabitants. Following the conquest, the indigenous population showed a sharp reduction from 1.3 million in 1519 to only 300 thousand in 1569 (Whitmore, 1992:156). McCaa (1995) hypothesises that between 20 and 30% of the indigenous population possibly died of smallpox. After reviewing Spanish chronicles and Nahuatl text written in the Roman alphabet, McCaa (1995) affirms that the Mexican smallpox epidemic of 1520 was the greatest sixteenth-century demographic catastrophe in the country. "There is agreement that a demographic catastrophe occurred and that epidemic disease was a factor in initiating a die-off, beginning, in Mexico, with smallpox in 1520 [...]. Killing with war and conquest was clearly a secondary factor" (McCaa, 1995).

Smallpox and the other diseases brought by the Spaniards, including measles, diphtheria, rubella, and mumps, were unexpected allies that decimated the indigenous population during the conquest of Mexico (Livi-Bacci, 2006). The persistence of these diseases, along with oppression and abuse by the colonisers, marked a 200-year process of inequalities and violence that put the survival of a culture at risk. Outbreaks of smallpox and other diseases continued to affect the indigenous population until the mid-seventeenth century when the population was reduced to its lowest level since the first encounter with the Spaniards (Márquez Morfín & Storey, 2016:193). The conquest of all this territory would have been much harder without the help of smallpox.

A cure was eventually found for smallpox but it came too late for the Aztecs. In 1796, an English physician, Edward Jenner, observed that women whose work was milking cows acquired cowpox pustules on their hands similar to the ones caused by smallpox but they were not infected by the smallpox virus (Henderson, 2011). In 1798, implanting cowpox in the skin became the first vaccine against smallpox. However, the last case of smallpox only occurred in 1977, in Somalia. The strategies used during the 1960s to fight smallpox were immunisation, contact tracing and isolation of cases (Committee on Infectious Diseases, 2002). It took 180 years from the discovery of the antibodies against smallpox to its eradication. The vaccination process took several decades to be implemented requiring the effort of 50 countries with the support of WHO and UNICEF. The fatality rate of smallpox was 30% for the common form of the disease (WHO, 2020a, b, c, d, e). To provide a reference, smallpox was nearly 10 times more deadly than COVID-19, which has a mortality rate of 3.8% based on the figures at the start of this chapter. Henderson (2011) states: "In fact, during the twentieth century alone, an estimated 300 million people died of the disease—more than twice the death toll of all the military wars of that century." Without a vaccine, the only resource against smallpox was the help of nurses to keep patients hydrated, which was the same resource and the only help that was eventually offered to the indigenous people in Mexico during the conquest 200 years before. Nowadays, without immunisation against COVID-19, humanity is still fighting viruses as it used to do centuries ago.

In Tenochtitlan, the decimation of the indigenous population was followed by a radical transformation in the built environment. In the seventeenth century, the lake around Tenochtitlan was drained and a new city was laid out and built on the top of the previous one. The material used in the Aztec monuments and buildings was reused to build the foundations for the current Mexico City. This had an impact on the passing of memories and the inheritance of the cultural legacy of the Aztecs. However, it is difficult to exterminate the whole history and traditions of a civilisation and remains can be seen in various museums in Mexico City.

Even though it could be argued that the socio-political structures of the Aztec Empire might not have been sustainable over time, would have they collapsed without the arrival of the Spaniards and their diseases? The fall of Tenochtitlan raises an interesting question about the boundaries of collapse and the difficulties of defining them. Understanding collapse as a process makes the definition of its beginning and end a subjective matter. In the case of Tenochtitlan when did the process of collapse start and end? Did it start when the Crusaders introduced smallpox to Spain, when the Spaniards brought it to America or when Tenochtitlan was conquered? Did the collapse end after the fall of Tenochtitlan or 200 years after when the indigenous population reached its lowest point? In the nineteenth century, the indigenous population started to grow again, showing possession of a level of resilience (Márquez Morfín & Storey, 2016:193). The fall of Tenochtitlan and the Mesoamerican settlements during the conquest may represent the collapse of the Aztec Empire but not the disappearance of the culture or the indigenous population. Nahuatl, the language of the Aztecs, survives in Mexico (Canger, 2006:433). In the movie *Roma*, which won the Golden Lion at the Venice film festival in 2018, the actresses playing the maids speak in Nahuatl.

Pandemics and Collapse

Epidemics and pandemics affect one factor related to the complexity of a society—its demography. Even though the complexity of a society is not necessarily defined by the quantity of people but by the diversity of functions and hierarchies (Tainter, 1988:91) without people there is no society and no built environment, only ruins. The last time there was a global fall in population was when the Black Death killed a third of the known world human population in the mid-fourteenth century (Hatcher, 1994). It was fast and deadly. In Europe, the disease moved at a pace of 4 km per day (Duncan & Scott, 2005). In 4 years, the Black Death, a bubonic plague in combination with a haemorrhagic plague, covered 300,000 km^2 and affected countries from Asia, to the Middle East and almost the whole of Europe (Christakos et al., 2007). Most of the people that caught the disease died between the third and sixth day after showing symptoms. The number of fatalities is still contested since calculations were done using data from clergy or estimations based on the number of houses. Speculation has the outbreak of Black Death occurring in Europe between 1347 and 1351 (Christakos et al., 2007), and reducing the population by 30–50% (about 20 million people) in that period. Until the eighteenth century in Europe, there were outbreaks of "the plague" almost every 15 years (Scheidel, 2017:304). The Black Death or plague is still infecting people and is found everywhere except in Oceania, with most cases since 1990 happening in Africa, although today it can be treated with antibiotics (WHO, 2020b). Despite this, in 2017 it caused the loss of 202 lives in Madagascar (WHO, 2020c), showing that our civilisation is still vulnerable to very old diseases. This should be a concern when thinking about the possibility of facing future pandemics or at least it should be a call to learn from past experiences and to be prepared.

The Black Death did not collapse human civilisations. All the affected cultures and major cities are still in the same place, at least from the point of view of built environment studies. Maybe this can be seen as proof that even a deadly pandemic by itself cannot collapse the built environment. The collapse of ancient civilisations was also not due to a single catastrophe like a pandemic (Tainter, 1988:52). In the case of Tenochtitlan, the smallpox epidemic was accompanied by war and political instability. However, the Black Death is a good example when it comes to analysing the impact of a single event on the stability of civilisation.

The Impact of the Black Death

Loss of life on the scale of the Black Death should have produced important changes, but its impacts remain inconclusive. The only thing that all the renowned scholars in medieval studies and related fields have agreed about is that it killed a lot of people everywhere. Apart from that everything else still seems debatable. What was the Black Death? What were the social, political and economic consequences?

Debatable issues regarding the Black Death cover whether it was a single disease stemming from the bubonic plague or more than one disease, given how quickly it spread in the fourteenth century, although modern DNA research has linked it to the bacterium Yersinia pestis (Bos et al., 2011). Its effects on the economy of the time have also been hard to determine. Fewer people made it harder to raise taxes. Cafferro (2018:84–112) describes the problems of financing war in Florence because the deaths caused by the Black Death had reduced the ability to raise taxes. The problem of financing wars in France had just produced an agreed system for raising taxes when the Black Death "...threw all into disarray" (Henneman, 1968).

Initially, it led to wages being raised and inequalities levelled but as Clark (2016) notes:

> "When it reduced population levels, real wages rose. When its severity declined in the sixteenth century real wages fell back to their earlier levels." Clark also investigated economic efficiency and found as the population fell economic efficiency increased but: "Once population starts to increase again...measured efficiency declines." The thesis that the Black Death led to the collapse of the manorial system has also been contested by proposing that the manorial system was in decay in many places before the Black Death, with the economy more reliant on payment by money than labour (Campbell, 2000).

Responses to the fourteenth-century global pandemic were also not homogeneous. In India and China, the disease spread far more slowly and infected fewer people than in Europe (Christakos et al., 2007; Nutton, 2008:8). The religious responses of Christian European societies and Muslim societies were different. In the former, the Black Death was understood as a punishment from God. Some of the responses included flight, praying, flagellation and prosecution of minorities. Jewish communities went through massacres unprecedented until the twentieth century. This attitude towards minorities was not found in Muslim communities but in both religious contexts, there was a prescriptive attitude towards regulating community behaviour that was not always aligned with the best practices for fighting a contagious disease (Dols, 1974:287). Communal prayers, processions and community attendance at mass funerals clustered people instead of encouraging social distancing.

Learning from the past is learning from a set of fragmented information and therefore it leads to interpretations that leave more questions than answers. We still know very little and therefore we are still learning from our own mistakes. This information should help us to be more cautious and less confident about the perception of progress.

The Impact of the Black Death on the Built Environment

After almost seven centuries, the impact of the Black Death on the built environment also remains inconclusive. Unlike earthquakes, fires, wars and "other shocks considered in the literature buildings and physical capital were not destroyed" (Jedwab et al., 2019:2). The Black Death is purely about people dying in their thousands regardless of whether they lived in rural or urban built environments, and regardless of their race, age or socio-economic levels. However, when a country loses half of its population in a couple of years, there might be critical changes that could be traced in the built environment.

There was a change in the complexity of the buildings, particularly big buildings like cathedrals. Probably due to the impact that the plague had on the building industry expensive and seemingly extravagant pre-plague buildings like Exeter Cathedral (remodelled 5 times between 1258 and 1369) gave way to more austere post-plague buildings, like Edington church in Wiltshire built by the Bishop of Winchester. While pre-plague churches were tall with heavy decoration and elegant windows, the church in Wiltshire no longer looked like a gothic cathedral. It was more horizontal, closed and fortress looking. Since many of the masons died during the plague those that survived were in high demand, so recruitment was a key factor that probably defined the final design and look of the buildings (Platt, 1996:138–139). Other churches started before the plague in England were not finished or they were finalised in alternative styles such as the perpendicular. In Italy, plans to double the size of Siena Cathedral were dropped (Telegraph Media Group, 2020) and in England, the tower of York Minster was completed in the perpendicular style rather than the decorated style of the nave.

Another challenge to the study of the impact of the Black Death on the built environment is the interpretation of the lack of archaeological evidence. If people were dying by the thousand, there should be evidence of the spatial impact of bodies in cemeteries. Even though bodies may have been incinerated or disposed of in large pits, such places have not been identified in the numbers expected considering the recurrence and impact of the plague in Europe (Antoine, 2008:104; Nutton, 2008:9).

Traditionally, the Black Death has been associated with a decline in urbanism with decay visible in terms of empty plots and buildings (Platt, 1996:25). However, more recent investigation has found "...evidence for resilience in towns and cities of the late fourteenth and fifteenth centuries, where some places gained in population, through in-migration and investment, while others waned" (Lilley, 2014). Cities with high mortality rates did not disappear and recovered their populations faster than rural settlements (Jedwab et al., 2019). Moreover, wealth rose faster in urban areas than in rural after the pandemic (Lilley, 2014). Rural to urban migrations during the plague and in its aftermath left some rural areas abandoned, which probably led to a natural regrowth of the forest. Information obtained from studies in the southeast of the Netherlands recorded changes in agriculture and deserted farmlands accompanied by regrowth of forest during and after the Black Death (AD 1350–1440). Changes in forest density could have had an effect in terms of carbon sequestration and contributed to the decline in carbon dioxide emissions between the thirteenth and fifteenth centuries (van Hoof et al., 2006).

This crisis in urban and rural landscapes turned into a recovery process. Although the plague caused many losses and the elimination of families, it was also an opportunity for others (see also Chap. 3). While some were leaving a village, others would be buying the land. Such recovery actions follow the framework of the adaptive cycle (see Chap. 4).

Containment Within the Built Environment: Quarantine

In the time of the Black Death, the only defence against the pest was avoiding contact with infected people. Quarantine was a response first used in Dubrovnik in 1377 that then spread throughout Italy. The main idea was to separate persons, animals and products that could have been exposed to the disease and isolate them for 40 days (Tognotti, 2013). The rationale behind the 40 days is unknown, although it is the root of the word quarantine. It could have been linked to the bible and the 40 days Jesus spent in the wilderness. Some places prevented people from outside entering the city, and such sanitary cordons were sometimes implemented with a death penalty for those that broke them (if the plague does not get you the authorities will). Other measures found in Venice like contact tracing, identification of the routes of imported goods to local stores, and the use of long coats, masks and gloves for those tending the sick, were also developed over time (Linkov et al., 2014). Quarantine in response to the Black Death marked the beginning of public health as an integrated system for cities. Since little was known about the disease, quarantine was relatively successful in fighting it.

Isolation from Urban Landscapes: Lazarettos

Before the Black Death, isolation had been used as precaution against the spread of leprosy. Lazarettos were buildings outside the city whose purpose was to isolate infected people. The name Lazaretto alludes to the biblical character Lazarus, who was raised from the dead by Jesus. He was also the patron of lepers. At one point, there were "19,000 lazarettos in Europe. There were 109 at one time in England and Ireland, the first of which was established in 1096, and the last in 1472" (Huber, 1926:124). During the Black Death, the existing lazarettos in Italy were used as quarantine stations. The intention was to protect the ports of the more important coastal cities in Italy that were also the most exposed to foreign contact. The lazaretto on the island Santa Maria di Nazareth was built to quarantine people arriving in Venice and became instrumental in the fight against the disease (Rosenberger et al., 2012:69). Considering that the causes of the plague were not known or misunderstood, quarantine and the use of lazarettos helped cities like Venice to survive the outbreak of the plague (Linkov et al., 2014).

Eyam: The Plague Village

After the Black Death, the next most serious epidemic was the 1665–1666 plague in London, which killed 15% of the city's population (The National Archives, n.d.). The plague spread beyond the city boundaries, with fatal consequences. Eyam is a small village in the Derbyshire Peak District in England with the 2018 population estimated at 1,000. In 1665, the year of the plague, the village had a population of 700 (Whittles & Didelot, 2016). The plague arrived in Eyam when some cloth was sent from London that reputedly contained fleas infected with plague. The visiting tailor who opened the parcel soon fell ill and was dead within less than a week (Holloway, 2017). At least this is the romantic tale. It is more probable that the visiting tailor was infected before he arrived in Eyam (Scott & Duncan, 2001:265). To stop the plague from spreading to the neighbouring town of Bakewell and beyond, the vicar organised that the village would close its borders. The Duke of Devonshire who resided in nearby Chatsworth provided food, which was left at the village boundary in exchange for coins soaked in vinegar. The villagers also practised social distancing with church services held in the open air with family groups standing apart from each other (Holloway, 2017). As a result of the villagers cutting themselves off, 275 people died of plague and 432 survived, ignoring deaths from other causes (Whittles & Didelot, 2016), but the plague did not spread locally. The closing of the village boundaries in this way has been viewed as an act of heroism and the story has been retold in various forms from historical novels (Brooks, 2001) to operas (Drummond, 1984). However, Eyam was not the only place to initiate quarantine procedures to stop the spread of plague as the same measures had been put in place in "…Penrith,

Carlisle and York...it is just that Eyam has had better public relations agents to promote its story" (Scott & Duncan, 2001:281).

Was this a collapse? For some families where every member died the answer is "yes" (Scott & Duncan, 2001:275) but the village survives to this day and includes buildings from the time of the plague, such as the church of St. Lawrence (Britain Express, n.d.) and Margaret Blackwall's house; she was a survivor although other members of her family died (Scott & Duncan, 2001:283). Eyam itself has hardly grown in terms of population from the time of the plague year although it is now a place of commuters to the nearby towns together with income from tourism rather than the lead mining village at the time of the plague, but such a transformation is common among many small English villages that have become desirable places to live rather than work, a situation made possible by the private car.

What Can Be Learned from Previous Pandemics?

Trying to compare a pandemic like the Black Death with COVID-19, or even the smallpox epidemic that was involved in the collapse of the Aztec empire, is difficult and perhaps makes no sense. Mortality and recurrence rates are strikingly different. The context in which both epidemics emerged is also very different even without the advances in medicine and science gained particularly after the nineteenth century. However, without a historical review, it would be hard to believe that responses to pandemics, separated by 700 years, could still share some similarities. It seems, however, that similar responses have occurred to earlier pandemics.

> The same language, the same observations, even the same recommendations were constantly repeated. Writers in eighteenth-century Norway described the manifestations of plague in their community in exactly the same words as a writer on plague four hundred years earlier (Nutton, 2008:2).

The Black Death challenged religious belief and stigmatised minorities. Religious beliefs are still playing a role in the current pandemic. The spread of coronavirus within the strict Haredi Jewish community, and the fact their behaviour has not always aligned with the advice from public health authorities, has raised criticism and discontent in Israel, the UK and the USA (Kasstan, 2020). It is understandable that the link between archaeological studies of the Black Death and possible changes to the built environment, or even its exact presence in the built environment at the time, are hard to find. Put in a different way, what would a future archaeologist find out about the present Coronavirus pandemic? If archaeologists in 2700 were to try to find evidence of the impact of COVID-19 on the built environment of New Zealand, they would probably come to the conclusion that it never happened.

Covid-19

Since the Black Death, the world has become more populated and its built environments more crowded. However, the more people we have, the smaller the spaces we must share and the shorter the distances between populations in cities and countries. At the same time, we rely more and more on, and dedicate more technology to, communication. Does this communication lead to a difference in the response to the COVID-19 pandemic compared to pandemics of the past?

Modern Communication

The increase in communication could be a good means of accelerating contract-tracing in times of a pandemic. Contact tracing is not a new thing. As noted above, it was used during the Black Death in Italy to track the movement and destination of people and goods arriving in ports. More recently, it has also played an important role during outbreaks of Ebola in Africa (WHO, 2020d). For centuries, manual contact tracing has been useful for identifying and isolating infected people. With the outbreak of the current pandemic, academics from Oxford were aware of the problems of contact tracing by stating that "Given the infectiousness of SARS-CoV-2 and the high proportion of transmissions from presymptomatic individuals, controlling the epidemic by manual contact tracing is infeasible" (Ferretti et al., 2020:1). Since this warning, currently several governments have put great faith in automated contact tracing systems.

Singapore was the first country to implement a nationwide Bluetooth contact tracing system. However, when Jason Bay (2020), who is Senior Director at the Government Technology Agency of Singapore and the product lead for TraceTogether (the app used in Singapore during the first outbreak of COVID-19) was asked if Bluetooth contact tracing could replace manual contact tracing, he answered "No. Not now and…not for the foreseeable future." He went on to note that even with the help of artificial intelligence and machine learning, automated algorithms would still generate false positives and false negatives. Bay also stated that human contact tracers are able to build a history including data that go beyond proximity. In the search conditions, people can include the environmental status, and exercise judgment to avoid biases, like acknowledging that a short encounter in an open-air public space (that would count as a contact in the Bluetooth system) should be treated differently from an encounter in a lift.

One of the difficulties of assessing the effectiveness of automated contact tracing is that countries that have presumably been more successful in their instrumentation, like China and Singapore, have not released official numbers about the percentage of the whole population that used their apps. In order to monitor the status of automated contact tracing apps, the MIT Technology Review is developing a database: the Covid Tracing Tracker (O'Neill et al., 2020). This database has collected information from

26 countries that have supported the use of an app to automatise contact tracing. It shows that in most of these countries information is centralised; Bluetooth is the technology commonly implemented; and the number of users is still unknown or has not been officially published for half of the countries ranked. The most surprising thing is that the penetration (percentage of users) is lower than 1% in some countries, like India, Malaysia and Poland, and only slightly higher than 20% in Australia, Norway and Bahrain. Iceland with the app Rakning C-19 has reached 38% of the population, becoming the leading country in percentage of users. Nonetheless, Gestur Pálmason, who was overseeing contact tracing efforts for the Icelandic Police Service declared: "The technology is more or less…I wouldn't say useless…but it's the integration of the two that gives you results. I would say [Rakning C-19] has proven useful in a few cases, but it wasn't a game changer for us" (Hadavas, 2020). When people have a choice, it seems they do not choose to use apps. In the present scenario, apps are only running in a small set of countries and monitoring a small portion of the population in each of these. These facts jeopardise one of the main intentions behind the automated contract tracing system, which is the ability to have instantaneous information that can help to isolate cases, reduce transmission and stop the spread of the virus (Ferretti et al., 2020:6). Moreover, Bluetooth works differently in smartphones with different operating systems so users can experience problems. Such problems again reduce the penetration in the population. There are two different ways of looking at this problem, either the technology produced was unnecessary or the stage of the present technological development is as rudimentary as medicine in the medieval age appears to us now.

Although even at low levels of take-up an app can be useful as a protective measure, an uptake of 60% is required for effective control of an epidemic "…in the absence of other strong interventions" (Fraser et al., 2020). Outside the technological efforts, and until a vaccine is made commercially available, manual contact tracing, quarantine, social distancing and hygiene are the only means to deal with the pandemic. Curiously, along with masks, gloves and protective clothing, these are the same strategies that humanity has been using since medieval times.

The present pandemic has also revealed deep inequalities in terms of internet accessibility. This is important in terms of receiving updated information, working from home, supermarket shopping online and keeping in touch in lockdown. In the USA, a country leader in technology, 62% of its counties do not have the minimum download speed for broadband Internet. The number of people without access to broadband in the USA has been estimated to be between 21 and 48 million, or 6.0–14.5% of the current population (Busby et al., 2020). The people most affected live in rural areas where the cost of extending the infrastructure is more expensive. The penetration of broadband in the country has not improved in the last 10 years (Stanton, 2011).

In a pandemic, the reliance on internet services is such that without them people could feel cut off from society. However, in July 2020, internet penetration worldwide is only 59% (Clement, 2020), so just under half of the world's population currently lives without the internet. In January 2020, broadband access in the UK was 96% (Kemp, 2020), making it a country with a very high broadband penetration rate. However, with this have come disadvantages in the pandemic. A group of vandals, based on delusional conspiracy theories, have been burning telecommunications towers in the United Kingdom, the Netherlands, Italy, Ireland, Belgium and Cyprus. Hundreds of acts have been reported in one month alone. In one of them, the vandals had the strange idea of burning the telephone mast of the Nightingale Hospital in Birmingham, even though the country is in the middle of a pandemic crisis. The conspiracy theories have gained momentum in social media through more than 400 Facebook communities, Instagram groups and Twitter accounts. Strangely, the groups have claimed that there is a link between the spread of the pandemic and the development of the ultrafast broadband network known as 5G. The number of baseless conspiracy theories is endless and there is no point in promoting them here. What is agreed between academics, governments and institutions is that there is no scientific evidence that 5G infrastructures can have a negative effect on health (PHE, 2019). The level of radiofrequency exposure of the 5G technologies is around 3.5 GHz, which is similar to the present mobile phone base stations that have no consequences for public health (WHO, 2020e). The Ministry of Health in New Zealand has been measuring radiofrequency exposure levels and found that in all cases measurements are below the standards, which are "50 times lower than the recognised threshold for established effects" (MH, 2020a). Nonetheless, this information has been ignored by those responsible for 15 fires around New Zealand since March 2020, most of them in areas where internet connectivity is most needed (Clent, 2020).

Curiously, the technology that these groups are fighting is the same technology that made possible the organisation of their conspiracy groups in social media and that has contributed to their diffusion. Research in the UK and Spain investigated the 5G conspiracy gossip using a social network analysis of Twitter data. The study found that a dedicated account was set up to spread the conspiracy theory and this created the first cluster of 408 Twitter users, which were expanded by people trying to mock the group, which only facilitated its diffusion. This group was supported by a commercial organisation, Infowars, which created content promoting their products, some of which just happened to be offering protection against electromagnetic fields. The findings showed that the spread of the conspiracy could have been avoided if any authority or policymaker had stopped it in time (Ahmed et al., 2020:6).

Such baseless fears and the ensuing acts of vandalism could be the subject of a useful psychological analysis about collective hysteria post lockdown. In the meantime, picturing a procession of fanatics trying to torch a telecommunication tower left no option other than thinking that there is a conspiracy to recreate the responses to the Black Death pandemic. Hopefully, they will not start a witch-hunt of programmers, engineers in computers and IT professionals. This is a clear example of how increments in the complexity of the technology involved in communications can eventually lead to unintended consequences. The self-organisation processes and the high connectivity that it allows can be used to entertain a hospitalised person with coronavirus or to organise the burning of a telecommunication tower in that same hospital. This should be a warning and a wake-up call to realise that complex systems, like social media, are very difficult to control but can make possible the collapse of something like 5G infrastructures.

The proliferation of the media in the pandemic has also contributed to polluting the environment with an overload of information, whose statements and predictions were forever changing and contradicting themselves. Since everyone looks convincing it is easy to follow the advice that most suits you. President Trump even suggested drinking disinfectant.

Where were the most precious achievements of the technology of our times, like artificial intelligence, virtual reality, machine learning, and robots, when we needed them? Since this pandemic could be understood as a trial in order to get ready for the next crisis, it gives an understanding of the extent to which societies can rely on these technologies.

COVID-19 and the Economy

A big difference between the Black Death that affected everybody but in a larger proportion the poor without the means to retreat from it (McKinley, 2020) and COVID-19 is that the latter seems to have hit developed countries harder during the first wave of the virus. Whether developed or developing countries will be more affected by COVID-19 is less evident at the moment of writing this chapter. If levels of human development (measured in the Human Development Index) are spread across six categories and compared with the global share (as a proportion of the total) of confirmed cases and deaths by thousands, the differences between more and less developed countries are striking (see Fig. 9.1). Countries that rank higher in the Human Development Index (HDI) have more confirmed cases and have suffered more deaths than countries that rank lower in the HDI. Less than 20% of all the countries affected account for almost half of the confirmed cases and deaths per thousand in the world. In the HDI, countries with an index above 0.7 are considered developed. If this parameter is used then 94% of all confirmed cases and 97% of the deaths per thousand occur in more developed countries. However, these numbers are very subjective. First, developed countries have been testing more and processing results faster than developing countries, which will affect the number of confirmed cases in a future count. Second, the timing of this measurement is also important, since developed countries have been hit by the virus first, not least because of the ease of travel between countries, while most of the developing countries have not yet reached the high "peak" of confirmed cases. The discrepancy in numbers between developed and developing countries should also take account of measurements of "severity," namely comparisons to pre-pandemic mortality patterns, so as to understand the impact of the virus in each country and to make comparisons between them (Schellekens & Sourrouille, 2020:6).

In the early stages in developing countries, the pandemic was mostly linked with the upper middle class and rich people who had the money to travel to Europe and the unfortunate fate of finding the virus before seeing La Gioconda. Even in a developed

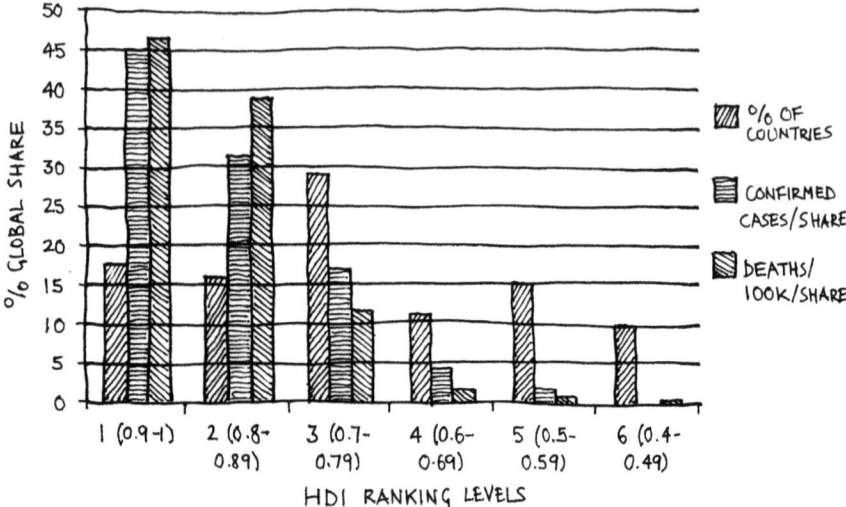

Fig. 9.1 Distribution of confirmed cases and deaths (by thousands) across development levels. Rank levels of HDI are organised from 1 to 6, namely, from more developed to less developed. (Based on data collected until April 2020). *Human Development Index (2019). **COVID-19 Dashboard (2020)

country like New Zealand, the majority of covid-cases have been linked to people who had travelled to Europe and other destinations affected by the virus. As of 6 August 2020, 41% of all NZ cases had travelled overseas and brought the disease back and a further 29% had caught the disease from someone who had travelled overseas (Ministry of Health, 2020a, b). Even though it is too soon to have a clear picture, the rising number of cases in developing countries and the lack of minimum resources and infrastructure, such as the lack of soap and a continuous water supply, could have severe implications for these countries. The fact that the virus has impacted countries with a higher development level and the most complex economies harder is a wake-up call to remind people that the perceived progress achieved by wealthy countries will not stop pandemics from spreading.

The final economic impact of COVID-19 will only be revealed in the future. In the meantime, it has induced economic uncertainty, which is the worst nightmare for any economist. One of the problems associated with economic uncertainty is that most of the world economies are built around the idea of continuous economic growth. The present disruption in the economic activities created by preventive responses to the pandemic has pushed countries into a situation where their economies depend on an estimated growth that is not real money in the present. The uncertainty created threatens the projected gains, which may well vanish. Under the present circumstances, projections of growth can be seen as what they are: abstractions.

"The COVID-19 pandemic is far more than a health crisis: it is affecting societies and economies at their core. While the impact of the pandemic will vary from country to country, it will most likely increase poverty and inequalities at a global scale"

(UNDP, 2020). Regardless of the uneven distribution of deaths caused in developed and developing countries, the pandemic has been affecting the economies of all countries. The International Monetary Fund (IMF) and the World Bank suggest that the economic crisis caused by the pandemic could be deeper than the 2008 global financial crash (see also Chap. 10). This is "the first recession to have been triggered solely by a pandemic during the past 150 years, and current forecasts suggest that it will be the most severe since the end of World War II" (The World Bank, 2020:20). In April 2020, a World Bank report (Lakner et al., 2020) stated that COVID-19 had pushed between 40 and 60 million people into extreme poverty (less than $1.90 US dollars per day), while another estimation suggested that the pandemic could raise the number of people living in extreme poverty from 434 to 922 million (OXFAM, 2020). The full panorama will be revealed once (if?) the pandemic has passed, but developing countries seem to be more vulnerable to poverty than to the disease.

The limiting of mobility because of global lockdown measures has had an impact on the demand for oil. The manager of a petrol station in my neighbourhood told me that before the pandemic, he was selling 9000 L of fuel per day but during the lockdown that went down to only 3000 L per day. Therefore, it is not surprising that the global economic crisis induced by COVID-19 has been accompanied by an unprecedented "collapse in oil demand and prices" (The World Bank, 2020:198). This highlights the dependency of our societies on non-renewable resources and the vulnerability that this creates. Perhaps the impact of the pandemic on the global economy could be a good excuse to rethink the diversification of economies in countries that depend on energy exporting and importing.

One of the outcomes of the initial impacts of the Black Death in Europe was the increase in wages and with this the softening of economic inequalities (Clark, 2016). It is uncertain but highly unlikely that the current pandemic will produce similar outcomes. In April 2020, as a result of the pandemic 74% of startups (companies in their early stages) have had a decline in their revenues with half of them affected significantly and 24% temporarily or permanently closed. If the situation continues 40% of them will be able to hold on for up to 3 months. Only 26% of the startups had no loss of full-time employees (Startup Genome, 2020:2). Of course, some companies have profited from the crisis and used the situation to grow their businesses. One of these is the online platform Zoom that became the preferred means around the world for working online because of its capacity to share a session with multiple users simultaneously. This has been one of the most downloaded applications. The owner was not ranked in the Forbes list in 2019 but appeared in the middle of the ranking of millionaires in 2020. However, these are exceptions. The trend for startups is very clear—it is a case of reinventing themselves or disappearing. The idea of reinventing yourself is not necessarily linked to being creative or innovative as it depends on having the money to "create" a second chance, as the capital owned will set concrete limits on the "creative capacity". People that have invested all their capital in a startup and lost it can be very creative people but without the money, they are just unemployed. After the lockdown, they will have to beg for a job in an institution or firm.

Businesses that depend on the physical flow of people like tourism, retail, hotels, sports events and airlines have been hit harder than other sectors. The construction industry has not remained untouched by the crisis, after all, designing and building remain human-driven activities. In April 2020, the Architecture Billings Index (ABI) showed a continuing downwards trend. The ABI is an economic indicator for non-residential construction activity supported by the American Institute of Architects (AIA). The focus is on the non-residential sector since the majority of buildings in this category are predominantly designed by architectural firms. When the index value is above 50, the industry is doing well but below 50 signals a decrease in billings and design contracts. In April 2020, the ABI index was 29.5 for billings and 27.6 for design contracts (AIA, 2020). This is the second month in a row that the ABI showed a score below 50, supporting the claim made by the National Bureau of Economic Research (NBER) that the USA is undergoing a recession (NBER, 2020). The pandemic has produced a new type of recession, which is shorter than in the past but with deeper impacts.

In the built environment, and particularly in the construction industry, the most profound economic impact has been taken by small companies (less than 50 employees) who are more likely to have stopped activities entirely (Schneider, 2020). Architecture and design firms related to health infrastructure did better due to the demands posed by the pandemic. The fact that small firms are being impacted harder is important because the industry is certainly divided. Firms of small size represent 97% of the total and account for almost 57% of the billing share, while big firms represent 3% of the total number of firms but account for 43% of the total share of billings. The economic impact of the pandemic has depleted small firms who are more numerous and therefore employ more people than a few big firms. The uneven distribution of the share of billings across firm sizes increases the inequality and therefore the complexity of the industry. If the distribution of billings was more equal each firm would have greater opportunities to buffer crisis and avoid cuts in income. The concentration of jobs around big firms means there is more chance of amplifying the economic impact of the pandemic.

The pandemic has pushed individuals and institutions to claim for a more human economy with "fairer taxation of rich individuals and corporations" (OXFAM, 2020:13). Even though this is still an ideal goal, the pandemic has revealed concrete ways in which governments, institutions and individuals can be more generous. The governments of many countries have offered economic aid to their people. The International Monetary Fund (IMF) through the Catastrophe Containment and Relief Trust (CCRT) has made available US$250 billion, which represents a quarter of its lending capacity, to help member countries. Between May and June 2020, 70 countries have benefited from the loans, most of them in Africa (IMF, 2020). Countries have also provided stimulus packages to deal with the economic impacts of the pandemic. The information collected by the economist Ceyhun Elgin (Elgin et al., 2020) was used to develop an index and a website (last updated in June 2020) with information that provides evidence of the economic efforts made by 160 countries to counter the effect of the pandemic. The data show that some countries like Luxembourg (22%), Japan (21.1%), Slovenia and Singapore (each 19.7%) have invested a considerable share of their GDPs to buffer the crisis. Wealthy countries have not been the only generous ones. Within the top 20 countries, both Iran and Greece have committed 14% of their GDP to battle the pandemic. A small country like New Zealand (10%) has double the stimulus of the UK (5%), a country that has been hit harder by the pandemic. In the case of New Zealand, the first stimulus package (NZ$12 billion) included a 6-month mortgage relief for negatively affected small business and mortgage holders, tax relief measures for commercial and industrial buildings, and support for companies facing insolvency. The wage subsidy scheme paid NZ$585 per week for full-time employees. Rent increases were frozen as well as tenancy terminations. Maori and Pacific communities were supported via funding to health services and Maori businesses (The Treasury, 2020). This set of actions looks more like the way some people would like to see a budget rather than a response to a critical situation. It has proved how generous governments can be when they are confronting a crisis. Why cannot some of this generosity be extended to normal situations? Even though it can be argued that these benefits are not sustainable under the present economic system, then why not treat this pandemic as an opportunity to rethink the system?

Environmental Impact of the Pandemic

The negative impact of COVID-19 on world economies and the health of people contrasts with the positive outcomes for the environment. The quarantine measures pushed changes in consumption behaviour and transportation globally. An analysis of energy demands until 14 April 2020 in 30 countries that consume two thirds of global energy showed that during this period, consumption was reduced by an average of 25% per week for countries in full lockdown and by 18% for those in partial lockdown (IEA, 2020:3). Global energy demand in the first quarter of 2020 declined by 3.8% (IEA, 2020:11) due to a decrease in oil use for road and air transportation and coal use, especially for electricity in China. Even though people spent more time at home, which could have increased the residential consumption of electricity, reductions in the demands from commercial and industrial activities resulted in an overall decline in the demand for electricity (IEA, 2020:22). As a result of these changes in energy demand during lockdown, global daily carbon dioxide emissions have been estimated to have declined by 17% during the lockdown (Le Quéré et al., 2020). The IEA (2020:4) states that the pandemic is likely to make global carbon dioxide emissions decline by 8% in 2020 to CO_2 levels of 10 years ago. However, these changes will probably be temporary since the global economic system, the transportation system and the energy system basically remain the same.

What might be the impact of this historical reduction in carbon dioxide emissions caused by the pandemic on the climate change crisis? The pandemic will not have washed away either the impact of climate change or its threat. This is because emissions and concentration of carbon dioxide are different things, as explained by Betts et al. (2020).

An analogy is filling a bath from a tap. If the tap represents CO_2 emissions, and the water level in the bath is CO_2 concentrations, while we have slightly turned the tap down temporarily, water is still flowing into the bath and so the level is still rising. To slow climate change, the tap needs to be turned right down—and permanently.

The concentrations of carbon dioxide are predicted to keep on increasing in 2020 regardless of the decline in emissions due to the pandemic. In the meantime, estimates show that due to the pandemic, concentrations of carbon dioxide emissions might rise 11% less than forecast in 2019 but they will keep on rising (Betts et al., 2020). For climate change, the impact of the pandemic on the environment represents what an aspirin is for a chronic migraine. However, the pandemic could be a sample of the scale and effort that it would take to reduce carbon dioxide emissions by 50% in the short term and 70% in the medium term. The scale of the problem and the failed attempts to solve it illustrate the increasing complexity of climate change, which might well lead to a collapse of the present development model.

Human-induced climate change could also be triggering new epidemics and pandemics (Altizer et al., 2013). The exploitation, destruction and fragmentation of species habitats due to urbanisation and climate change impact negatively on biodiversity. High biodiversity can be helpful in buffering transmissions of certain infectious diseases (Ostfeld & Keesing, 2017). Therefore, battling against climate change cannot be put on hold, not even during a pandemic. The link between climate change and the transmission of diseases is still a matter under study that perhaps deserves more attention considering the lives lost during the present pandemic. However, it is vital to recall that the pandemic induced reduction in carbon emissions caused the suffering of and the loss of thousands of jobs for people. Relying on pandemics as environmental levellers and a way to reduce carbon emissions would be a very ungraceful way of collapsing. However, if climate change issues are not tackled in time, we may well need to start looking at the current pandemic as a warm-up for the next global crisis.

The Built Environment

One thing that has traditionally been observed when it comes to the relationship between epidemics, pandemics and the built environment is the issue of density. The reason for this is very simple, as proximity and prolonged direct contact with someone infected create a scenario that increases the chances of contracting an infectious disease. Without getting too deep into the many ways of understanding density, the proximity induced by the built environment can be produced by sharing a big space with many people or by being in a very small space like a lift with only one person. The more important clusters of COVID-19 in New Zealand were related to a school event, a wedding, a conference, and a funeral. All happened in spaces that are bigger than any room in an average house but that were shared with a lot of people. Even big spaces, partially open, like stadia, were epicentres for the dispersion of the virus. Following this line of thought, it would be easy to assume that built environments with denser populations would have been hit harder by the pandemic. Nonetheless, the relationship between the pandemic and the density of the built environment is not that simple. Density (see Chap. 8) is not such an easy concept from which straightforward conclusions can be simply assumed. It can, however, add complexity to the built environment, especially when built and population densities are accompanied by multiple or mixed uses.

Most countries are clustered around the 100 people/km^2 mark, however looking at the figures in detail, within that density countries like Macedonia, France and Spain have very different deaths per million. Canada that has fewer than 10 people/km^2 has more deaths per million than Singapore and Monaco, with population densities above 10,000 people/km^2.

Figure 9.2 shows that across countries, the population density does not seem to be related in a linear way with deaths caused by the pandemic. However, these data do not show the link between the pandemic and specific types of built densities. It would seem more appropriate to look at the relationship between m^2 per person across countries to explore the relationship if any between built density, population density and COVID-19 (Table 9.1).

Table 9.1, represents a snapshot mid-2020 that was taken before there is any sign of an end to the pandemic, but by grouping countries with similar infection rates it shows no relationship between COVID-19 cases and density in terms of built area

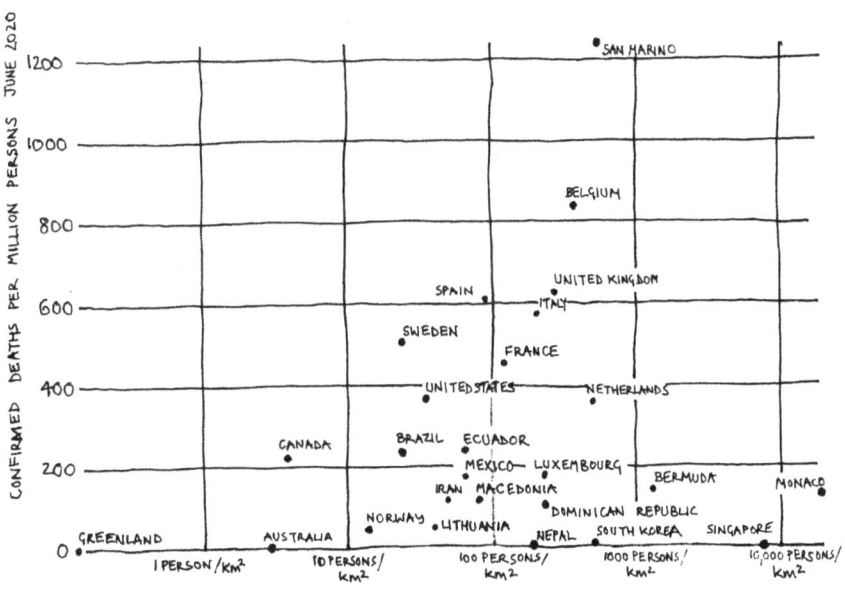

Fig. 9.2 Deaths per million as of 23 June 2020 against the population density of selected countries (based on Our World in Data, 2020a, b, c)

Table 9.1 European Union, COVID-19 and built area per capita

Country (European Union only)	Built up area[a] m²/capita (2014)	COVID-19 reported[b] cases as if 19 August 2020	Population 2020[c]	% of population infected
Luxembourg	354.39	7,499	625,978	1.2
Belgium	399.19	78,804	11,589,623	1.2
Spain	179.20	364,196	46,754,778	0.8
Sweden	288.97	85,219	10,099,265	0.8
Ireland	210.38	27,499	4,937,786	0.6
Portugal	323.80	54,448	10,196,709	0.5
Romania	N/A	72,208	19,237,691	0.4
Italy	233.67	254,636	60,461,826	0.4
Netherlands	285.16	63,911	17,134,872	0.4
Malta	N/A	1,322	441,543	0.3
Austria	268.56	23,875	9,006,398	0.3
Germany	287.98	226,914	83,783,942	0.3
France	369.16	221,267	65,273,511	0.3
Denmark	416.45	15,855	5,792,202	0.3
Bulgaria	N/A	14,669	6,948,445	0.2
Croatia	N/A	6,855	4,105,267	0.2
Estonia	195.45	2,200	1,326,535	0.2
Poland	227.01	57,876	37,846,611	0.2
Czech Republic	278.95	20,483	10,708,981	0.2
Republic of Cyprus	N/A	1,359	1,207,359	0.1
Lithuania	204.45	2,474	2,722,289	0.1
Finland	213.71	7,776	5,540,720	0.1
Greece	228.97	7,427	10,423,054	0.1
Slovenia	241.85	2,456	2,078,938	0.1
Latvia	156.65	1,323	1,886,198	0.07
Slovakia	282.27	2,922	5,459,642	0.05
Hungary	320.36	4,970	9,660,351	0.05

[a]OECD.Stat (2020)
[b]ECDPC (2020)
[c]Worldometer (2020)

per capita. The group of Austria, France, Germany and Denmark has similar rates of infection but the built environment per capita varies by a factor of over 1.5. The attitude taken to controlling the infection and the speed at which governments acted will probably have had much more to do with the infection rate than density.

COVID-19 and Cities

More might be learned of COVID-19 and the built environment by looking closer in scale, namely, by comparing cities within countries. New York is the city with the densest population and has also been the one most affected by the virus in the USA. The same is applicable to Madrid in Spain and Buenos Aires in Argentina, which are the most populated cities within their countries and the most affected by the pandemic. However, this assumption is challenged when Seoul and Hong Kong are considered. Both cities have a high population density but to date, they have had a mild impact from the pandemic in comparison with the cities previously mentioned. In China, when plotting the number of confirmed cases and the population density, most of the confirmed cases happened in cities with a density between 5000 and 10,000 people/km² and cities with higher population densities did not show more cases (Fig. 9.3). Even at the city scale, density seems to refuse to follow a single pattern globally, at least at this stage of the pandemic.

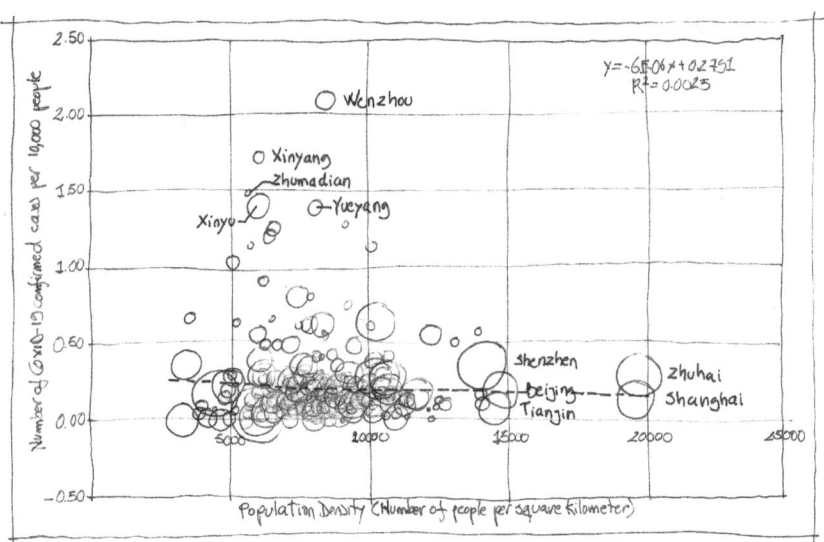

Fig. 9.3 Number of COVID-19 cases and population density (adapted from https://blogs.worldbank.org/sustainablecities/urban-density-not-enemy-coronavirus-fight-evidence-china)

COVID-19 and Housing

The top-down analysis of the role of urban density in the pandemic contrasts with what happens when the same relationship is analysed bottom-up, namely, from the population and built density at the household scale to the neighbourhood scale. Since most countries progressively moved to lockdown in order to secure social distancing, dwellings tended to be used in a more intensive way. This change in the way the built environment was used induced by the lockdown can be seen in data related to the change in time spent at home. This number is available for both countries and cities. The change in behaviour at the small scale is an interesting variable because it cascades up to be visible at the global scale. Figures 9.4 and 9.5 show the change in the percentage of time spent at home from the beginning of the pandemic. In April 2020 (Fig. 9.4), there were more countries whose populations spent more time at home (between 20 and 50%) than in June (Fig. 9.5) when some of the countries had left lockdown, reducing the change in the time spent at home to between 10 and 20% higher than before the lockdown.

The pandemic revealed many inequalities in household conditions. This is an obvious situation in developing countries where inequality is high, producing contrasting housing conditions between rich and poor people. However, the pandemic has also increased the visibility of hidden inequalities in developed countries. In Canada, the disparity in size between single-detached houses and "condos" (apartments) has been growing wider in the last few decades. In Vancouver, "for new

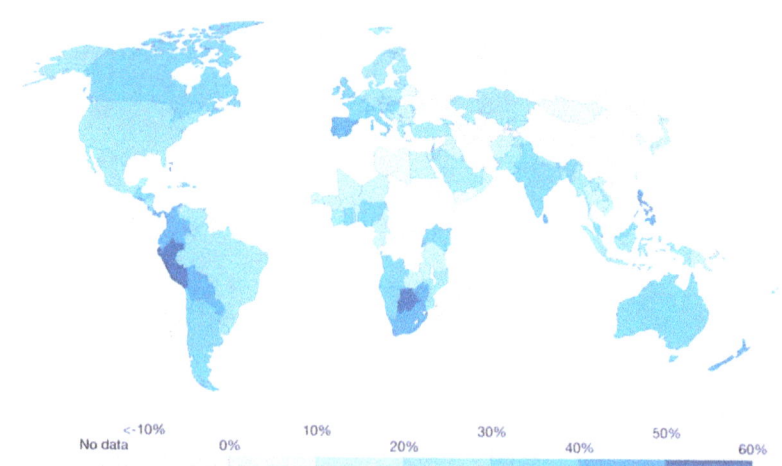

Fig. 9.4 Change in time spent at home. 11 April 2020 (Our World in Data, 2020b)

Residential areas: How did the time spent at home change since the beginning of the pandemic?, Jun 20, 2020
This data shows how the number of visitors to residential areas has changed relative to the period before the pandemic.

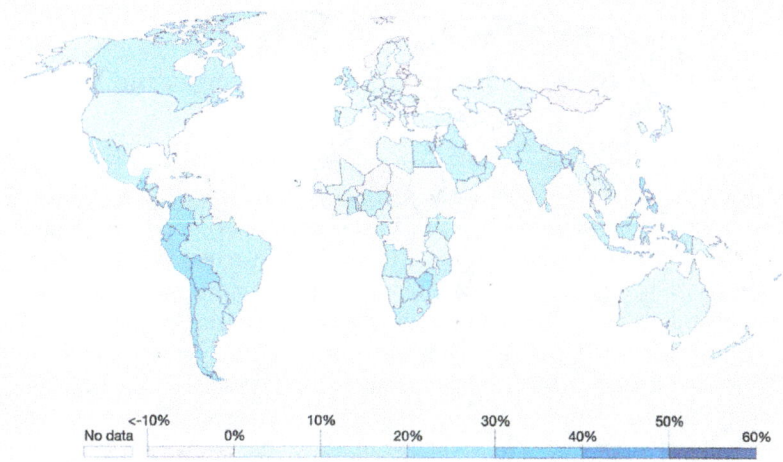

Source: Google COVID-19 Community Mobility Trends – Last updated 20 February, 15:02 (London time)
Note: It's not recommended to compare levels across countries; local differences in categories could be misleading.
OurWorldInData.org/coronavirus • CC BY

Fig. 9.5 Change in time spent at home. 20 June 2020 (Our World in Data, 2020c)

dwellings built between 2016 and 2017, the average new house had an area of 3,820 sq. ft., and condo, 644 sq. ft." (Gold, 2020). This situation has pushed one council to allow the construction of windowless bedrooms in condos. With more people working from home in the pandemic, it is easy to imagine how difficult it would be to divide up space to cope with different functions in these small condos. In the end, families will do what they can with what they have but this cannot be an excuse to accept what are unsatisfactory living conditions as normal.

The increased time spent at home produced discrete results depending on the number of family members within a house, their age distribution (more or fewer children or elderly people in need of assistance), the space available to work online, workload per capita within households, socio-economic level and income, available m^2 per person, and having a garden or balcony and other amenities. For wealthy families with steady incomes, flexible jobs, big houses with gardens and accessibility to green areas within their neighbourhood, the lockdown looked more like a strange vacation when you had the time to learn how to bake sourdough bread. For the most vulnerable socio-economic sectors, it was totally the opposite. The pandemic put families in a state of complete uncertainty, increased stress and anxiety, creating the perfect environment for depression. This situation was worsened by limited accessibility to outdoor spaces. In some countries like Spain or France, the lockdown measures were so restrictive that walking and running outside were not encouraged. Nonetheless, the imagination, temper and resilience of people are unstoppable.

Social media and the news documented people running marathons in their back yards (Farzan, 2020) and a man walking the dog from a balcony (Salo, 2020).

The Resilience of the Built Environment to Changes Induced by COVID-19

The old-fashioned concept of quarantine used during the Black Death defined the core strategy for dealing with COVID-19. With different levels of permitted mobility, people around the world were asked to stay at home for most of the time. Retail outlets, shops and everything else with the exception of some chemists, supermarkets and petrol stations were closed. The circulation of cars in the streets was almost nil. The single "Living in a ghost town," a song released by The Rolling Stones during the lockdown, shows the empty streets of London like a dead ghost town, something impossible to imagine a few months ago. What were the adaptations made by people within the built environment?

The lockdown did something that links to the hypothesis of collapse: it simplified everyday life to buffer the risk of contagion, but what do we mean by simplified and how did that happen? As discussed in Chaps. 3 and 4, the ecological resilience approach in urban landscapes implies understanding the built environment as an "urban panarchy." This means seeing it as a complex adaptive system, with many scales and discrete changes happening within and across these. Persistence at big scales, like the city scale, can happen because other changes happen at small scales, like the domestic scale. Changes in the built environment during the pandemic can be analysed using a multi-scale approach like that of the urban panarchy to understand what adjustments made the adaptation possible. Figures 9.6, 9.7 and 9.8 present simple diagrams to illustrate changes at different scales using a bottom-up approach, namely, using domestic life as the unit of analysis because, as shown earlier (see Chap. 4), some changes are more elusive when viewed at large scales.

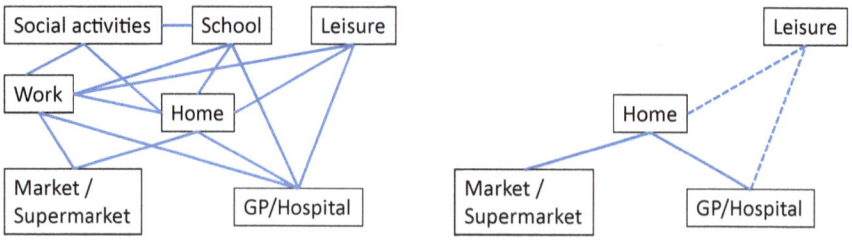

Fig. 9.6 Reduction in complexity (city scale)

The Resilience of the Built Environment to Changes … 255

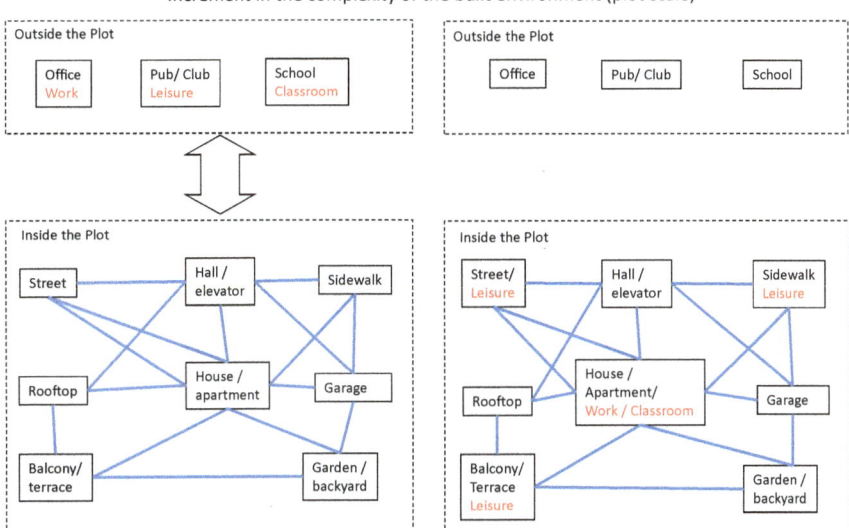

Fig. 9.7 Increment in complexity (plot scale)

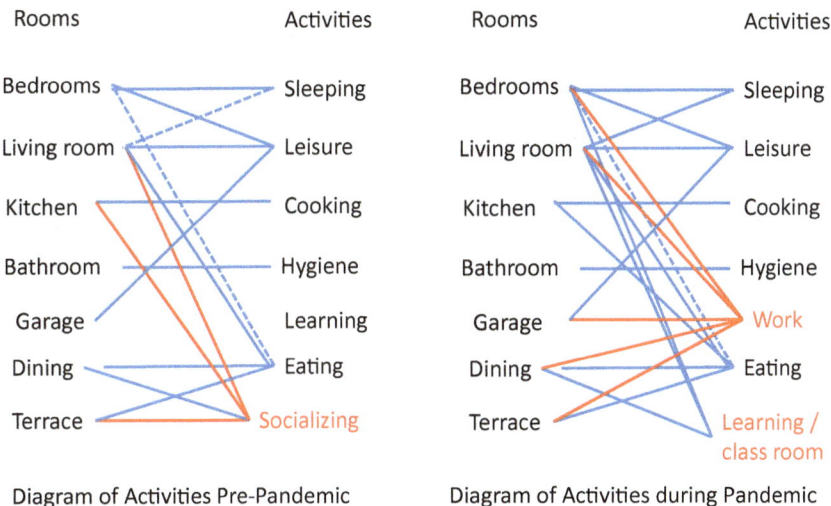

Fig. 9.8 Increment in complexity (building scale)

At the city scale (Fig. 9.6), the complexity of life was reduced by limiting movement to basically one destination, the supermarket or closest chemist. Depending on the country's restrictions and level of the lockdown, leisure time was more or less confined to outside the house but in the immediate neighbourhood. Houses, supermarkets and eventually hospitals and parks were the only buildings and spaces within which all city life was concentrated. For example, In the USA, normal mobility patterns were reduced by 36–63% (Badr et al., 2020). It is amazing to realise that people could keep on living in cities by using only two buildings and a very small portion of the whole city. This is possible because we live in cities that have redundant spaces, buildings only for work, others for domestic life and others for complementary activities. This creates obsolescence during normal times but it can create opportunities during a pandemic. In cases where the health infrastructure was saturated, some hotels and office buildings were used to house people affected by the virus (The Guardian, 2020).

The reduction of complexity at the city scale was also possible because that complexity was absorbed by the plot and building scale. As shown by Garcia and Vale (2017:185–197) in a morphological analysis of Auckland, small-scale changes buffer the impact of disturbances. The difference, in this case, is that the disturbance (the pandemic) was buffered by decreasing complexity and diversity instead of increasing diversity and complexity at the city scale. The plot scale (Fig. 9.7) shows fewer changes in terms of the number of spaces in use but it shows more activities happening in some of them due to the inclusion of work and much increased leisure at home.

At the domestic scale, and assuming that households did not transform the infrastructure of the dwelling by adding new rooms or walls to subdivide spaces, changes mostly happened in the way of using and sharing spaces. Figure 9.8 shows in a diagrammatic way the differences between the relationships and activities before and during the pandemic. The biggest change was the loss of social activities (with other people outside the dwelling) and the inclusion of work as an activity that demanded its own space. The dining or living room became spaces that acquired multiple uses from a workspace to a classroom, and sometimes both things at the same time.

The introduction of work as an activity within the dwelling could be welcomed or not depending on the pre-existing conditions, availability of spaces and their size, structure of the family (age diversity, occupations, expectations), and job stability and income. If more than one person was working in a family, the situation could shift from moving the workspace to home to turning home into a workspace. The second change could also be problematic. Teaching online, parenting and preparing food at the same time can be as challenging as having to look at your children in the bath while answering emails. It is not only a matter of space availability but also how many things someone can do at the same time.

The analysis of changes pre and during the pandemic and the exploration of how adaptation happened are useful for understanding the resilience of the built environment to work from home. This, in turn, might be useful for forecasting scenarios in which working from home and living with limited mobility become the norm. As discussed in the section above on the environmental impact of the pandemic, limiting mobility, shopping, and tourism have decreased carbon dioxide emissions, although this has happened at the economic cost of many people losing their jobs. Would it be possible to live in lockdown for a couple of years to achieve the goals of the Paris agreement? For how long would this be needed? Would it even be possible for the majority of people to work from home in such a situation? Such an idea would never be acceptable as the risk from climate change is not perceived by the majority of people as being as severe as the risk from COVID-19.

Conclusion

Some of our worst fears, like a deadly pandemic, can be emotionally and economically harmful but still not enough to collapse the built environment of cities or a civilisation. The smallpox epidemic during the Spanish conquest of Mexico was a key factor in the collapse of the Aztec civilisation. However, it was the combination of the epidemic with certain political conditions, social inequalities, and war that created the perfect environment for conquest. As noted before, the collapse of ancient civilisations happened mainly due to internal problems, through processes that were raising the cost of sustaining an increasing complexity. The Black Death and the present pandemic are good examples of what a pandemic can do by itself. However, not even the mortality rate of the Black Death was enough to cause the collapse of a civilisation, and the same is valid for the present pandemic. The fact that nine months since COVID-19 first appeared, technology has not found a way of stopping it, is useful for realising that our civilisation has transformed the expectations generated by technological development and economic growth into a cult, whereby we expect these two together can solve all problems—immediately. In the same way that the Black Death led to questioning of religious affiliations, this pandemic can be used to question this faith in economic growth and technological development.

If current responses to a pandemic still share similarities with responses that occurred centuries ago, we should question the benefits of our investments in the present complexity, particularly when we realise that the pandemic will be managed sooner or later but climate change issues will still be there. The current pandemic has also shown that abrupt and radical changes can be made, even under the present economic and political interests. Who would have believed that all the casinos in Las Vegas could be closed, Time Square empty and the streets of London deserted?

As noted, technological advances such as tracing apps have so far been relatively useless to combat the spread of the pandemic. There is the belief that, in times of crisis, somebody will have a genius idea and things will be solved. However, nobody came out with a genius idea this time. At the time of writing this chapter, there is still no commercial vaccine against COVID-19, though trials are underway. Contrary to the expectation generated by the ideals of development and progress, medicine and science still need time to deal with the unpredictable. From an economic point of view, the pandemic has shown that it is possible to stop the commercial wheel almost completely and the system will not suddenly collapse.

The pandemic, through the lockdown responses, has been useful for experiencing the tight links between economic interests, consumption and climate change. This development model is what makes the problem of the latter so complex. The current development model is effectively fuelled by and dependent on carbon dioxide emissions. These grow continuously and produce diminishing marginal returns via the concentration of carbon dioxide that impacts the climate. Solutions to this problem become more expensive and more difficult to solve when more development is the goal. The pandemic lockdown can serve as an example of what happens when the current economic system stops and consumption is reduced. Carbon emissions were reduced but carbon concentration will keep on increasing. If it is assumed that consuming less could be a simple and cheap solution to reducing carbon emissions, the pandemic has shown that it is a very expensive measure when the economic damage produced to millions of families is taken into consideration.

The type of work that can be done from home and the spaces needed should be the starting point in imagining an alternative development model for the built environment. The pandemic clearly separated the types of work deemed essential in built environments from the rest. Producing and distributing food, health services, street cleaners, garbage collectors and postmen were among the essential workers that were allowed to keep working. The rest of the work suddenly became non-essential and could or could not be done from home. Critical times are a good test for recognising what matters the most. Indirectly, the lockdown challenged the significance and relevance for society of most of the work we do for a living.

COVID-19 has not led to a collapse of human society at either the global or national scale. Many countries have borrowed money to cope with the crisis that will eventually have to be paid back, but so far there has been no bankrupt country. Cities have also survived but some households have had to transform themselves in order to survive as former jobs have been lost. The problem with climate change is that it is a big scale problem that will lead to transformation at every scale. COVID-19 has supplied a small taste of what these changes might mean.

References

Ahmed, W., Vidal-Alaball, J., Downing, J., & López Seguí, F. (2020). COVID-19 and the 5G conspiracy theory: Social network analysis of twitter data. *Journal of Medical Internet Research, 22*(5), 9.

AIA (American Institute of Architects). (2020). *Architecture Billings Index (ABI)*. Retrieved May 12, 2020, from https://www.aia.org/resources/10046-the-architecture-billings-index.

Altizer, S., Ostfeld, R. S., Johnson, P. T. J., Kutz, S., & Harvell, C. D. (2013). Climate change and infectious diseases: From evidence to a predictive framework. *Science, 341*(6145), 514–519.

Antoine, D. (2008). The archaeology of "plague." *Medical History. Supplement, 27*, 101–114.

Badr, H. S., Du, H., Marshall, M. Dong, E., & Squire, M. M. (2020). Association between mobility patterns and COVID-19 transmission in the USA: A mathematical modelling study. *The Lancet on-line* (9pp.) Retrieved August 7, 2020, from https://www.thelancet.com/action/showPdf?pii=S1473-3099%2820%2930553-3.

References

Bay, J. (2020). *Automated contact tracing is not a coronavirus panacea.* Retrieved April 20, 2020, from https://blog.gds-gov.tech/automated-contact-tracing-is-not-a-coronavirus-panacea-57fb3ce61d98.

Betts, R., Jones, C., Jin, Y., Keeling, R., Kennedy, J., Knight, J., & Scaife, A. (2020).*Analysis: What impact will the coronavirus pandemic have on atmospheric CO_2?* Retrieved May 12, 2020, from https://www.carbonbrief.org/analysis-what-impact-will-the-coronavirus-pandemic-have-on-atmospheric-co2.

Bos, K., Schuenemann, V., Golding, G., Burbano, H., Waglechner, N., Coombes, B., Mcphee, J., Dewitte, S., Meyer, M., Schmedes, S., Wood, J., Earn, D., Herring, D., Bauer, P., Poinar, H., & Krause, J. (2011). A draft genome of Yersinia pestis from victims of the Black Death. *Nature, 478*(7370), 506–510.

Britain Express. (n.d.). *Eyam St. Lawrence Church—The Plague Church.* Retrieved April 24, 2020, from https://www.britainexpress.com/counties/derbyshire/churches/eyam.htm.

Brooks, G. (2001). *Year of Wonders.* Penguin Books.

Busby, J., Tanberk, J., & BroadbandNow Team. (2020). *FCC underestimates Americans unserved by broadband internet by 50%.* Retrieved April 28, 2020, from https://broadbandnow.com/research/fcc-underestimates-unserved-by-50-percent.

Cafferro, W. (2018). *Plutarch's war: Florence and the Black Death in context.* Cambridge University Press.

Campbell, B. (2000). Britain 1300. *History Today, 50*(6), 10–17.

Canger, U. (2006). Nahuatl. In *Encyclopedia of language and linguistics, 14 volume set.* Elsevier ScienceDirect Books.

CDC (Centers for Disease Control and Prevention). (2019). *1968 pandemic (H3N2 virus).* Retrieved August 4, 2020, from https://www.cdc.gov/flu/pandemic-resources/1968-pandemic.html.

Christakos, G., Olea, R. A., & Yu, H. (2007). Recent results on the spatiotemporal modelling and comparative analysis of Black Death and bubonic plague epidemics. *Public Health, 121*(9), 700–720.

Clark, G. (2016). Microbes and markets: Was the Black Death an economic revolution? *Journal of Demographic Economics, 82,* 139–165.

Clement, J. (2020). *Worldwide digital population as of July 2020.* Retrieved August 5, 2020, from https://www.statista.com/statistics/617136/digital-population-worldwide/.

Clent, D. (2020). *Why setting cell phone towers on fire won't stop coronavirus.* Retrieved May 25, 2020, from https://www.stuff.co.nz/national/crime/121537885/why-setting-cell-phone-towers-on-fire-wont-stop-coronavirus.

Committee on Infectious Diseases. (2002). Smallpox vaccine. *Pediatrics, 110*(4), 841–845.

COVID-19 Dashboard by the Center For Systems Science and Engineering (CSSE) at Johns Hopkins University (JHU). (2020). Retrieved May 5, 2020, from https://coronavirus.jhu.edu/map.html.

Dols, M. W. (1974). The comparative communal responses to the Black Death in Muslim and Christian societies. *Viator, 5,* 269–288.

Drummond, J. D. (1984). *Plague upon Eyam: An opera in three acts.* University of Otago Press.

Duncan, C. J., & Scott, S. (2005). What caused the Black Death? *Postgraduate Medical Journal, 81*(955), 315.

ECDC (European Centre for Disease Control). (2020). *COVID-19 situation update worldwide, as of 3 August 2020.* Retrieved August 4, 2020, from https://www.ecdc.europa.eu/en/geographical-distribution-2019-ncov-cases.

Elgin, C., Basbug, G., & Yalaman, A. (2020). Economic policy responses to a pandemic: Developing the COVID-19 economic stimulus index. In *Covid economics: Vetted and real time papers* (Vol. 3, pp. 40–54). CEPR Press

Encylopaedia Britannica. (2020). *Tenochtitlán.* Retrieved August 4, 2020, from https://www.britannica.com/place/Tenochtitlan.

ECDPC (European Centre for Disease Prevention and Control). (2020). *COVID-19 situation update for the EU/EEA and the UK, as of 19 August 2020.* Retrieved August 20, 2020, from https://www.ecdc.europa.eu/en/cases-2019-ncov-eueea.

Farzan, A. N. (2020) *A British man ran a marathon in his 20-foot backyard during the coronavirus lockdown—and thousands tuned in.* Retrieved August 7, 2020, from https://www.washingtonpost.com/nation/2020/04/02/backyard-marathon-coronavirus/.

Ferretti, L., Wymant, C., Kendall, M., Zhao, L., Nurtay, A., Abeler-Dörner, L., Parker, M., Bonsall, D., & Fraser, C. (2020). Quantifying SARS-CoV-2 transmission suggests epidemic control with digital contact tracing. *Science (New York, N.Y.), 368*(6491), 9.

Fraser, C., Abeler-Dörner, L., Ferretti, L., Parker, M., Kendall, M., & Bonsall, D. (2020). *Digital contact tracing*, Retrieved August 5, 2020, from https://github.com/BDI-pathogens/covid-19_instant_tracing/blob/master/Centralised%20and%20decentralised%20systems%20for%20contact%20tracing.pdf.

Garcia, E. J., & Vale. B. (2017). *Unravelling sustainability and resilience in the built environment.* Routledge.

Gold, K. (2020, April 17). COVID-19 puts urban density to the test; OPINION. *Globe & Mail* [Toronto, Canada] (p. 3). Retrieved May 20, 2020, from https://link-gale-com.ezproxy.auckland.ac.nz/apps/doc/A621083142/AONE?u=learn&sid=AONE&xid=5549af80.

Hadavas, C. (2020). *How effective are contact tracing apps?* Retrieved May 20, 2020, from https://slate.com/technology/2020/05/contact-tracing-apps-less-effective-iceland.html.

Hatcher, J. (1994). England in the aftermath of the Black Death. *Past & Present 144,* 3–35.

Henderson, D. A. (2011). The eradication of smallpox—An overview of the past, present, and future. *Vaccine, 29*(4), 7–9.

Henneman, J. B. (1968). The Black Death and royal taxation in France, 1347–1351. *Speculum, 43*(3), 405–428.

Holloway, J. (2017). Resounding the landscape: The sonic impress of and the story of Eyam, plague village. *Landscape Research, 42*(6), 601–615.

Huber, E. (1926). The control of communicable diseases prevalent in Massachusetts. *The Boston Medical and Surgical Journal, 195*(3), 122–127.

Human Development Index and Its Components. (2019). Retrieved September 29, 2020, from http://hdr.undp.org/en/content/table-1-human-development-index-and-its-components-1.

IMF (International Monetary Fund). (2020). *Factsheet.* Retrieved May 12, 2020, from https://www.imf.org/en/About/Factsheets/Sheets/2016/08/01/16/49/Catastrophe-Containment-and-Relief-Trust.

International Energy Agency (IEA). (2020). *Global energy review 2020. The impacts of the covid-19 crisis on global energy demand and CO_2 emissions.* IEA Publications. Retrieved August 20, 2020, from https://webstore.iea.org/download/direct/2995.

Jedwab, R., Johnson, N., & Koyama, M. (2019). Pandemics, places, and populations: Evidence from the Black Death. *IDEAS Working Paper Series from RePEc,* CEPR Discussion Papers 13523.

Kasstan, B. (2020). Angry at ultra-orthodox Jews for 'Defying' coronavirus rules? it's more complicated than that. *Haaretz* Retrieved April 10, 2020, from https://www.haaretz.com/israel-news/EXT.premium.EXT-STATIC-netanyahu-s-plane-israel-s-controversial-new-air-force-one-1.8993542.

Kemp, S. (2020). *Digital 2020: the United Kingdom,* Retrieved August 5, 2020, from https://datareportal.com/reports/digital-2020-united-kingdom#:~:text=There%20were%2065.00%20million%20internet,at%2096%25%20in%20January%202020.

Lakner, C., Mahler, D. G., Negre, M., & Prydz, E. (2020). *How much does reducing inequality matter for global poverty?* World Bank Group. Retrieved May 12, 2020, from http://documents.worldbank.org/curated/en/765601591733806023/pdf/How-Much-Does-Reducing-Inequality-Matter-for-Global-Poverty.pdf.

Le Quéré, C., Jackson, R. B., Jones, M. W., Smith, A. J., Abernethy, S., Andrew, R. M., De-Gol, A. J., Willis, D. R., Shan, Y., Friedlingstein, P., Creutzig, F., Peters, G. P., & Canadell, J. G. (2020). Temporary reduction in daily global CO_2 emissions during the COVID-19 forced confinement. *Nature Climate Change,* 1–7.

Lilley, K. D. (2014). Urban planning after the Black Death: Townscape transformation in later medieval England (1350–1530). *Urban History, 42*(1), 22–42.

Linkov, I., Fox-Lent, C., Keisler, J., Sala, S., & Sieweke, J. (2014). Risk and resilience lessons from Venice. *Environment Systems and Decisions; Formerly the Environmentalist, 34*(3), 378–382.

Livi-Bacci, M. (2006). The depopulation of Hispanic America after the conquest. *Population and Development Review, 32*(2), 199–232.

McKinley, K. (2020). *How the rich reacted to the bubonic plague has eerie similarities to today's pandemic.* Retrieved August 6, 2020, from https://theconversation.com/how-the-rich-reacted-to-the-bubonic-plague-has-eerie-similarities-to-todays-pandemic-135925.

Márquez Morfín, L., & Storey, R. (2016). Population history in precolumbian and colonial times. In D. Nichols & E. Rodríguez-Alegría (Eds.), *The Oxford handbook of the Aztecs.* Oxford University Press. Retrieved September 29, 2020, from https://www.oxfordhandbooks.com/view/10.1093/oxfordhb/9780199341962.001.0001/oxfordhb-9780199341962-e-35.

McCaa, R. (1995). Spanish and Nahuatl views on smallpox and demographic catastrophe in Mexico. *The Journal of Interdisciplinary History, 25*(3), 397–431.

MH (Ministry of Health New Zealand). (2020a). *Cellsites and 5G.* Retrieved August 5, 2020, from https://www.health.govt.nz/your-health/healthy-living/environmental-health/radiation-environment/cellsites-and-5g.

Ministry of Health New Zealand. (2020b). *COVID-19—Current cases.* Retrieved August 6, 2020, from https://www.health.govt.nz/our-work/diseases-and-conditions/covid-19-novel-coronavirus/covid-19-current-situation/covid-19-current-cases.

National Bureau of Economic Research. (2020). *Determination of the February 2020 peak in US economic activity.* Retrieved August 20, 2020, from https://www.nber.org/cycles/june2020.html.

Nutton, V. (2008). Introduction. *Medical History. Supplement, 27*, 1–16.

O'Neill, P. H., Ryan-Mosley, T., & Johnson, B. (2020). *A flood of coronavirus apps are tracking us. Now it's time to keep track of them.* Retrieved May 10, 2020, from https://www.technologyreview.com/2020/05/07/1000961/launching-mittr-covid-tracing-tracker/.

OECD.Stat (2020). *Built-up area and built-up area change in countries and regions.* Retrieved August 20, 2020, from https://stats.oecd.org/Index.aspx?DataSetCode=BUILT_UP.

Ostfeld, R. S., & Keesing, F. (2017). Is biodiversity bad for your health? *Ecosphere, 8*(3), 1–12.

OXFAM. (2020). *Dignity not destitution.* Retrieved May 12, 2020, from https://oxfamilibrary.openrepository.com/bitstream/handle/10546/620976/mb-dignity%20not%20destitution-an-economic-rescue-plan-for-all-090420-en.pdf.

Our World in Data. (2020a). *COVID-19 death rate vs. population density.* Retrieved June 20, 2020, from https://ourworldindata.org/grapher/covid-19-death-rate-vs-population-density.

Our World in Data. (2020b). *Residential areas: How did the time spent at home change since the beginning of the pandemic?* Retrieved April 11, 2020, from https://ourworldindata.org/grapher/changes-residential-duration-covid?stackMode=absolute&time=earliest..latest&region=World.

Our World in Data. (2020c). *Residential areas: How did the time spent at home change since the beginning of the pandemic?* Retrieved June 20, 2020, from https://ourworldindata.org/grapher/changes-residential-duration-covid?stackMode=absolute&time=earliest..latest&region=World.

Platt, C. (1996). *King death: The Black Death and its aftermath in late-medieval England.* UCL Press.

PHE (Public Health England). (2019). *5G technologies: Radio waves and health.* Retrieved May 10, 2020, from https://www.gov.uk/government/publications/5g-technologies-radio-waves-and-health/5g-technologies-radio-waves-and-health.

Rosenberger, L. H., Riccio, L. M., Campbell, K. T., Politano, A. D., & Sawyer, R. G. (2012). Quarantine, isolation, and cohorting: From cholera to klebsiella. *Surgical Infections, 13*(2), 69–73.

Salo, J. (2020). *Chinese resident on coronavirus lockdown 'walks' dog from balcony with leash.* Retrieved August 7, 2020, from https://nypost.com/2020/02/19/chinese-resident-on-coronavirus-lockdown-walks-dog-from-first-floor-balcony-with-leash/.

Scheidel, W. (2017). *The great leveler: Violence and the history of inequality from the stone age to the twenty-first century.* University Press.

Schneider, A. (2020). *Measuring the fallout of COVID-19 for the design industry.* Retrieved May 12, 2020, from https://www.metropolismag.com/interiors/thinklab-industry-impact-survey/.

Schellekens, P., & Sourrouille, D. (2020). *Tracking COVID-19 as cause of death: Global estimates of relative severity*. Retrieved May 12, 2020, from https://www.brookings.edu/wp-content/uploads/2020/05/Tracking_COVID-19_as_-Cause_of_Death-Global_Estimates_of_Severity.pdf.

Scott, S., & Duncan, C. J. (2001). *Biology of plagues: Evidence from historical populations*. Cambridge University Press.

Stanton, L. (2011). FCC 706 report finds broadband deployment too slow for second year; McDowell Dissents. *Telecommunications Reports, 77*(11), 1–45.

Startup Genome. (2020). *The impact of COVID-19 on global startup ecosystems*. Retrieved June 21, 2020, from https://startupgenome.com/reports/impact-covid19-global-startup-ecosystems-startup-survey.

Tainter, J. A. (1988). *The Collapse of Complex Societies*. Cambridge University Press.

Telegraph Media Group. (2020). *Amazing buildings you didn't know were unfinished*. Retrieved August 5, 2020, from https://www.telegraph.co.uk/travel/galleries/spectacular-buildings-you-would-never-guess-are-unfinished/siena-cathedral/.

The Guardian. (2020). *Delhi to transform 25 luxury hotels into Covid-19 care centres*. Retrieved August 7, 2020, from https://www.theguardian.com/global-development/2020/jun/22/delhi-to-transform-25-luxury-hotels-into-covid-19-care-centres.

The National Archives. (n.d.). *Great plague of 1665–1666*. Retrieved August 5, 2020, from https://www.nationalarchives.gov.uk/education/resources/great-plague/.

The Treasury. (2020, April). *COVID-19 economic package at a glance: He waka eke noa: We are all working together*. Retrieved May 12, 2020, from https://treasury.govt.nz/publications/glance/covid-19-economic-package-glance-he-waka-eke-noa-we-are-all-working-together-april-2020.

The World Bank. (2020). *Global economic prospects*. The World Bank. Retrieved May 12, 2020, from https://www.worldbank.org/en/publication/global-economic-prospects.

Tognotti, E. (2013). Lessons from the history of quarantine, from plague to influenza A. *Emerging Infectious Diseases, 19*(2), 254–259.

UNDP. (2020). *Socio-economic impact of COVID-19*. Retrieved August 5, 2020, from https://www.undp.org/content/undp/en/home/coronavirus/socio-economic-impact-of-covid-19.html.

van Hoof, T. B., Bunnik, F. P. M., Waucomont, J. G. M., Kürschner, W. M., & Visscher, H. (2006). Forest re-growth on medieval farmland after the Black Death pandemic—Implications for atmospheric CO_2 levels. *Palaeogeography, Palaeoclimatology, Palaeoecology, 237*(2), 396–409.

Whittles, L. K., & Didelot, X. (2016). Epidemiological analysis of the Eyam plague outbreak 1665–1666. *Proceedings of the Royal Society B, 283*, 1–9.

Whitmore, T. M. (1992). *Disease and death in early colonial Mexico: Simulating Amerindian depopulation*. Westview Press.

Worldometer. (2020). *Countries in the EU by population (2020)*. Retrieved August 20, 2020, from https://www.worldometers.info/population/countries-in-the-eu-by-population/.

WHO (World Health Organisation). (2020a). *Smallpox*. Retrieved August 4, 2020, from https://www.who.int/biologicals/vaccines/smallpox/en/#:~:text=Two%20forms%20of%20the%20disease,most%20prominent%20on%20the%20face.

WHO. (2020b). *Plague*. Retrieved August 4, 2020, from https://www.who.int/health-topics/plague#tab=tab_1.

WHO. (2020c). *Plague—Madagascar*. Retrieved August 4, 2020, from https://www.who.int/csr/don/27-november-2017-plague-madagascar/en/.

WHO. (2020d). *Contact tracing*. Retrieved August 5, 2020, from https://www.who.int/csr/disease/ebola/training/contact-tracing/en/.

WHO. (2020e). *5G mobile networks and health*. Retrieved May 10, 2020, from https://www.who.int/news-room/q-a-detail/5g-mobile-networks-and-health.

Chapter 10
The Architecture of Wealth

All that glitters is not gold
Proverb

Introduction

It may seem odd to have a chapter on wealth in a book on collapse and the built environment, but as noted in Chap. 8, wealth could be viewed as a way of measuring environmental impact, and one of the big problems in terms of potential change facing both the man-made and natural environments is the impact of human-induced climate change. The relationship between wealth and climate change has already been noted (Chap. 8). There is also a potential link between wealth and complexity. The medical care system in the USA is the most expensive in the world and one of the reasons given for this is its complexity. Papanicolas et al. (2018) note that the USA spends on medical care double what 11 other high-income countries spend, while "life expectancy in the USA was the lowest of the 11 countries." The value was an average 78.8 years, which curiously is the same average lifespan as Communist Cuba (see Table 8.3, Chap. 8). This increase in complexity for what appears to be no gain, certainly in terms of life expectancy, has come about because of high costs linked to the privatised US system.

> Behind the scenes is a huge army of people who are managing how the hospital and doctors are going to be paid for those services. And that's extremely complicated, because exactly how they get paid depends on our particular insurance. And my insurance is different from your insurance. (Harvard T. H. Cahn, 2020).

This example suggests that complexity may also make things more expensive. The example of the additional cost of high-rise buildings discussed in Chap. 8 can also be linked to the fact they are more complex in terms of both structure and additional services such as lifts than the equivalent floor area in a single storey building. This suggests that they may also be a link between wealth and a propensity to collapse.

Expression of Wealth

As much as clothes and the car, where there is sufficient money available, the home is treated as an extension of the personality of those who dwell within. The greater the sum of available money, the greater the efforts that go into building an edifice to reflect that very wealth and status. Perhaps the ultimate architecture of wealth is the folly—a building with no purpose. "Follies are built for pleasure, and pleasure is personal, difficult to define. Follies are fashionable or frantic, built to keep up with the neighbours, or built from obsession. They are at once cheerful and morbid, both an ornament for a gentleman's grounds and a mirror for his mind" (Jones, 1974:1).

A prime example of such folly is the 40-foot high stone pineapple near Airth, in Stirlingshire, Scotland (Fig. 10.1). To grow pineapples in a climate like that of the UK needed heated glass houses, so the fruits were themselves a symbol of luxury, so to also create a pineapple in stone above the hothouses was hammering home the point that John Murray, 4th Earl of Dunmore and then owner of Dunmore Park, was a wealthy man.

The stone pineapple is only one example of the many buildings that have deliberately been designed to express the wealth of their owners, ranging from the great pyramid at Giza began c.2550 BC for Pharaoh Khufu to the 27-storey private house in Mumbai reported to have cost over US$1 billion (Arch20, 2020).

Fig. 10.1 The 1761 stone pineapple sits on a seven-sided drum in Dunmore Park, forming a summerhouse above the walls that formed part of the walled garden (adapted from author's photograph)

Measuring Built Assets and Wealth

The pineapple is an expression of the wealth of a single person or family but equally, buildings can express the wealth of a society or nation. The global built assets wealth index has been proposed as an alternative way of measuring "…economic health and prospects for growth" (Tottathil, 2013). However, this raises another issue as some built assets such as factories contribute to economic growth in a way that other built assets might not.

> Typically, middle income emerging economies tend to rely more heavily on their built assets to generate economic returns as they have heavier manufacturing industries. More developed countries, on the other hand, will have relatively lower GDP from the built environment as their economies have become more diversified into services and people orientated industries. (Arcadis, 2016).

This also ignores the contribution to growth that comes from non-built asset intensification activities such as forestry and farming, especially given the latter is a vital part of human environmental impact. That suggests that the global built assets wealth index may not be useful in exploring the architecture of wealth.

Dent (2018) argues that in the past the landed estate, with its grand house that could be viewed as an architecture of wealth with its battery of servants required to run the place, was rather focused on land ownership that had a community role as it involved "…a conscious identity with its environment, both social and natural " (Dent, 2018:30). This ownership of land was essentially rurally based, and there is an implied duty of care in maintaining the land in good order so that it would continue to be productive so the estate could be bequeathed to one's heir. In this case, ownership is also stewardship. With industrialisation and urbanisation, land is now owned both for manufacturing purposes and also for investment. The latter means that land is held in order to gain a financial reward, whether from rent income or rise in capital value (see also Chap. 6). "Of course, such holdings will still have some acknowledgement of the community, but this tends to be the broader, more disparate community of shareholders, policy holders, stakeholders and so forth" (Dent, 2018:30). This list does not include any reference to the environmental community role, and also suggests that those outside the disparate community are not considered at all. So, does the building created as an "investment" rather than for immediate use by its creator look different from the architecture of wealth of the past, such as the grand mansion set in its park?

Dwellings as Investment

The first thing to note is that all building is an asset for someone. The distinction being made here is between someone investing in a building for a specific purpose, such as buying a dwelling to live in, and those investing in a building or buildings as a way of using surplus money such that it will grow and make more money. This has led to the practice of "flipping" as discussed later in this chapter (see also Chap. 6). In recent years in many developed societies, such as the UK, Australia and New Zealand, house prices have tended to increase. A rise happens when the demand for houses exceeds the supply, so those who can afford to pay more, whether cash or through borrowing money, can push up the price. The gap between supply and demand has been attributed to a variety of reasons. In the UK, the Conservative policy of selling off social housing reduced properties for rent, leading to one type of shortage as renters tried to enter at the low end of the housing market, but shortage of land and lack of money to provide infrastructure to unbuilt land can also be a factor, as can the willingness of banks to lend money. Lending freely, which of course makes money for banks, means more people are in search of a house, also increasing pressure on the supply.

However, if this is your only house and you have to move, you may not necessarily get any richer from house price increases as you have to buy another house, which might or might not be cheaper, depending on its size and where it is. If you need a bigger house to accommodate a growing family you are going to need to find more money to pay for the difference between what you can get for your smaller house and what you need to pay for the larger one than you would have needed if house prices had not risen. This usually means having a higher mortgage, which comes with higher interest costs, so it is the lender of this money (the bank?) who benefits from the increase in house prices. The money contained in a house may be useful for the owners towards the end of life, as a means of providing for any care required but most people do not invest in a single-family home as a means of getting richer but rather as engendering a sense of security through having more control over their housing—at least while they can still afford to pay the mortgage. For these reasons, home ownership had been viewed as "a good thing." However, this has not always been the case.

The 2008 Global Financial Crisis and Near Collapse

As the global financial systems found, extending the idea of home ownership in the USA to many more people so they could partake in this "good thing" led to a situation where "…financial intermediaries invented programs whereby individuals who were not responsible enough to save up a down payment for a house could get credit and purchase a home with no equity of their own invested in the home" (Klock, 2013). With nothing to lose, there was also nothing to keep these putative owners maintaining their mortgage payments. The eventual outcome was the 2008 financial crisis and the housing market crisis, leading to estimates of 8–13 million foreclosures (Klock, 2013). The effect spread way beyond the housing market. "This was the scariest week he had ever experienced, one trader recounts. It was the week when the global economy almost had cardiac arrest" (Norberg, 2012:72). The pressure to increase home ownership meant that rather than dwellings being sound investments, through reckless lending they caused a major crisis that almost grounded the global economy.

The series of events that led to the 2008 crisis also illustrates the problem of increased complexity. The history of borrowing money to pay for a house through the means of a mortgage appears to originate in the mutual or self-help societies of Victorian England. Samy (2016:1) states "Demand for building societies arose out of a growing need for a place to invest the surplus incomes of the working and middle classes, and for a lending institution to fund the construction of suitable housing for their accommodation." The first such societies were groups of people who paid into a fund to buy the materials to build a house. When the house was finished a ballot decided who would live in it. Members continued to pay into the find until all members were housed when the society was wound up. Permanent building societies came into existence later. Members paid into the building society and received interest on their savings to save up for a deposit on a house. The society lent additional money in the form of a mortgage, which the new house owner paid back over time with interest. This system could continue, providing there were both lenders and borrowers but the connection between them—the building society—was relatively simple. The situation that led to the 2008 financial crisis was not. Norberg (2012:46–63) sets out what happened in detail, which is summarised here. A bank might go and buy a substantial number of mortgages (thousands), thus spreading the risk as not everyone will default and also by buying from different geographical locations in case anything happens to a particular local housing market. These mortgages were then bundled into securities, some of which would carry more risk than others. The interest of those repaying the loans is then passed on to those holding the securities. Other investors would buy and rebundle these securities into a new kind, and this would be done again, and even again, so that the investor in one of these new bundles is a long way from knowing whose mortgages he or she is holding. The next step involves banks. Banks take in money and lend out money but if everyone suddenly wants to take out their money at the same time the banks may not be able to get in the money from whoever or whatever they have loaned it to fast enough. Central or government banks are there to fill this gap in bank liquidity. However, this meant that banks could take risks in this complicated mortgage market since the central bank was there to bail them out if they ran into trouble. To cut a long and complicated story short, while house prices rose everything was good but once the housing bubble burst the crisis followed. As Norberg (2012:61) comments some houses were never occupied. "They [the mortgage holders] had just bought the houses so they could resell them at a higher price in a market that could only go up. When house prices started to fall, they just quit paying their loans, staying in their old homes or buying a cheaper one somewhere else." Although the starting point of the crisis was encouraging people to buy houses without the security of a deposit, the complexity of what followed was what dragged the issue into a global panic.

The Landlord

There is another type of home owner—the person who puts their surplus money into property, which is then rented out. This is different from the person buying a house in order to sell it on at a profit in the short term in a rising housing market. However, again the situation of the investor in rented housing is not one of equity, at least in New Zealand, where the following is extracted from a letter to the Sunday paper (Little, 2020).

> When a young family is saving for a home deposit via a bank savings account the interest is taxed, including the portion lost to inflation. With inflation currently greater than the interest banks pay, this young family's home deposit savings are going backward. However, we have the ludicrous situation where an investor can borrow this money plus more from the bank, buy a house and rent it to this young family, who then pays off the investor's mortgage.
>
> The investor can also claim the mortgage interest plus repair and upkeep costs against tax due on the rental income, something a live-in homeowner cannot do.

This situation contributes to the lack of affordable housing as often these investors are looking for houses to rent out at the cheaper end of the market, and there is no control over who buys a house, whether a family that wants to buy it and live in it or a putative landlord. This increases the gap between owners and renters (see also Chap. 6). When it comes to buildings and wealth in the present day it is not so much the buildings as the system which can make some people wealthy, although as the 2008 financial crisis revealed, buildings for all their physical presence are not always a safe investment.

That said, the tradition of using surplus funds to buy houses for rent perhaps stems from the landed gentry with their estates building cottages for the estate workers, often designed in order to enhance the beauty of the estate. Examples range from the *cottages ornés* at Edensor on the Duke of Devonshire's Chatsworth estate in Derbyshire (Donner, 1944), through to the industrial revolution ideal of 'five percent philanthropy'. The latter refers to the Victorians who wished to find a solution to housing the poor, particularly the urban poor, in decent housing but at the same time receive a 5% return on their financial investment (Tarn, 1973:43). This led to ventures such as the Peabody estates in London, designed by Henry Darbishire. "His blocks were austere, minimally Italianate, with solid brick walls of yellow London stocks relieved by bands of hard white Suffolks and cornices and lintels of a tough terracotta. It was the aesthetic of the sublime rather than the beautiful. Cleanliness, sanitation and ventilation were key concerns" (Stamp, 2016). However, this is perhaps less the architecture of the wealthy than the architecture for the respectable (and deserving) poor. The architecture of wealth was very visible at this period. If you were landed and rich you lived in a big house attended by many servants, for some of whom you might build cottages as an aesthetic adventure as much as to keep the workers healthy and working. Advice on what to build came through organisations like the Country Gentlemen's Association in the UK, founded in 1893 as a "…society of landowners, land agents, farmers and others interested in the land, numbering many thousands, and residing in all parts of the kingdom" (The CGA, 2020). The problem was always

Fig. 10.2 Cottages at Bournville (adapted from https://upload.wikimedia.org/wikipedia/commons/0/09/Bournville._Cottages_in_Linden_Road_%28front_view%29.jpg)

whether a decent cottage could be built such that a labourer "…earning less than 20s. a week for six days' labour" could afford to rent it (CGA, 1914:11). Nor was building cottages for their workers confined to the landowners. Industrialists also demonstrated both their success and their thoughtfulness by building settlements of workers' housing such as New Earswick in York for Rowntree, Bournville in Birmingham for Cadbury, and the Krupp Colonies around Essen, Germany (Meakin, 1905:364–372). This can be considered as an architecture of wealth precisely because having healthy workers meant that the factories would continue to make money for the owners, while at the same time demonstrating the philanthropic leanings of these same owners through artistic design (Fig. 10.2).

Flipping

Flipping is the process of buying a house at a low cost, having the capital to refurbish it, and then selling it on in the hope of making a profit once all fees are paid (Boardman, 2013:9). This works when house prices are rising and can become a phenomenon when they rise fast. A report of flipping in the USA states "Nearly 54,000 houses were flipped during the second quarter of 2017, with investors making an average profit of $67,516 per property and 48.4% return on investment" (Anon, 2017). However, fees and taxes, including income tax, mean that house flipping is not always so lucrative. As one commentator noted, "…there are renovation costs, holding costs to cover the mortgage, legal costs and selling costs. Even if you sell well, the profits are nothing like the popular estimates" (Bell, 2019). Because houses are renovated and sold on, flipping perhaps improves the housing stock, but flipping is for those that have money to invest with the intention of making a profit from the short-term rise in house prices in a rising housing market. The traditional landlord who holds on to a property will receive not only a regular rent but also any long-term rise in the value of the dwelling when it is eventually sold. However, both methods of house ownership are for those who have money, which they use to make more money. Given a growing population and the need to balance land for housing against land for growing food to feed the human population has meant that the consequent increase in house prices still makes house ownership the more attractive option than renting, as with ownership comes greater control over the dwelling. The problem is that in many places fewer and fewer people can see house ownership as an option (see Chap. 6).

My House—My Castle

J. M. Richards began his 1940 book on modern architecture by stating "…the Man in the Street only sees in the new architecture another bewildering addition to the variety of architectural styles he is already offered" (Richards, 1940:10). However, in his post WWII book *Castles on the Ground,* which was an exploration of suburbia and the fact that badly designed buildings—from an architectural aesthetic perspective—do not always result in poor overall environments he was far more generous to design in the suburbs. More pertinently for this chapter, in describing the suburbs he stated "…the need to identify ourselves, through the patterns into which we ourselves have moulded it, with the place we regularly return to—grows greater" (Richards, 1973:7–8). Rather than needing to be guided through why a new architecture looked as it did, the "Man in the Street" was now recognised as contributing to his suburban environment through the personalisation that went with creating home. This visible aspect of personalisation has more to do with how an individual views their social roles and relationships, or how they identify themselves within a group, rather than a private view of self (Cheek & Briggs, 1982). If the interior of the dwelling has more to do with the latter, then what Richards observed was how the exterior of the dwelling was altered so as to create an identity that related to the social group of neighbours and near community. Perhaps the extreme example of this is the Best Kept Village competition that since the 1970s has been run by the Campaign for Rural England (CPRE) and that involves making sure all community spaces such as playing fields, footpaths, and public toilets are clean and in good order and the village appears as a tidy and cohesive whole.

Personalisation of the dwelling exterior is neither always so restrained nor necessarily only the prerogative of the traditionally wealthy. In Peru, the word "huachafo," which is associated with informal settlements, is used to describe in a pejorative way something kitsch, an aesthetic associated with the taste of the "nouveau riche" or someone "who unsuccessfully aspires to a social rank higher than that to which he or she belongs" (Parker, 1991:77). Huachafo is also used to describe someone who has no money or standing and tries to look like the elite of society through imitating aristocratic fashions. The origin of the word Huachafo has been linked to the Quechua Peruvian word "waqcha," which means orphan. It has also been associated with the English expression from the eighteenth-century "whitechaps" (a shortened form indicating "chaps from Whitechapel"). This term referred to elite workers from the Whitechapel neighbourhood in London in the UK who, during the Industrial Revolution, acquired a good economic situation and dressed in an ostentatious way. In Lima, Peru, huachafo has become more than a slang word and is a social phenomenon (Schwab, 1940). The built environment is the perfect medium through which to express the aspirations of people that were discriminated against by the elite for having humble origins in poor socio-economic levels (Parker, 1991:78). It can be also the way that one family or household competes with or detaches themselves from other neighbours to clearly state that they are a little less poor than them, or that they represent the top tier of an informal settlement. Personalisation of the exterior can be seen in houses constructed with Greek architectural elements, or where ornamentation has been added to the façade of a house to make it look like a French palace or an Italian villa. As the fashions and trends of the elite change, so does the huachafo style. Sometimes, these changes can be traced to the same house that has been built in one style for the ground floor and a different one for the upper floor.

The Huachafo, as an architectural expression of informal dwelling, is not only about expressing aspirations of greatness but is also part of an honest desire to personalise houses and make them look beautiful using very limited resources. Some of the characteristic elements used are fake gable facades and the ornamentation of windows and balconies with fake arcades and other historic motifs (Fig. 10.3). Fake façades are also used temporarily to conceal a new level that is still a work in progress. All these elements introduce a greater complexity since the diversity of the façade increases from the elements needed to the elements added but not needed. For example, the gable roof is not necessary in Lima since rainfall is low but is copied because it is linked to the aesthetics of expensive villas (Dreifuss-Serrano, 2019:304).

Huachafo is a wider social phenomenon that appears in many Latin American countries and is the result of a colonial process that in turn has created economies that have produced noticeable inequalities. In Bolivia, the term "cholets" is used to refer to a type of vernacular architecture with aspirational intentions (Fig. 10.4).

The phenomenon appears in other parts of the world. Since the 1990s when Vietnam opened up to the global economy, some of the informal "cake houses" began to use the French colonial style (Fig. 10.5) to express modernity, cosmopolitan life and a certain standard of quality (Herbelin, 2013).

The point here is not about judging the personal taste of the owners, after all the entire education of students of architecture is based on copying buildings and elements of buildings conceived by other architects (along with learning to despise anything that non-architects like). The stylistic difference between an architect's work and the owners of these houses lies in the sources and the ways of using them. What is common in this category of vernacular architecture are the aspirations of wealth as a driver. This can become problematic when the complexity added by the ornamentation involves using more materials and resources, all of which have

Fig. 10.3 Example of Huachafo architecture (adapted from https://www.archdaily.mx/mx/786643/que-es-lo-huachafo-en-la-arquitectura)

Fig. 10.4 Cholets (adapted from http://arquitecturahuachafa.blogspot.com/)

Fig. 10.5 House in the "New French Style" near Hanoi (adapted from Herbelin, 2013)

an environmental impact. Using large windows in Peru or Vietnam will also have a negative impact on the indoor comfort of the house, raising temperatures, and if the windows have limited openings, precluding useful natural ventilation, which in turn encourages the use of air conditioning. Large glazed areas in the façade will demand shading devices, which adds another layer of material to a façade that is already using materials that strictly it does not need. There can also be social impacts. This aspirational architecture could be the sign of an ongoing gentrification process as well as a way of supporting inequalities within neighbourhoods. These inequalities can harm one of the main strengths of informal settlements, which is the capacity for sharing and relying on each other when things do not work as expected. The architecture of wealth has no socio-economic boundaries but can also create complications at all socio-economic levels.

Large Houses

In a study of housing in Malaysia Omar et al. (2012) noted that it was the outside of the house that inhabitants modified, especially the entrance façade. Townsend (2010:51) states in reference to the home "...the marking of human territory can be seen in environmental symbols such as fences, hedges and nameplates." However, rather than working together for a common goal, as in the Best Kept Village competition, the exterior of the dwelling can also become the place for the display of wealth, and in some countries like Australia and New Zealand, where there is sufficient land, this has led to a significant growth in house size, which can then command a greater presence on the street. As an example, Khajehzadeh and Vale (2017a) show the average floor area of a new house in New Zealand has increased by 72% from 108.7 m^2 in 1974 to 187.1 m^2 in 2014. This increase in floor area is seen in larger rooms, and more rooms with a specialised function such as a home office and en-suite bathrooms. A perhaps unintended consequence of larger houses is an increase in circulation space. As Khajehzadeh and Vale (2017a) comment, "More circulation does not give more flexibility to occupants and will impact on the energy needed for maintenance and cooling/heating of the house." Large houses thus consume more resources for their operation as well as their construction. In the large mansions of the past where the circulation route from the servants' quarters, where food was cooked, to the grand dining room where food was consumed, was often both long and tortuous, special vessels had to be designed to keep the food warm. These warming dishes were in two parts, with a plate on top and dish underneath to hold hot water. Obviously, houses are not so large today and there are other ways of keeping food warm, but the point of the example is that growing in size can create additional problems.

The advent of these modern large houses has led to terms like McMansions (Fig. 10.6), normally used pejoratively to describe large houses that stand out from

Fig. 10.6 A McMansion in Wellington, NZ (adapted from author's photograph)

their surroundings. In his study of the term in the USA, Miller (2012) found four meanings "...a home that is large; a home that is large in comparison to its surrounding neighbourhood; a new home with a particular architectural or design style, often invoking criticism; and homes serving as symbols for broader concepts including sprawl and excessive consumption." However, whatever the meaning, a McMansion is going to cost more to build than a more modest house and is, therefore, an architectural expression of wealth.

In a study of housing in New Zealand, Khajehzadeh and Vale (2017b) found that, similar to other studies, New Zealanders spend on average 69% of their time at home indoors. Of this those in small four-room houses (living room, kitchen, dining room, bedroom) spent the least time (15.04 h/day) and those living in the standard six-room or three-bedroom house the most time at home indoors (16.16 h/day). Of more significance, however, was that as houses became even larger less time was spent at home indoors, dropping to 15.40 h/day for those living in houses with nine or more habitable rooms. Given that on average, these same New Zealanders were spending 57% of their time at home in some sort of bedroom, this suggests that as houses get bigger, they also become underused, at least in New Zealand. This is a pity as it takes both energy and resources in the form of materials to make larger houses. Nor do most people leave these large houses empty of furnishings. In a separate study of the life-cycle energy bound up in furniture, appliances and tools (termed FATs) (Khajehzadeh & Vale, 2017c), the following emerged. "For instance, comparing the energy embodied in FATs in a ≤5 and an 8 room house shows that in year 0 the latter uses 168 GJ more energy for furnishing while this difference increases to 730 and 1328 GJ after 50 and 100 years." This is because furniture and appliances are replaced many times in the life of a house. The most sustainable thing to do when it comes to housing is to live in a small house (Vale & Vale, 2009:189).

Large houses tend to go with wealthier societies or with the wealthy in poorer societies. Table 10.1 sets out some examples of dwellings and the average space occupied per person in various countries (Khajehzadeh, 2017) alongside GDP/person (Worldometer, n.d.) and GNI/person (World Bank Group, 2020). Here GDP is the total value of all goods and services within a country in a certain time, usually a calendar year, while GNI is all the income the country gets from its registered nationals and businesses within a set time. Income from overseas nationals is included and income generated at home from non-nationals is deducted.

Although the dates do not correspond in the table, with some exceptions, average floor area per person tends to increase with increasing GDP and GNI. Available land and culture also play a part in the figures in places such as Hong Kong and Japan. Tsenkova (2009:115) also points out that Romania is unusual in having less floor space per person than other south-eastern European countries. In a study of the EF of families from around the world, Vale et al. (2018:213) found a wide difference in m^2/person in the dwellings, ranging from a high 144 in the USA to 1 for the UNMISS refugee camp in South Sudan. In warmer climates like South Sudan space outside the dwelling will also be used, but again this study found that wealth tended to go with an increase in dwelling size, since more money means that more land and more building materials can be purchased to go into each dwelling.

Table 10.1 Average per person dwelling area, GDP and GNI for selected countries

Country	Dwelling floor area per person m^2	Date	GDP $US/capita 2017	GNI $US/capita 2018
USA	77.0	2009	59,939	63,170
Denmark	51.9	2009	57,545	60,170
Australia	89.0	2009	53,831	53,250
Austria	42.9	2009	47,261	49,350
Hong Kong	15.0	2009	46,733	50,300
New Zealand	56.2	2010	43,415	41,020
Japan	35.0	2009	38,214	41,310
Spain	33.0	2008	28,175	29,300
Estonia	29.7	2009	20,170	21,130
Latvia	27.0	2008	15,613	16,500
Poland	24.2	2008	13,871	14,100
Romania	15.0	2008	10,781	11,300
Bulgaria	25.2	2008	8,197	8,860

Plot Size

Another aspect of reading wealth into a residential environment is house frontage, which is related to plot size although not dictated by it. Long thin plots that give a narrow frontage are characteristic of terraced housing, which can be modest. In the UK, Victorian terraced houses could have a plot width as small as 3.0 m (Muthesius, 1982:107, 121). Long thin plots have also produced certain housing typologies, such as the tube house of Hanoi, where a plot width of 3.5 m with a length up to 35 m was traditional (Garcia & Vale, 2017:110–111). However, the distance between the front door and the street is also indicative of wealth. In the least expensive houses in a particular society, such as that of the UK, the front door opens directly onto the street, whilst some area of land between the front door and the street not only gives separation for privacy but also offers the opportunity for personalisation (Fig. 10.7).

The car has also had an impact on street frontage. Not only have many existing front gardens been paved over to accommodate one or two cars, thus increasing the area of impervious surfaces (see Chap. 5), but having an integral garage means that at a minimum the house has to be wide enough to accommodate a car and some form of entrance, leading to the classic three or more storey townhouse. Where land is plentiful, as in New Zealand, a wide frontage can accommodate a car and a front garden for display in even relatively modest houses (Fig. 10.8).

With an even longer frontage, an in and out drive way can be achieved, which is an echo of the carriage sweep of the large houses of the rich, which was an area of land sufficient to turn a carriage and horses, though this was normally some distance from the road edge, and the greater the distance the greater the probable wealth of the owners.

Dwellings as Investment

Fig. 10.7 Three terraced houses, Northampton, UK: in the left hand one the door opens to the street, the centre has a bay window, and the right hand one has an area with railings to give light to a basement room (adapted from author's photograph)

Fig. 10.8 House with garage and front garden, Hawkes Bay New Zealand (adapted from author's photograph)

Manufacturing and Commerce

Wealth can also be read in commercial and institutional buildings. The headquarters of a firm have often been a vehicle for an architecture expressive of the firm's aspirations, and an expensive-looking building could also express the solidity and credit worthiness of the enterprise. Bank buildings form a good example of such expression, since the purpose of a bank is to ensure the wealth entrusted to it is in safe hands. "Even as artistic styles change—from neoclassical pillars to abstract modern art—the message presented by a bank's physical appearance speaks volumes" (Anon, 2014). Classical architecture was the means used to express the trustworthiness of banks in new countries like New Zealand and Australia (Skinner, 2020:348–349). However, as the nature of banking has changed from one of storing money to one of selling banking and other services (Harris, 2007), so have bank buildings. These have moved from a solid structure with few windows, to the all glass shop windows, making the bank just another 'shop' in the high street. The bank as shop is even more important in its presence on the street as money is now largely virtual and transactions can be dealt with without ever setting foot inside the bank building. The all glass bank headquarters building has followed this trend, with examples such as Foster's 1986 Hong Kong and Shanghai Bank headquarters in Hong Kong, and the 1984 twin glass towers that form the headquarters of the Deutsche Bank in Frankfurt, dubbed by the locals as debit and credit (Tourismus+Congress, n.d.). However, the glass bank allowing a view in from outside can also be seen as a return to early banking in thirteenth-century Florence, where transparency of transactions was key (Jacobi, 2019:36–38). The point here, however, is not so much what the style of architecture might symbolise but how much money has been lavished on the building as an expression of the wealth and perhaps forward-thinking of the business enterprise on the client. From this perspective, the HKSB headquarters in Hong Kong look far more impressive than the rather mundane glass towers of the Deutsche Bank in Frankfurt.

Governments, whether national or local, also have a vested interest in architecture that expresses both trustworthiness and the promise of wealth. The townhall was not just a place of work for those running local government but also a symbol of the prosperity of the town. This meant that such buildings were often the focus of avant-garde architecture, such as Dudok's 1928–31 Hilversum town hall (Richardson, 1996), and the 1937 townhall at Aarhus in Denmark won in a competition by Arne Jacobsen and Eric Møller (Hawkes, 2014). The same idea of looking towards the avant-garde is seen in the New Zealand parliament buildings. The original 1876 Old Government Buildings, that for 56 years housed the "…Ministers' offices, the Cabinet room and all Wellington-based civil servants" (Heritage New Zealand, 2020) are the largest timber buildings in the southern hemisphere but were constructed to look like a stone building (Fig. 10.9), suggesting government was associated with permanency. A fireproof library in the gothic style was added to the set of government buildings in 1899, followed by the 1918 Edwardian classical parliament building in stone that was part of a scheme that was never completed. The final much later

Manufacturing and Commerce

Fig. 10.9 Old Government Buildings, New Zealand (adapted from https://en.wikipedia.org/wiki/Old_Government_Buildings,_Wellington#/media/File:Old_Government_Buildings_-_whole.JPG)

building (1969–79) by Basil Spence is a circular office building locally known as the Beehive (New Zealand Parliament, 2016).

This need to express not so much wealth as the forward architectural thinking of the time applies equally to places of manufacture. The Fagus Factory by Gropius and Meyer has become an iconic building of the Modern Movement (Pevsner, 1968:213), although it is the administration wing, and not the factory where shoe lasts were and still are made, that has been given a place in the icons of modernism. The 1930–32 Boots Factory at Nottingham in the UK by Owen Williams, in reinforced concrete and glass, celebrated the success of the high street pharmacy company with a five-storey atrium in the packing hall (Anon, 1932). In the USA, Louis Kahn was commissioned in 1966 to design the Olivetti Underwood factory, completed in 1970, with its internal clear span to allow for change, and exposed services (Archeyes, 2020). However, the competition in creating buildings that formed a memorable image for a company came to a head with the skyscraper, as here it was not only the form of the building but its height that could be used to express corporate ambition.

The Quest for the Tallest Building

Constructed in 1930–31, the Empire State Building with its 102 storeys held the title of the world's tallest building for 40 years, only being surpassed by the World Trade Centre in 1970. Helsely and Strange (2008) in describing the race in New

York to make the city's tallest building, which was finally won by fixing an airship mooring mast to the Empire State Building, make the point that the early pioneers of skyscrapers "…assigned value to being tallest that was independent of the narrow value of a skyscraper as a piece of real estate." Since then the race for the sky has, as noted in Chap. 8, led to competition to be both the country and the commercial enterprise with the tallest building. However, there was a time when the tall buildings in New York were seen as being "the tombstones of capitalism" given the economic problems of that time (Schleier, 2008). A somewhat similar issue has been raised about modern skyscrapers, in that they are often taller than necessary for maximising the profit from building them, making them "…potentially a misuse of resources" (Barr et al., 2015). Lawrence (Brass & Lawrence, 2012) realised that the completion of a very tall building tended to coincide with the start of an economic recession and in 1999 he developed what he called the skyscraper index. Lawrence noted "…that skyscrapers seem to mark a very large economic boom that typically ends in large recession" (Brass & Lawrence, 2012). This is because easy money is available during a boom to support building big and tall, with such buildings only being completed as the economic cycle moves from its boom to recession phase. This has led to the idea that announcing the construction of the tallest building in a country or region could mean the local economy was reaching its peak (Jadevicius, 2016).

Although there has been criticism of the skyscraper index (Barr et al., 2015) building tall takes money and such buildings express wealth. Moreover, height is not the only factor in play, and imageability has become a further aspect of tall building. A skyscraper that is given a name that reflects its shape, such as the Cheesegrater (the Leadenhall Building) or the Gherkin (30 St Mary Axe) in London immediately disguises whose money went into building or renting the structure, so may be counterproductive in the building as embodying a corporate identity. Imageability can also work the other way with the NatWest Tower, completed in 1980 and also in London, designed to look like the NatWest Bank logo when viewed from above. It is now known rather prosaically as Tower 42 and is no longer the headquarters of the NatWest Bank, making the symbolic footprint meaningless.

Another suggestion put forward has been the use of tall buildings as vertical farms using hydroponic methods (Wagner, 2010), thus bringing food production into cities. Such ideas are not without problems. Al-Chalabi's (2015) life-cycle study found "…that vertically grown produce has a carbon footprint that is much higher than conventionally grown produce." In resource terms such ventures do not make sense.

Very tall buildings are expressive of wealth simply because they are in terms of cost per floor area more expensive to build than lower rise structures. The economics of skyscrapers have never been good (Helsely & Strange, 2008). Additionally, by their very nature, they also change the way the indoor environment is managed since opening windows are a low rise rather than high rise solution to ventilation, and this aspect is explored in the next section.

"Green" Buildings as Investment

One aspect of corporate architecture that can be viewed as an expression of wealth is the creation of an internal atmosphere sealed from the vagaries of the outside weather. This idea can be traced back to the 1902–06 Larkin Building by Frank Lloyd Wright where a fresh air supply was organised as the surrounding neighbourhood was somewhat polluted (Banham, 1969:86–92). It developed into the fully air-conditioned office space, with examples such as the 1935 office building for the Hershey Chocolate Corporation and the 1936–39 Johnson Wax building (Siry, 2013). The air-conditioned office building has since become a standard.

> The corporate image is often linked to an image of environmental seclusion facilitated by advancements in building heating and cooling technologies, lighting and the advent of lifts. These technologies underpinned the possibility of spreading design philosophies and an export of the 'Modern' movement globally (Hamza, 2014:93–94).

Given this, when the idea of natural ventilation and daylight was proposed as part of the greening of buildings, it should be no surprise that there was an initial reluctance to invest in green or sustainable buildings as these were not necessarily demonstrating that a company was wealthy enough to afford to run an all air-conditioned building. Arguments were made that green buildings would probably cost more to build, even if they cost less to run, although this may now be changing. "There is growing belief that over time more environmentally conscious property will have higher income growth, be viewed as less risky and deliver higher returns" (McNamara, 2010). To demonstrate that a commercial building is really green, it can be assessed and awarded, for example, a LEED or BREEAM rating. For many years, rating only happened at the design stage before any occupants had moved in, leading to a discrepancy between what was claimed and what was achieved. A further study (Scofield, 2009) found that in primary energy terms "…LEED certification is not yielding any significant reduction in GHG emission by commercial buildings." Since the development of the Australian NABERS in 2000, which only rated buildings on performance once built and occupied, other rating systems, such as LEED, have followed suit with LEED version 4 issued in 2015. However, this makes it more difficult for a developer to claim green credentials for a building as a status symbol in order to attract tenants since its green performance is dependent on those very tenants.

Just as the all air-conditioned office building was once seen a status symbol, firms are starting to realise that a green headquarters building is part of the green credentials when it comes to marketing a company's products. An example of the green building as a reflection of the company image is the 2017 headquarters for Apple in Cupertino, California, designed by Foster and Partners. This circular doughnut of a building has a diameter measuring from the outer wall slightly larger than the height of the Empire State Building, so amply expresses the wealth of the company. Although claimed to be the greenest building in the world in terms of energy which comes from renewables in the form of PVs and biogas fuel cells (Apple Inc., 2020), the suburban location of the headquarters means the workers will drive there, with 11,000 car parking spaces being provided for 14,000 workers. These cars in turn occupy 325,000 m^2 as against the 318,000 m² built space for the office and laboratory workers (The Economist Newspaper, 2020). Additionally, many mature trees were to be transferred as part of the landscaping plan but the small amount of carbon they could sequester (Bernstein, n.d.) will in no way offset the emissions generated from the commute to the new headquarters. The Apple headquarters is as much an example of a company using money to sell an image as was the race between the New York City building for Woolworth and the Metropolitan Life Building, to be the tallest occupied building in the world at that time (Helsely & Strange, 2008).

Empty Buildings

Whatever the perceived merits of the Apple headquarters, at least the building is occupied, even if by many cars. Another aspect of the built environment and wealth is the phenomenon of empty buildings. Buildings without people are like monuments—effectively structures without use. Empty buildings occur as a result of recession as companies can no longer trade but these are less a reflection of wealth than of the opposite. In 2010 in a USA trying to come out of recession (Miller, 2009), "The national office vacancy rate rose to 17.4% as office buildings across the country lost 1.8 million square feet of occupied space in the second quarter compared with the first quarter" (Tuohy, 2010). However, empty buildings do not necessarily mean that the resources invested in the built environment are not being used, as they may be forming part of wealth generation. This aspect becomes magnified when people are wealthy enough to have a building that is either not used or only used a little. The following sections look into different types of empty building, starting with those associated with tourism.

Partial Occupancy

Tourism brings its own issues of buildings that are not fully used. An important performance indicator for hotels is the occupancy rate (Tugores & Valle, 2016) but tourism is a seasonal activity, meaning that occupancy rates fluctuate, being high in peak periods and lower in the off-season. Often room rates drop in the latter in an effort to attract local people (Singh et al., 2010). The EU publishes occupancy rates for "bed places in hotels and similar establishments" and for 2019, occupancy rates averaged from the high of 67% in August to a low 37% in January (Eurostat, 2020). Occupancy is only one measure used to predict revenue as the room rate is also important. However, these figures show that hotels and similar establishments will for much of the time be partially occupied in terms of the number of bed spaces. This should perhaps be a concern given the growth in tourism and the dependence of some economies on it. More tourism will mean more hotels being built to be occupied only some of the time. In 2007, tourism was estimated to be the sixth largest sector in the global economy, coming after "...trade in fossil fuels, telecommunications and computer equipment, automotive products, and agriculture" (Lew, 2011). It is also a sector that has continued to grow, "1.5 billion international tourist arrivals were recorded in 2019, globally. A 4% increase on the previous year is also forecast for 2020" (World Tourism Organisation, 2020). This UNWTO prediction from January 2020 was before the COVID-19 pandemic, which in turn will lead to more empty tourist beds while borders remain closed as international tourist arrivals have been estimated as dropping by 78% in 2020 (Sigala, 2020).

Under-occupancy has an effect on the local built environment in the parts of the world where the climate means tourism is seasonal. This is because the number of people in a locality at certain times of the year may be dwarfed by the number of buildings, many of which are waiting near empty for the arrival of the summer season in coastal resorts or the winter season in winter sports areas. The opposite is true in the height of the season. In a study of a Norwegian ski-resort, Henningsen et al., (2015:123) noted the following.

> During the winter season, the small town becomes a hub connecting different mobilities: tourists who come for the skiing experience, people looking for work, and some people who defy this distinction. At that time of year, tourists decisively outnumber locals and approximately half of these are foreigners.

Tourist accommodation is not the only building type subject to under-occupancy. Modern life has separated work from home, meaning that when home is empty, the workplace is occupied and vice versa. School buildings are much less used during holiday periods and churches are another building type with intermittent use. Perhaps most noticeable in the built environment are the large unused car parking areas after the end of work, mirrored by the empty garages and car parks at home during the day, and all to accommodate a vehicle that is used simply to get from home to work. The more cars there are the wealthier the society (Law et al., 2015) so the number of cars and area of car parks in a built environment could be an indicator of wealth.

Second Homes

Hall and Müller (2004:3) describe second homes as both a boon to tourism and as a problem because they can increase competition for dwellings so prices increase, which can make it harder for those living in attractive places to find permanent homes. Societies as far apart as Sweden and New Zealand have a tradition of second homes. Larsson and Müller (2019) state Sweden with the other Nordic countries globally have the highest proportion of second home owners. This has come about because of the pull of the city in the nineteenth century leaving rural dwellings vacant and because of government programmes whereby "...new second home settlements were built not least on the outskirts of the cities and in amenity-rich areas." Some 20% of Swedes own a summer house and 50% have access to one through family and friends (Swedish Institute, 2020). The romance of escaping from the city to the countryside in the summer and living there in a small wooden house where life is simpler is considered part of being Swedish, through finding contact with past generations who did the same thing. At the same time, there is a boost to the local rural economy when the visitors arrive, both in the hospitality and construction sectors, although local services such as health and the police may be stretched (Larsson & Müller, 2019). If the use of second homes in Sweden means holidays are local rather than overseas, especially if air travel is avoided, then the resources that go into them may be offset by the resources that are saved in terms of their share of air travel and overseas tourism. The money spent also goes back into the national economy. A study of Finland (Tran, 2014:144–145) noted that one in five Finns owned a summer cottage, the average floor area of which was 50m^2. More importantly, the study looked at energy consumption and found that energy was still being used in the main dwelling, whether to keep the fridge running or through the use of district heating in seasons when the summer cottage was occupied for spring, autumn and winter holidays. Having a summer cottage in 2011 raised the average energy use per person in the city of Oulu by 18% from 13,345 to 15,810 kWh.

In a similar way, the bach (pronounced "batch", not like the composer) or crib in New Zealand was a simple dwelling for summer occupation. Lowe (1999) states the bach in New Zealand dates from the 1920s with another boom occurring after WWII. Data on the current number of baches are difficult to obtain because of the way census data are collected. The last available data show that, in 1981, 4.3% of all households had a second home (Keen & Hall, 2004:174). Like the Swedish summer cottage, these were simple structures built of materials to hand, kitted out with old furniture, and situated near the sea or lakeside, and were passed down through the family. Lloyd Jenkins (2004:290–291) describes the true New Zealand bach as a "…crude seaside shelter through which a key generation of New Zealanders had staked out their claim to the new Valhalla—beachfront property." Going to the bach, or camping, was the New Zealand summer holiday before air travel became more accessible. However, for some, the bach as a second home was the opportunity to explore the design of a beachfront dwelling that also represented how New Zealanders saw themselves. "The ultimate extension of the new casual lifestyle into the environment came with the architect-designed beach houses of the 1950s. These were not traditional baches, nor did they borrow anything of architectural consequence from the bach form" (Lloyd Jenkins, 2004:128). In their study, Walters and Carr (2015) concluded that the simple bach was a myth, stating the "…New Zealand second home is, and always has been, a site for the consumption of luxury," noting this was evident from the 1930s, and more obvious from the 1980s onwards. Unfortunately, this luxury can remain unused for large parts of the year. A luxury bach is a bit like having a top of the range BMW, parking it on the street and only driving it to the supermarket once a year. The only difference is the BMW will depreciate as it sits there while the luxury bach will appreciate in value.

Mention has been made in Chap. 6 of the phenomenon of absentee owners in cities like London and New York, in particular the owner who "…bought the property for capital appreciation and sees no reason to live in it" (Farwaji, 2020). This is a little different from the second home owner if the property has been left vacant in the hope that it will appreciate in value and can then be sold off. However, like unoccupied second homes in rural areas, unoccupied property—ghost mansions—reaffirms that wealth can do what it likes, while those without money struggle to find somewhere to live (Booth & Laadam, 2017). Through digital platforms like Airbnb, some empty houses are now rented out but this has brought its own problems. Rather than just renting out a house whilst the owner is on holiday, which was how Airbnb started, in a study of Melbourne, which was early to take up Airbnb, it seems dwellings are being bought up for use as Airbnbs rather than for permanent renting not least because, "Short-term letting via sites like Airbnb allows investors to earn up to three times the amount they'd receive in rent" (Alexander, 2016). Following a literature review of the Airbnb effect, Garay et al. (2020) note that:

> …various recent studies have made a connection between the spatial distribution of Airbnb properties, housing crises and the generation of socio-spatial inequalities in different cities. Such inequalities are generally manifest as commercial and residential displacement, widening rent gaps and deterioration in the quality of everyday life.

If you can make more money renting to tourists then without some form of control this is what will happen.

Ghost Cities

Because an urban population is viewed as the way to create national wealth in some parts of the world, China being an example, ghost cities—or areas of the built environment that are not fully occupied—have been created. Sorace and Hurst (2016) see this happening because "China's urbanisation of land and creation of infrastructure often far outpace the urbanisation of its people." As a result, Harvey (2013:60) notes "whole new cities, with hardly any residents or real activities as yet, can now be found in the Chinese interior." Eco-cities have also received the same fate, such as Tianjin's Binhai New Area which in 2013 was "…another ghostly hollow urban landscape (Sorace & Hurst, 2016).

Although ghost cities may appear to sit uncomfortably with the architecture of wealth, there is a connection. Economic booms have in the past produced settlements that then become ghost towns, such as Waiuta on New Zealand's West Coast. From 1906 to 1951, it served as the largest gold mine in the South Island but was abandoned with the closing of the mine (New Zealand History, n.d.). Ghost towns can also be awarded World Heritage status, such as the Japanese former mining settlement of Hashima Island (Hansen, 2018). Such settlements were abandoned because they no longer had any economic meaning. As Wade (2015:39) notes "Throughout history ghost towns have started out as boom towns." However, the new ghost cities of China are created not through traditional rural and urban migration but through a process of urbanisation aimed at creating a middle-class keen on consumption. In this way, China is aiming to create an economy based on production and domestic consumption, rather than relying on exports and the global market (Wade, 2015:197).

The phenomenon in China may well be temporary, particularly for the cities that are still under construction. The idea of an economy that serves itself rather than relying on trade to generate wealth has appeal. Perhaps the ghost cities of China are not so much the architecture of wealth but rather as Wade (2015:199–200) suggests the result of crushing the process of urbanisation into a short timescale.

The 2008 financial crisis has also led to ghost developments where investors hoping to make money in the short-term failed, leaving banks the owners of deserted property. This situation was particularly noticeable in Spain, where in the 2011 census 3.4 million homes were unoccupied (Sevillano & Marcos, 2018), representing 13.7% of total dwellings (Instituto Nacional de Estadística, n.d.). Some of this number must be holiday homes rather than real estate speculation gone wrong, but whatever the cause empty property is indicative of wealth and the desire to accumulate more. In this, Spain seems to have been particularly affected amongst European countries (Neate, 2014). According to The Economist (Anon, 2008), "Spain's problem is not that it is suffering more than other European countries, but that it was doing so much better than them before. Bloated by cheap credit and a property bubble, it became an easy place to make money."

Cities of the Dead

It is true that in the words of Shakespeare's song from *Cymbeline*, "Golden lads and girls all must, as chimney-sweepers, come to dust" but wealth can still leave an impression on the built environment after death. *Cymbeline* was written in 1610 when the theatres reopened after the shut down for the plague (Royal Shakespeare Company, 2020) and maybe Shakespeare had this in mind, since disease is no respecter of wealth (see Chap. 9). Inequalities may virtually disappear on the beach when everyone is wearing swimming gear, but the cemeteries of some cultures remind that there is inequality even in death. Noting that not everyone can afford or needs a monument, Jones (1967:203) continues "…but still there are millions of monuments all over the earth, setting out grandeur or faith or conformity, made of granite, wood, gold, cast-iron, plastic, clay, whatever was available or in fashion." Examples can be found in the 700-year-old Muslim necropolis the City of the Dead in Cairo, which has now become home to tens of thousands of people in the crowded city (Hoffman, 2011), the 1873 Willesden Jewish Cemetery in London, described as the "Roll-Royce" of London Jewish cemeteries (Historic England, 2017), and the Christian La Recoleta Cemetery, Buenos Aires sitting within an expensive neighbourhood, and with its more than "…6,400 statues, sarcophagi, coffins and crypts" (turismo.beonosaires, n.d.). Such cemeteries are also famous because of the people buried there—Eva Peron in Recoleta and Julius Vogel, one-time prime minister of New Zealand, in Willesden—but many less well-known people are also honoured with graves and mausoleums that indicate the wealth of the families who paid for them.

This introduces the point of the gravestone or memorial, which is a focus for honouring or remembering the dead. While those close to us who have died are never forgotten Graham (2017:55) states "While acknowledging that objects are not essential to remembering, it is argued that material items play a fundamental role in defining persons in remembrance and this is achieved through the senses and the emotions." Different cultures also have different attitudes to the artefacts that commemorate the dead. "In Chinese culture, it is important for a deceased family member to be buried close to his or her native place, so he or she can watch over descendants and they, in turn, can visit the grave or permanent resting place regularly to pay their respects" (Blundy & Davis, 2017). In Hong Kong where land is scarce cremation has virtually replaced burial but it is becoming increasingly difficult to find a public site where ashes can be housed so these cultural traditions can be upheld. In Mexico, cemeteries are the focus of the Day of the Dead, from the end of October to the start of November, which is the time when the dead return to earth to live with their families and loved ones (Haley & Fukuda, 2014:1). During this time, the graves and memorials in cemeteries are highly decorated and the whole is a huge festival (Figs. 10.10, 10.11 and 10.12).

In some societies where an expanding population led to overflowing cemeteries and graveyards around churches, bones were removed and placed in catacombs, such as in the converted Roman quarries under Paris, a process that began in 1784 (Jones,

Fig. 10.10 Cemetery in Missions de Sierra Gorda (Querétaro, Mexico) before the celebrations for Dia de Muertos (Day of the Dead) (adapted from author's photograph)

Fig. 10.11 "Ofrenda" made with flowers in the main square of the town. Sierra Gorda, Queretaro, Mexico (adapted from author's photograph)

1967:178). Catacombs are levellers as all that remains are the bones and in Paris some of these have been arranged for decorative effect (Les Catacombs de Paris, 2018). War graves, with their simple crosses or headstones, are also levellers. In the establishment of war graves after WWI, the Imperial War Graves Commission felt it was essential that every grave was treated equally as the sacrifice of each soldier of his life was also equal (Morris, 1997). This approach was even more poignant in the comment from Morris (1997) that "The Somme battlefield itself was strewn with 50,000 scattered white wooden crosses, temporarily marking graves regardless of nationality."

The reverse is obvious in some graves that have become global icons, such as the pyramids of Egypt. The Great Pyramid at Giza, built in 2789–2767 BC with an original height of 485 ft (148 m) (Fonte, 2007:9) remained the world's tallest structure until overtaken by the spires of a number of European gothic cathedrals, such those of Lincoln and Beauvais, both since lost. If you are a pharaoh then you have the means to build a tomb that is so big it will last, even if most of the facing stone has been removed. However, many cemeteries reveal the fact that wealth is transitory and some mausoleums and graves fall into disrepair because there is no longer money for their upkeep.

Fig. 10.12 Celebration during the Day of the Dead in Mixquic, Mexico (adapted from author's photograph)

Architects and Wealth: Do Architects Only Work for the Wealthy?

In an interview with 2020 Priztker Architecture Prize winners Yvonne Farrell and Shelly McNamara, from the Irish firm Grafton Architects, McNamara stated that: "people think of architecture as being for the rich, and nothing to do with them. And that's a pity" (Wallace, 2016). The architects, whose comments are usually directed to other architects, argued that part of the responsibility for the lack of affordability of architectural services is due to the complexity of the approval processes and building costs that end up as a source of struggle for people. Does this mean that people are wrong to believe that architecture is a privilege of the wealthy? Parvin (2013:92) noted: "The uncomfortable reality is that almost everything we know as 'architecture' today is, in truth, design for the 1%: individuals, organisations, governments and corporations with the financial resources and capital to build, and to hire an architect."

Most people do not even know what architects do. A UK survey commissioned by InBuilding.org (Thompson, 2012) found that people did not know that architects prepared plans, applied for permissions and managed contractors. Of those asked, 15% did not know that architects designed buildings. Given this, it is no surprise that people are not prepared to spend money on hiring architects. The results of the survey can be read in many ways. There is definitely a gap between people and architects. Since architectural education was put into universities, thus finally losing the system whereby prospective architects became articled pupils in architectural firms, the contact between future architects and real clients is rare. The programmes of schools of architecture and the egos of their tutors make things worse by demanding bigger and bigger projects from students as they progress through their education. Most of these projects have no concern for the real problems that people have. These students will rarely be involved in the design of a new neighbourhood and they have almost no chance of designing a city. After finally graduating, students will be lucky to have clients for whom they can design bathrooms, stairs and extensions to small houses. Before this, they undergo a very demanding and expensive education, often without sleep, to find approved solutions for problems that nobody has.

The expectation is that after 5 or 6 years of sleep deprivation and eating instant noodles, life will be better. Salaries of architects are usually higher than the minimum wage, which also impacts on the demand for their services. However, this does not make architecture a career where you can expect to find many billionaires. With the exception of Sir Norman Foster, the only architect in the Sunday Times Rich List (Hopkirk, 2019), the majority of architectural firms is small; they struggle to find clients and they work long hours. Small firms are very sensitive to the ups and downs produced by the boom and bust cycle of the construction industry (Baker, 2018). As noted in Chap. 9, architectural firms were soon affected by the pandemic. Architects struggle to find wealthy clients while being too expensive for the entire potential market of non-wealthy people. One factor that makes the services of architects expensive is that their fees are usually linked to a percentage of the total building cost. Therefore, it is not difficult to infer that architectural services are out of the reach of middle and low-income households. What makes the gap between people and architects even wider is the popular perception that architects tend to provide expensive solutions that are not very functional and more linked to the personal taste of the architect than that of the client (Michelson, 1968).

The Argentinian architect Rodolfo Livingston proposed that instead of making "architecture for buildings," architects should try to do "architecture for families." In his book *"Cirugia de casas"* (Surgery of Houses), Livingston (2012) stated that the role of architects should be very similar to that of general practitioners, architects should be doctors of houses. Livingston realised that two thirds of the houses in Argentina are renovated at least once, and 95% of these renovations are made without architects (Estudio Livingston, 2018). This represents an immense number of square metres of building. According to Livingston's philosophy, the renovation of a house is the moment when it needs surgery and a good doctor can make the difference. The reality is that many mistakes are done during renovations, errors that could have been avoided if families had an architect that they could trust, someone keen to listen to every individual problem. Livingston (1995) developed "El Metodo" (The method), a manual for discussing problems with clients, and making them participate in the design process of their houses by playing role games. The method introduced a way of interviewing clients and creating habitats that were human and that make the most out of very little. Livingston's method was used in Cuba during the 1990s after he and Selma Diaz developed architectural workshops in Habana, Holguin and Guantanamo. The method was spread in Cuba and served as the base for the first Architects' Group of the Cuban Community. A key aspect of Livingston's method is that the architect charges "reasonable fixed fees" for every stage of the design process instead of a percentage of the total budget for the house or renovation. This made the architectural services more accessible but there was also more work for all architects. From 1994, these groups of architects proliferated and by 1996, there were 74 collectives. Over 2 years, they had offered architectural services to 11,846 families. At 89 years old, Livingston is still practising in Argentina and claims to have personally offered his services to 6000 clients. His firm, Estudio Livingston (2018) has a website with a friendly interface where consultations can be made. The firm offers a 50% discount to low-income families. Livingston's successful career shows that architecture is not necessarily just for the wealthy.

Monuments, Wealth and Collapse

The built environment is often what remains when there is some type of collapse in a society that leads to a reordering and possibly a change in location. However, the type of built environment can also be a signal of an impending change, which might not be for the better. When Lawrence coined the skyscraper index, he was trying to link an aspect of the built environment with a change, in this case the swing from economic boom to economic recession. Although others have disputed such a clear connection there is the simple fact that resources invested in creating monuments, whether ancient pyramids or modern super tall buildings, also require resources both human and material to maintain these. Tainter (1988:118) points to the problem of monuments in his discussion of the effect of diminishing marginal returns.

> Consider, for example, the plight of a society that must simultaneously face declining marginal productivity in any combination of the following: agriculture, minerals and energy extraction, science, education and information processing, size and costs of civil and military organisations, upkeep of capital stock (such as monumental architecture, or more practical things like aqueducts and bridges), etc. As each of these spheres requires an increased proportion of the society's budget, the portion available for investment in future growth must decline.

Another link between wealth as manifested in the built environment and the theory of collapse is the fact that seeing the built environment as a way of making money introduces problems, such as ghost mansions and house flipping. To combat problems, new legislation is then enacted to try to tax the wealth created. For the rich, this just means employing people to minimise the tax bill, and so the whole complexity of the system increases, again creating a spiral of diminishing returns. The more people are taxed the more they will find loopholes, which then have to be filled in, and so the cycle goes on (Tainter, 1988:116).

Ultimately, does it matter if a pharaoh builds a pyramid or a wealthy New Zealander buys a second home large enough to house a family of eight? The answer is "yes" in terms of inequality and environmental impact to the latter and "yes" in terms of economic stability and environmental impact to the former. Looking first at the pyramid, a pharaoh was a pharaoh and so both a ruler and a mediator between the gods and the people. It was no use aspiring to be a pharaoh as this was determined by dynastic succession. Assembling the materials and manpower to build a pyramid had an environmental impact and could have an economic effect as building projects were funded by the taxes raised on the Egyptian estates (Winkler, 2018). When it comes to the large second home, the additional resources that go into making and running large houses have already been discussed. The bigger issue with the large house is the owner is not a pharaoh but an ordinary person who has acquired wealth. The aspiration, therefore, is that wealth is something we could all have if we just found the way to get it. This leads to seeing the built environment as a means of wealth creation and concentration, with all the problems that follow. History contains examples of what happens when inequalities seem just too unjust—the 1787–89 French Revolution, the 1917 social revolution in Russia. Although pointing to the manifest inequalities in both cases is a simplistic view of complex societal problems (Skocpol, 1979:47–51), the problem with the built environment is that the inequalities are all too visible.

References

Al-Chalabi, M. (2015). Vertical farming: Skyscraper sustainability? *Sustainable Cities and Society, 18*, 74–77.
Alexander, J. (2016). *How Airbnb is reshaping our cities*. Retrieved August 9, 2020, from https://theconversation.com/how-airbnb-is-reshaping-our-cities-63932.
Anon. (1932). Boots' factory at Beeston. *The Architectural Review, 72*(430), 86–88.
Anon. (2008). Europe: Vacant lot; Spain's bust. *The Economist, 388*(8593), 1.
Anon. (2014). Bank buildings create a lasting impression in the community. *Northwestern Financial Review, 199*(4), 44.
Anon. (2017). 5 best and worst cities for house flipping ROI. *National Mortgage News, 42*(3), 38.
Apple inc. (2020). *Apple now globally powered by 100% renewable energy*. Retrieved July 6, 2020, from https://www.apple.com/newsroom/2018/04/apple-now-globally-powered-by-100-percent-renewable-energy/.
Arcadis. (2016). *Global built asset performance index 2016*. Retrieved June 6, 2020, from https://images.arcadis.com/media/B/6/6/%7BB6614982-7686-41AA-861F-FCEA08A48DE6%7DGlobal%20Built%20Asset%20Performance%20Index%202016_002.pdf?_ga=2.142961267.774914447.1591407610-749030968.1591407610.
Arch20. (2020). *Twelve of the most expensive houses in the world*. Retrieved August 8, 2020, from https://www.arch2o.com/most-expensive-house-in-the-world/.
Archeyes. (2020). *Olivetti underwood factory in Pennsylvania/Louis Khan*. Retrieved July 1, 2020, from https://archeyes.com/olivetti-underwood-factory-in-pennsylvania-louis-kahn/.

References

Banham, R. (1969). *The architecture of the well-tempered environment*. The Architectural Press.
Baker, K. (2018). *How many architects does our economy need?* Retrieved August 8, 2020, from https://www.architectmagazine.com/aia-architect/aiafeature/how-many-architects-does-our-economy-need_o.
Barr, J., Mizrach, B., & Mundra, K. (2015). Skyscraper height and the business cycle: Separating myth from reality. *Applied Economics, 47*(2), 148–160.
Bell, M. (2019). *The truth about flipping*. Retrieved June 18, 2020, from https://www.landlords.co.nz/article/976515588/the-truth-about-flipping.
Bernstein, (n.d.). *How sustainable is Apple Park's tree-covered landscape, really?* Retrieved July 6, 2020, from https://www.archdaily.com/875782/how-sustainable-is-apple-parks-tree-covered-landscape-really.
Blundy, R., & Davis, H. (2017), *Why dying in Hong Kong is getting more complicated...and expensive*. Retrieved July 8, 2020, from https://www.scmp.com/news/hong-kong/education-community/article/2102544/why-dying-hong-kong-getting-more-complicated-and.
Boardman, M. (2013). *Fixing and flipping: Real estate strategies for the Post-Boom Era*. Apress Publishing, e-book
Booth, W., & Laadam, K. (2017). London struggles with 'ghost mansions'; empty properties hot button amid housing shortage. *National Post (Index-Only)*. Retrieved July 7, 2020, from https://search-proquest-com.helicon.vuw.ac.nz/docview/1942461925?accountidL-=14782.
Brass, K., & Lawrence, A. (2012). Talking tall: the Skyscraper index. *CTBUH (Council on Tall Buildings and Urban Habitat) Journal, 2*, 42–44.
Cheek, J. M., & Briggs, S. R. (1982). Self-consciousness and aspects of identity. *Journal of Research in Personality, 16*, 401–408.
Dent, P. (2018). Introduction. In D. Lorenz, P. Dent, & T. Kauko (Eds.), *Value in a changing built environment* (pp. 29–33). Wiley Blackwell.
Donner, P. F. R. (1944). Edensor or Brown comes true. *Architectural Review, 95*(566), 39–43.
Dreifuss-Serrano, C. (2019). Huachafo as a reading key for self-building housing: Study on the formal and social aspects of informal architecture in metropolitan lima (PERÉ); el huachafo como clave de lectura para la vivienda autoconstruida: Estudio sobre los aspectos formales y sociales en la arquitectura informal de lima metropolitana (PERÚ). *Arquitetura Revista, 15*(2), 291–311. https://doi.org/10.4013/arq.2019.152.05
Estudio Livingston. (2018). *Un arquitecto, ¿para qué?* Retrieved from https://estudiolivingston.com.ar/un-arquitecto-para-que/.
Eurostat (2020). *Tourism statistics—Occupancy rates in hotels and similar establishments*. Retrieved July 13, 2020, from https://ec.europa.eu/eurostat/statistics-explained/index.php/Tourism_statistics_-_occupancy_rates_in_hotels_and_similar_establishments.
Farwaji, S. (2020). *What is an absentee owner in real estate and how to find one?* Retrieved July 7, 2020, from https://www.mashvisor.com/blog/absentee-owner/#:~:text=An%20absentee%20owner%20is%20someone,let%20their%20property%20sit%20vacant.
Fonte, G. (2007). *Building the great pyramid in one year: An engineer's report*. Algora Publishing.
Garay, L., Morales, S., & Wilson, J. (2020). Tweeting the right to the city: Digital protest and resistance surrounding the Airbnb effect. *Scandinavian Journal of Hospitality and Tourism, 20*(3), 246–267.
Garcia, E. J., & Vale, B. (2017). *Unravelling sustainability and resilience in the built environment*. Routledge.
Graham, B. (2017). *Death, materiality and mediation: An ethnography of remembrance in Ireland*. Berghahn Books.
Haley, S. D., & Fukuda, C. (2014). *Day of the dead: When two worlds meet in Oaxaca*. Berghahn Books.
Hall, C. M., & Müller, D. K. (2004). Introduction: Second homes. Curse or Blessing? Revisited. In C. M. Hall & D. K. Müller (Eds.), *Tourism, mobility and second homes: Between elite landscape and common ground* (pp. 3–14). Channel View Publications.

Hamza, N. (2014). The sustainable corporate image and renewables: From technique to the sensory experience. In A. Sayigh (Ed.), *Sustainability, energy and architecture: Case studies in realizing green buildings* (pp. 93–112). Elsevier Science & Technology.

Hansen, K. (2018). *30 of the most stunning abandoned towns around the world.* Retrieved July 7, 2020, from https://www.architecturaldigest.com/gallery/most-stunning-abandoned-towns-around-world.

Harris, D. (2007). Branch lines. *Estates Gazette 67.* Retrieved June 28, 2020, from https://link-gale-com.helicon.vuw.ac.nz/apps/doc/A164555249/ITOF?u=vuw&sid=ITOF&xid=7e6bd9e7.

Harvard T. H. Chan (2020). *The most expensive health care system in the world.* Retrieved August 8, 2020, from https://www.hsph.harvard.edu/news/hsph-in-the-news/the-most-expensive-health-care-system-in-the-world/.

Harvey, D. (2013). *Rebel cities: From the right to the city to the urban revolution.* Verso.

Hawkes, D. (2014). Aarhus Town Hall and the 'other' environmental tradition. *Architectural Research Quarterly, 18*(3), 273–282.

Helsely, R. W., & Strange, W. C. (2008). A game-theoretic analysis of skyscrapers. *Journal of Urban Economics, 64,* 49–64.

Henningsen, E., Jordhus-Lier, D., & Underthun, A. (2015). The resort as a workplace: Seasonal workers in a Norwegian mountain municipality. In D. Jordhus-Lier & A. Underthun (Eds.), *A hospitable world?: Organising work and workers in hotels and tourist resorts* (pp. 121–136). Routledge.

Herbelin, C. (2013). What is "French style"? Questioning genealogies of "western looking" buildings in Vietnam. *ABE Journal, 3*(3). https://doi.org/10.4000/abe.392.

Heritage New Zealand. (2020). *Old government buildings.* Retrieved July 1, 2020, from https://www.heritage.org.nz/places/places-to-visit/wellington-region/old-government-buildings.

Historic England. (2017). *New listings announced to celebrate 70 years of protecting England's historic buildings.* Retrieved July 8, 2020, from https://historicengland.org.uk/whats-new/news/70-years-of-listing/.

Hoffman, C. (2011). *Cairo's City of the Dead.* Retrieved July 8, 2020, from https://www.nationalgeographic.com/travel/egypt/cairo-city-of-the-dead/.

Hopkirk, E. (2019). *Foster is only architect on rich list—Again.* Retrieved from https://www.bdonline.co.uk/news/foster-is-only-architect-on-rich-list-again/5099443.article.

Instituto Nacional de Estadística. (n.d.). *Population and housing census.* Retrieved July 8, 2020, from https://www.ine.es/en/censos2011_datos/cen11_datos_inicio_en.htm.

Jacobi, L. (2019). *The architecture of banking in Renaissance Italy: Constructing the spaces of money.* Cambridge University Press.

Jadevicius, A. (2016). Skyscraper indicator and its application in the UK. *Entrepreneurial Business and Economics Review, 49*(2), 37–48.

Jones, B. (1967). *Design for death.* Andre Deutsch.

Jones, B. (1974). *Follies and Grottoes* (2nd ed.). Constable.

Keen, D., & Hall, C. M. (2004). Second homes in New Zealand. In C. M. Hall & D. K. Müller (Eds.), *Tourism, mobility and second homes: Between elite landscape and common ground* (pp. 174–195). Channel View Publications.

Khajehzadeh, I. (2017). *An investigation of the effects of large houses on occupant behaviour and resource use in New Zealand.* Ph.D. thesis, Victoria University of Wellington.

Khajehzadeh, I., & Vale, B. (2017). How house size impacts type, combination and size of rooms: A floor plan study of New Zealand houses. *Architectural Engineering and Design Management, 13*(4), 291–307.

Khajehzadeh, I., & Vale, B. (2017b). How New Zealanders distribute their daily time between home indoors, home outdoors and out of home. *Kōtuitui: New Zealand Journal of Social Sciences Online, 12*(1), 17–31.

Khajehzadeh, I., & Vale, B. (2017). How house size impacts on the number of furniture, appliance and tool items (FATs) in a house: An embodied energy study of New Zealand houses. *Energy Procedia, 142,* 3170–3175.

Klock, M. (2013). The virtue of home ownership and the vice of poorly secured lending: The great financial crisis of 2008 as an unintended consequence of warm-hearted and bone-headed ideas. *Arizona State Law Journal, 45*(1), 135–182.

Larsson, L., & Müller, D. K. (2019). Coping with second home tourism: Responses and strategies of private and public service providers in western Sweden. *Current Issues in Tourism, 22*(16), 1958–1974.

Law, T. H., Hamid, H., & Goh, C. N. (2015). The motorcycle to passenger car ownership ratio and economic growth: A cross-country analysis. *Journal of Transport Geography, 46*, 122–128.

Les Catacombs de Paris. (2018). *L'ossuaire*. Retrieved July 8, 2020, from https://www.catacombes.paris.fr/lhistoire/lossuaire.

Lew, A. A. (2011). Tourism's role in the global economy. *Tourism Geographies, 13*(1), 148–151.

Little, D. F. (2020). Tax system fails. *The Sunday Star Times June 14*, 20.

Livingston, R. (1995). *El método*. Ediciones de la Urraca.

Livingston, R. (2012). *Cirugía de casas* (16a ed.). Editorial Nobuko.

Lloyd Jenkins, D. (2004). *At home: A century of New Zealand design*. Godwit Books.

Lowe, P. (1999). *Survival of an icon: The great New Zealand bach*. Bachelor of Architecture Thesis, Victoria University of Wellington.

McNamara, P. (2010). Two takes on…green property investment. *Property Week, 75*(38), 79.

Meakin, B. (1905). *Model factories and villages: Ideal conditions of labour and housing*. T Fisher Unwin.

Michelson, W. (1968). Most people don't want what architects want. *Transaction, 5*, 37–43.

Miller, B. J. (2012). Competing visions of the American single-family home: Defining McMansions in the *New York Times* and *Dallas Morning News*, 2000–2009. *Journal of Urban History, 38*(6), 1094–1113.

Miller, D. (2009). *Stagflation in 2010 may look like reruns of the 1970s*. Retrieved July 6, 2020, from https://moneymorning.com/2009/12/21/stagflation-2/.

Morris, M. S. (1997). Gardens 'for ever England': Landscape, identity and the First World War British cemeteries on the Western Front. *Ecumeme, 4*(4), 410–434.

Muthesius, S. (1982). *The English terraced house*. Yale University Press.

Neate, R. (2014). *Scandal of Europe's 11m empty homes*. Retrieved July 8, 2020, from https://www.theguardian.com/society/2014/feb/23/europe-11m-empty-properties-enough-house-homeless-continent-twice.

New Zealand History. (n.d.). *Goldfields*. Retrieved July 7, 2020, from https://nzhistory.govt.nz/keyword/goldfields#:~:text=Today%20Waiuta%20is%20a%20West,of%20Waiuta%20in%20its%20heyday.

New Zealand Parliament. (2016). *Buildings and grounds*. Retrieved July 1, 2020, from https://www.parliament.nz/en/visit-and-learn/history-and-buildings/buildings-and-grounds/.

Norberg, J. (2012). *Financial Fiasco: How America's infatuation with homeownership and easy money created the economic crisis*. Cato Institute.

Omar, E. O., Endut, E., & Saruwono, M. (2012). Personalisation of the home. *Procedia—Social and Behavioral Sciences, 49*, 328–340.

Papanicolas, I., Woskie, L. R., & Jha, A. K. (2018). Health care spending in the United States and other high-income countries. *The Journal of the American Medical Association, 319*(10), 1024–1039.

Parker, D. (1991). *The rise of the Peruvian middle class: A social and political history of white-collar employees in Lima, 1900–1950*. Retrieved from https://search-proquest-com.ezproxy.auckland.ac.nz/docview/303940120/?pq-origsite=primo.

Parvin, A. (2013). Architecture (and the other 99 percent): Open-source architecture and the design commons. *Architectural Design, 83*(6), 90–95.

Pevsner, N. (1968). (originally published 1936). *Pioneers of modern design*. Penguin Books Ltd.

Richards, J. M. (1940). *An Introduction to modern architecture*. Penguin Books Ltd.

Richards, J. M. (1973). *Castles on the ground: the anatomy of suburbia* (2nd ed.). John Murray.

Richardson, M. (1996). Restoring a building and a reputation (Dudok Festival, Town Hall, Hilversum, Netherlands). *Architects' Journal, 203*(24), 54.
Royal Shakespeare Company. (2020). *Dates and sources*. Retrieved July 8, 2020, from https://www.rsc.org.uk/cymbeline/about-the-play/dates-and-sources#:~:text=license%20our%20images-,Dates,closure%20due%20to%20the%20plague.
Samy, A. (2016). *The building society promise: Access, risk and efficiency 1880–1939*. Oxford University Press.
Schleier, M. (2008). The Empire State Building, working-class masculinity, and King Kong. *Mosaic, 41*(2), 29–54.
Schwab, F. (1940). Lo huachafo como fenómeno social. *Revista, 3*(4), 16–22.
Scofield, J. H. (2009). *A Re-examination of the NBI LEED Building Energy Consumption Study, Energy Programme Evaluation Conference, Portland* (pp. 764–777). Retrieved July 6, 2020, from https://www2.oberlin.edu/physics/Scofield/pdf_files/Scofield%20IEPEC%20paper.pdf.
Sevillano, E. G., & Marcos, J. (2018). *Spanish government planning measures against owners of empty homes*. Retrieved July 8, 2020, from https://english.elpais.com/elpais/2018/09/18/inenglish/1537258176_649049.html.
Sigala, M. (2020). Tourism and Covid-19: Impacts and implications for advancing and resetting industry and research. *Journal of Business Research, 117*, 312–321.
Singh, D. R., Wright, A., & Hayle, C. (2010). Factors influencing hotel occupancy in Jamaica: The role of events 1991–2008. *Tourism Analysis, 15*(3), 357–365.
Siry, J. M. (2013). Frank Lloyd Wright's innovative approach to environmental control in his buildings for the S.C. Johnson Company. *Construction History, 28*(1), 141–164.
Skinner, R. (2020). A pretty true reflection of our civilisation: Classical architecture in nineteenth-century New Zealand. In N. Temple, A. Piotrowski, & J. M. Heredia (Eds.), *The Routledge handbook on the reception of classical architecture* (pp. 343–355). Routledge.
Skocpol, T. (1979). *States and social revolution: A comparative analysis of France, Russia, and China*. Cambridge University Press.
Sorace, C., & Hurst, W. (2016). China's phantom urbanisation and the pathology of ghost cities. *Journal of Contemporary Asia, 46*(2), 304–322.
Stamp, G. (2016). Long may George Peabody's legacy continue to flourish. *Apollo, 183*(640), 48–49.
Swedish Institute. (2020). *The Swedish summer house love affair*. Retrieved July 6, 2020, from https://sweden.se/culture-traditions/the-swedish-summer-house-a-love-affair/.
Tainter, J. A. (1988). *The collapse of complex societies*. Cambridge University Press.
Tarn, J. N. (1973). *Five percent philanthropy: An account of housing in urban areas between 1840 and 1914*. Cambridge University Press.
The CGA (Country Gentlemen's Association). (1914). *Artistic country building 1913–14*. The CGA Ltd.
The CGA. (2020). *History of the CGA*. Retrieved June 15, 2020, from https://www.thecga.co.uk/pages/history.
The Economist Newspaper. (2020). *How not the create traffic jams, pollution and urban sprawl*. Retrieved July 6, 2020, from https://www.economist.com/briefing/2017/04/08/how-not-to-create-traffic-jams-pollution-and-urban-sprawl.
Thompson, M. (2012). It's true: People don't know what architects do. *Architects' Journal, 236*(3), 9–11.
Tottathil, M. (2013). *Global built environment wealth = $163 Trillion [infographic]*. Retrieved June 6, 2020, from https://arabiangazette.com/global-built-environment-wealth-tops-193-trillion-infographic-20130904/.
Tuohy, C. (2010). Filling the vacancy vacuum: Empty building mean opportunity for a new class of hybrid product in commercial insurance. *Risk and Insurance, 21*(7), 8.
turismo.beonosaires. (n.d.). *Recoleta Cemetery*. Retrieved July 8, 2020, from https://turismo.buenosaires.gob.ar/en/otros-establecimientos/recoleta-cemetery.

Tourismus+Congress Gmbh. (n.d.). *Deutsche Bank Twin Towers*. Retrieved June 28, 2020, from https://www.frankfurt-tourismus.de/en/Media/Attractions/Skyscraper/Deutsche-Bank-Twin-Towers.

Townsend, L. (2010). *Home making and identity: A psychology of personality processes in north east Scotland*. Ph.D. thesis, Robert Gordon University.

Tran, T. H. (2014). *Sustainable patterns of living based on an investigation of footprint on Hanoi-Vietnam, Wellington-New Zealand and Oulu-Finland*. Ph.D. thesis. Victoria University of Wellington.

Tsenkova, S. (2009). *Housing policy reforms in post socialist Europe*. Physica-Verlag.

Tugores, M., & Valle, E. (2016). Innovation, hotel occupancy and regional growth. *Tourism Economics, 22*(4), 749–762.

Vale, R., & Vale, B. (2009). *Time to eat the dog?: The real guide to sustainable living*. Thames and Hudson.

Vale, B., Vale, R., & Chicca, F. (2018). Dwelling. In F. Chicca, B. Vale, & R. Vale (Eds.), *Everyday Lifestyles and Sustainability: the environmental impact of doing the same things differently* (pp. 208–217). Routledge.

Wade, S. (2015). *Ghost cities of China: The story of cities without people in the world's most populated country*. Zed Books.

Wagner, C. (2010). Vertical farming: An idea whose time has come back. *The Futurist, 44*(2), 68–69.

Walters, T., & Carr, N. (2015). Second homes as sites for the consumption of luxury. *Tourism and Hospitality Research, 15*(2), 130–141.

Wallace, A. (2016). *Architects—They're not just for the rich*. Retrieved July 8, 2020, from https://www.irishtimes.com/culture/art-and-design/architects-they-re-not-just-for-the-rich-1.2789530.

Winkler, A. (2018). *How the ancient Egyptian economy laid the groundwork for building the pyramids*. Retrieved July 8, 2020, from https://theconversation.com/how-the-ancient-egyptian-economy-laid-the-groundwork-for-building-the-pyramids-107026.

World Bank Group. (2020). GNI per capita, Atlas method (current US$). Retrieved June 19, 2020, from https://data.worldbank.org/indicator/NY.GNP.PCAP.CD.

World Tourism Organisation (UNTWO). (2020). *International tourism growth continues to outpace the global economy*. Retrieved July 13, 2020, from https://www.unwto.org/international-tourism-growth-continues-to-outpace-the-economy.

Worldometer. (n.d.). *GPD per Capita*. Retrieved June 19, 2020, from https://www.worldometers.info/gdp/gdp-per-capita/.

Chapter 11
What Should We Do?

> *"Exactly!" said Deep Thought. "So once you do know what the question actually **is**, you'll know what the answer means."*
> Douglas Adams

Introduction

It is clear there are many serious problems awaiting us. Issues of climate change and its effects, inequality and pandemics have been examined in this book, and it appears a collapse of some sort seems likely. As we attempt to collapse gracefully, it is all very well to point out the problems, what is needed are suggestions for what could be done to solve the problems, or at least to reduce their impact. This is a book about the built environment so these suggestions need to be related to what type of built environment we should be creating now. What kind of built environment might best serve as an inheritance for our descendants, its occupants?

The first big problem to be tackled is that of climate change. Unless this is dealt with, other problems, like inequalities, are only going to get worse.

The Cost of Climate Change

There are several ways of looking at the likely cost of climate change. In a world based largely on the financial point of view, it probably makes sense to consider it first solely from the perspective of money, rather than the "emotional cost" of losing your home to rising sea level or the "environmental cost" of the extinction of koalas.

In a 2009 article in the magazine *Foreign Policy*, the Danish political scientist Bjorn Lomborg, who gained fame as the "skeptical environmentalist" (Lomborg, 2001), wrote the following with a co-author (Lomborg & Rubin, 2009) when considering the costs of the effects of climate change: "The total costs of global warming for the next 100 years are estimated at $5 trillion, which should be compared to the total income of $800 trillion to $900 trillion expected to be generated in the same period." They went on to comment "As such, the sheer scale of the problem makes it unreasonable to talk about a limit to growth." Presumably, they meant that the cost of dealing with the problems created by global warming was so small compared with total income that it was not worth bothering about.

But this is only one point of view and although it is an economic prediction, it does not come from an economist. It always pays to check the figures. The authoritative *Stern Review*, commissioned by the UK government, reached a different conclusion two years earlier, so Lomborg and his colleague could have read it before writing their own article.

> Using the results from formal economic models, the Review estimates that if we don't act, the overall costs and risks of climate change will be equivalent to losing at least 5% of global GDP each year, now and forever. If a wider range of risks and impacts is taken into account, the estimates of damage could rise to 20% of GDP or more.
>
> In contrast, the costs of action—reducing greenhouse gas emissions to avoid the worst impacts of climate change—can be limited to around 1% of global GDP each year (Stern, 2007:xv–xx).

Global GDP in 2007 was just over $US 53 trillion (IMF, 2020) so Stern was saying that the cost of climate change would be at least $2.65 trillion *a year*, which is 5% of annual global GDP. Using Stern's higher estimate of 20% of global GDP, the annual cost would be $10.6 trillion. On the basis of Stern's lower estimate of 5%, the cost of climate change over "the next 100 years," the timeframe given by Lomborg and Rubin, would be at least $132.5 trillion, assuming annual GDP did not increase at all over the same period. This is a lot more than Lomborg's $5 trillion. Stern's lowest figure for 1 year's cost of climate change is more than half Lomborg and Rubin's estimate for the cost over 100 years.

Paying to Fix It

Stern's figure that the cost of reducing emissions to avoid climate change would require 1% of global GDP would mean a cost in 2007 of $530 billion a year. This sounds like quite a lot of money although compared with the world's military expenditure, which was more than twice this at $1.3 trillion in the same year (The World Bank Group, 2019a), it does not seem so bad. A report in 2007 from a military-focused research group stated "Projected climate change poses a serious threat to America's national security" (CNA, 2007:6) so it might be argued that military spending would need to rise to counter it. It is interesting to note that it has been claimed by the Union of Concerned Scientists that the US military is the largest institutional user of oil in the world (UCS, 2014), although it is taking steps to try to reduce this dependence.

The world's population in 2007 was 6.6 billion (PRB, 2020), which means the annual cost of climate change back then would have been about $400 per person, while the "cost of action" to fix it would have been around $80 per head per year and the annual cost of military expenditure then was around $200 per person. Military expenditure in 2018, the latest year for which data are available, was $1.8 trillion (SIPRI, 2019). By 2018, well before COVID-19, the world's GDP was $86.4 trillion (IMF, 2020) and the population was 7.6 billion (Worldometers, 2020a) so military expenditure was around $240 per person. Assuming that Stern's figure of 1% of GDP would still be enough to avoid the worst of climate change, the cost would be $864 billion or $114 per person, more than it was in 2007 because of more people and more wealth. This suggests delays in dealing with climate change are going to cost humanity more, just as delays in dealing with minor building maintenance can lead to much higher costs of fixing a major problem some years later.

It might be argued, particularly by the armaments industry, that military expenditure is essential, so we should not attempt to get the money to solve climate change by spending less on weapons. If we are not able to face giving up our military forces to save the world, other comparisons can be made with things that we choose to buy, to give another sense of the affordability or otherwise of dealing with climate change. In 2018, the global market in cosmetics was estimated to be worth around $508 billion, rising to $758 billion by 2025 (Shahbandeh, 2019). The money we spent on cosmetics in 2018, $508 billion, could have gone two thirds of the way to solving climate change, which would have cost $864 billion in 2018. As consumers, we do not often buy a bomber but we do probably buy makeup, deodorant and all the other things that make us feel lovelier. Would we be prepared to forego our beauty to save the world?

Choosing not to buy cosmetics might be a step too far for most people. Perhaps there are other costs that are more directly connected with climate change that could be used to fund solutions? According to work by the International Monetary Fund (Coady et al., 2015) for example, the global cost of subsidies on fossil fuels was around $5 trillion in 2015 alone, far greater than would be required, using Stern's estimate, to have a chance of avoiding climate change. This IMF figure for fossil fuel subsidies includes both direct subsidies to the fossil fuel industry and to consumers, as well as the costs borne by society in the form of ill-health, environmental damage and the like. With a world population in 2015 of 7.3 billion people (PRB, 2015) the IMF figure gives a cost of $685 per person for subsidising the use of fossil fuels at that time, making solving the climate problem look like a very cheap option.

Of course, the problem has worsened in the years between the publication of the *Stern Review* and now. Population in 2020 was around 7.8 billion (Worldometers, 2020b) compared with 6.6 billion in 2007, and the CO_2 in the atmosphere had risen from 380 ppm in 2007 to 413.4 ppm in January 2020 (NOAA, 2020). This is a very rapid rise. Recent measurements of 23 million-year-old mummified leaves found in the South Island of New Zealand revealed that the CO_2 level in the atmosphere at that time varied between 450 and 550 ppm and average global temperatures were between 5 and 6 degrees Celsius higher than today (Reichgelt et al., 2020). The change in CO_2 concentration in the last 13 years means that roughly we will reach the 450 ppm level in another 13 years. We do not have very long in which to act. Because our economy and population also grow, we may well get to this level of CO_2

concentration in about 10 years. If this means that we will also see a temperature rise of 5–6 degrees, it may well prove to be more than an inconvenience.

The figures about the cost of stopping climate change make clear that it is not that the world cannot afford to stop climate change, it simply chooses not to. In a cost-driven world economy, it seems bizarre that we are choosing to pay to ruin the climate although it would be cheaper not to. It appears that the politicians we elect to represent us are preferring to pay more of the global budget, which in the end comes out of the pockets of the people via taxation, in order for us to have climate change. One wonders why we put up with it and why we continue to vote for it?

It is no surprise that when Greta Thunberg, who created the idea of climate strikes, addressed world leaders at the United Nations' General Assembly in 2019, she said "…all you can talk about is money and fairy tales of eternal economic growth. How dare you?" (Rowlatt, 2020). The figures show that she has every right to be angry.

Who Should Pay to Fix It?

Reasonably, it could be argued that the countries that have both caused climate change and have benefited from causing it should be the ones to pay for it, but what would that mean for people living in those countries?

The members of the OECD which in 2007, the year of the *Stern Review*, comprised "30 member countries committed to democracy and the market economy" (OECD, 2007:7) could be used to represent these relatively richer countries. In 2007 (OECD, 2007:6), the organisation represented 14% of the world population, 84% of world trade and 49% of world CO_2 emissions. It also had a 76% share of world Gross National Income (GNI) in "current US dollars." The 2007 Gross National Income for the whole world per capita was $11,677.50 in "current international dollars" (The World Bank Group, 2019b). Knowing the world population, the total GNI was roughly $77 trillion and we can work out that the OECD's share of this was just over $58 trillion. Using the GNI figures and applying Stern's 1% value, the cost of reducing emissions would have been $770 billion. If only the OECD countries paid for this it would have been only 1.3% of the OECD's GNI. It does not seem like an impossibly large amount. If spread equally among the OECD's population of 924 million it would be $833 each per year, or $2.30 a day.

Even in the relatively wealthy OECD, there are plenty of poor people for whom $2.30 a day would be a lot of money to have to find to pay for avoiding climate change. It has been shown that the majority of the emissions causing climate change is produced by the wealthy, who tend to buy more stuff, use more energy and do more flying. Oxfam, for example, has asserted that the richest 10% of people worldwide are responsible for 50% of global emissions (Oxfam, 2015), which is similar to the OECD's 14% of the world population causing 49% of emissions.

A more recent study came to a similar conclusion. Oswald et al. (2020) in a study of 86 countries using data from the EU and the World Bank found that the richest 10% of people consume roughly 20 times more energy overall than the poorest 10%.

As people become richer, they spend more on holiday travel and cars meaning that the richest 10% use 187 times more vehicle fuel energy than the poorest 10%. If incomes rise, consumption will rise.

This can already be seen in the data reported by the International Energy Agency (Cozzi & Petropoulos, 2019), which shows that the increasing world demand for large and inefficient SUVs, which consume around 25% more energy than average size cars, has overtaken any emission reductions resulting from improved efficiency in other vehicles. The researchers found that SUVs alone were the second-largest contributor to the increase in global CO_2 emissions since 2010 (the largest increase came from electricity generation). The European Environment Agency has found that the average emissions from new cars have been increasing in the past 3 years due to the increase in SUVs (EEA, 2020). This has happened even in Norway, which has a policy of 100% zero-emission car sales by 2025 and where electric cars formed 25% of sales in 2017 (Elvestuen, 2018).

The rich are the problem. What if the rich paid to fix the problem for everyone else? Otto et al. (2019) state that, in 2017, the world had 36 million "high net worth individuals" meaning each had net assets of more than $US 1 million. In 2017 the world GDP was $81.306 trillion (World Bank, 2021). Using Stern's figure of 1% of GDP to solve climate change, it would cost $813 billion a year. If the $813 billion were shared among the high net worth people, each person would have to pay out around $22,600 each per year. It would mean that the people who Otto et al. call the "super-rich" might have to hang onto last year's Porsche another year before trading it in but would that be a total catastrophe?

It should not be beyond the ability of politicians to adjust taxation so that the burden of payment would fall on those who cause the emissions, but in countries "committed to democracy and the market economy" like the OECD members, this has not happened yet and it seems unlikely that it ever will.

The Cost of Giving up Fossil Fuels

Another way to consider the problem is to look at the cost and the likely effect of solving a large part of climate change. One of the major causes of carbon emissions is the use of fossil fuels, which make up about three quarters of total emissions (USEPA, 2019). If these fuels could be replaced by zero-carbon energy sources, it would go a long way to reducing total emissions and staving off climate change. Some time ago, researchers at Stanford University looked at what would need to be done to replace fossil fuels worldwide with "WWS"—wind, water and solar energy—using a variety of available technologies (Delucchi & Jacobson, 2011a, b). They found that in a global energy system based on 100% WWS, not just for electricity generation but for all uses that are supplied currently from fossil fuels, the cost of energy would be similar to the cost of energy at present and energy supply would be at least as reliable as today. One of the interesting aspects of this study is that the authors listed all the equipment that would be needed for the transition, such as nearly two billion

3 kW solar photovoltaic systems on people's roofs, and they also worked out that the world has sufficient materials to build all the necessary equipment and systems. It is also worth considering that the project would provide a huge amount of employment. Delucchi and Jacobson concluded "that barriers to a 100% conversion to WWS power worldwide are primarily social and political, not technological or even economic." A fully renewable zero-carbon global energy system could provide reliable power at no more cost than the current system. There seem no reasons not to do it, other than social and political reasons, of course.

Another estimate (Bullis, 2014) based on the annual report of the International Energy Agency states "Switching from fossil fuels to low-carbon sources of energy will cost $44 trillion between now and 2050". This spreads the cost over 36 years, making it just over $1.2 trillion a year. The report concludes "the costs of switching will be paid for in fuel savings between now and 2050" so the conclusion is very similar to that reached by Delucchi and Jacobson.

Time to Eat the Rich?

The "back of the envelope" calculations made above are not intended to be definitive but are intended to be an exploration of the situation. What these crude calculations and the far more detailed studies by Delucchi and Jacobson make clear is that it is unreasonable to say that the costs of avoiding the worst effects of climate change will be too high, indeed we could even get the wealthy to pay for it. This might be in their best interests because as the philosopher Jean Jacques Rousseau is supposed to have said "When the people shall have nothing more to eat, they will eat the rich." (Thiers, 1850:359).

The Built Environment

The arguments above show that it is possible to change to a mode of energy supply at a global scale that could fulfil current needs without carbon emissions. The aim here is to summarise the effect this might have on the built environment, based on the issues raised throughout the book. To do this, it seems useful to adopt the resilience approach and look at different scales within the modern built environment.

Cities

Much has been written about the increasing urbanisation of the global population with little discussion as to whether this is a good thing to happen. Ecological resilience theory would suggest that diversity is one strategy for persistence of a system, since the collapse of a small part does not affect the survival of the whole. This would suggest that labelling desirable cities as "smart" or "floating" misses the point. A city might contain elements of both ideas as a survival strategy. In part, this comes from the notion that it is possible to design a city but this almost never happens in reality. The process of city creation is much more complicated than a designer's dream and must first rely on a source of economic activity that brings people together, who then create the built environment to support more economic activities. This means urbanisation happens because of economic growth but it does not cause it, as it is more of a chicken and egg relationship.

This raises another issue about modern cities. Currently, they are predicated on economic growth and as the economy grows so does the city through pulling people in to take advantage of the situation. The growth of cities causes problems in terms of trying to ensure people can move around them and that a reasonable quality of life is attainable within the concentration of people. In 2003, Henderson studied the relationship between economic growth and optimum city size. His conclusion is relevant.

> For any country size and level of development, there is a best degree of urban concentration, which balances the gains from enhanced concentration such as local knowledge accumulation against the losses such as shoring up the quality of life in crowded mega cities. That best degree of concentration declines with country size and level of development.

A somewhat more recent study of the same issue (Grossman, 2013) came to the conclusion that "...urban congestion associated with structural change may leave economic growth unsustainable in the long-run."

What is useful in both studies from the point of view of collapse is that gains from growth have to be balanced against losses—as in ecological resilience, there is no win–win situation. Tainter's analysis of the collapse of civilisations concluded that collapse occurred when the efforts and resources needed to solve problems outweighed the gains from problem-solving. The seemingly never-ending growth of cities needs to be investigated in terms of whether the benefits gained from expansion show a marginal return. This is the question that governments and authorities should be trying to answer. This means tackling the currently accepted situation that people can make money out of urban growth and so cities grow.

Neighbourhoods

Diversity is also useful at the neighbourhood scale. The concept of zoning within cities can be seen in Tony Garnier's design for the 1917 Cité Industrielle, where the city was split into industrial, residential, public offices and agricultural areas with circulation systems between, which meant there was no diversity at the neighbourhood scale. Within medieval cities, there had been zoning by trades, with areas of the city housing the same trade, on the modern idea of knowledge sharing, such as the Shambles in York in the UK. This street was full of butchers and, in 1830, 28%

of all butchers in York were still trading from the Shambles, although there are none there now (York Civic Trust, 2020). However, the Shambles was also home to people living above the shop and one zoned street is a very different scale from separating industrial or service areas from residential, and thus introducing commuting between home and work.

The COVID-19 pandemic proved to many people the importance of the neighbourhood in providing residential areas with local services and space for recreation, largely because work for many people was restricted to home. With driving discouraged, many people walking for exercise discovered previously unknown areas of their neighbourhood. Others living in much less diverse neighbourhoods were less fortunate. What this shows is that diversity is useful in a crisis and urban design at this scale should be based on diversity. However, this very diversity is a design issue as research has shown the importance of designing spaces that people want to use so that there is a chance of meeting other people and hence encouraging a sense of community (Liu et al., 2019). There is no point in providing green areas and play spaces if there is no safe way to cross a busy road to get to them.

This raises the issue of moving around at the neighbourhood scale. Before the mass use of the private car, this was a simple issue. Long distance travel was by rail and perhaps by ship or plane for the very wealthy, and local travel was bus, cycling or walking. This made the neighbourhood a much safer place, just as people found during lockdown. With the need to move to non-fossil fuels as discussed above, there is also the opportunity to rethink neighbourhood travel and whether the move from petrol to electric private vehicles is really an answer to the problem or just another example of diminishing marginal returns.

Inequalities can also be visible at the neighbourhood scale. Although, as suggested earlier, the only way to try and balance inequalities is to transfer money from the rich to the poor via taxation (or robbery for the modern Robin Hood), as long as the built environment is seen as a wealth generator there will be inequalities. If housing were nationalised and everyone had to rent off the government a very different situation might result.

Plots and Buildings

Mention was made above of the need to fit PVs to building roofs in the move to renewable energy supply and climate change mitigation. This suggests that roofs have to be suitable both in terms of orientation and form. The Ancient Greeks in their solar cities were well aware of design for solar orientation but such simple matters are normally overlooked in design where the view may be much more important than solar access. A study of an Auckland suburb by Ghosh and Vale (2006) found that moving from the common hipped roof to a gable roof would increase the potential of the roof to collect solar energy from 73 to 82.5% of all domestic energy demands excluding space conditioning. This is one aspect of design seldom considered but one that should be an essential issue even if PVs are not yet fitted, given the inherited nature of the built environment. The issue here is really about building now for future needs when all energy has to come from renewables, but doing this is also a step to climate change mitigation.

On the other scale of building, monuments should also be treated carefully when it comes to a potential societal collapse. Modern monuments are the new generation of super-tall buildings, with the thought that these will be able to house many more people on a smaller area of land. Since these same people need to be fed and watered and their wastes dealt with in a safe way, this may be another example of marginal returns when all resources going into the construction and running of such buildings are evaluated. Such evaluations should be mandatory so that those who benefit from them pay the full cost, rather than passing on many costs to the community at present.

What Should Designers Do?

To move to a sustainable society implies a change in behaviour since technology alone cannot fix global problems. However, who has to change their lifestyle is also critical because the impacts of the poor and rich are completely different. If poor people follow the steps of the rich, the world ends miserably and there is no hope of a graceful collapse. If only 50% of the population make a radical change towards sustainability, it perhaps does not matter or change anything. The only hope is to change the top 10 or 20% radically. This is important for designers because these are the people most designers work for as they have the money for new projects. The remaining 80% might not have that much money and what they have is normally invested in the built environment in the form of the family dwelling. If inequality keeps on growing, fewer people will have more money. The lucky few designers will be commissioned by this elite but the real future of architecture is much more likely to be architecture for the much less wealthy. Future designers will need to learn how to design for the poor and working with many clients trying to solve small complex situations with the idea of reducing and avoiding future problems. This work will attract much lower fees but there could be plenty of such work. This will also change everything at the academic level as the training will need to be focussed on the architect as local practitioner rather than prima donna designer.

Future design will also need to recognise that though there will be more people in the future, and hence potential clients, there will not be more land to put them on. They will also need air, food and water and the land to provide this. The challenge to design is to recognise that this means much more than putting more people on less land, as design has to include how these vital resources will be sourced in a way that can be sustained.

Final Thoughts

What the previous chapters have shown is that humanity faces the two problems of an environmental crisis and a crisis of inequality. Simultaneously, through the built environment, human ingenuity and technology have the ability to make those problems smaller or bigger. History tends to show that the more technological and complicated solutions only lead to even more complicated problems. Why spend efforts to 3D print buildings when we know human hands are very good at making them? Is it better that machines put builders out of a job, leading to the need to find increased means of supporting the unemployed builders through taxation, rather than letting people do what they know how to do? This is an example of design ideas leading to marginal returns through substituting a problem that really does not exist—how to make buildings—with the problem of how to make as much money as possible making buildings. Along with always looking at the marginal returns from innovations that appear to solve problems, we also need to examine the problem to see if it is real. Collapse may come closer just because we are solving the wrong problems. If we wish to collapse gracefully then radical change is needed. The degree of change also has a relationship with the gracefulness of collapse, since if the problems continue to grow the collapse is probably going to be less than graceful.

The last point is that since we are not all equally responsible for this mess, we should contribute to solving it according to the level of responsibility we had in creating it. A two-year old is not responsible for the consumption rates and problems created by his/her parents or grandparents, but unless we act quickly now he/she will be the person who has to solve these problems as an adult.

Civilisations have collapsed before and we have some understanding as to why such things happened—increasing complexities, inequalities, diseases, external hazards, all of which eventually led to a situation where human ingenuity could no longer solve the problems in front of it in a way that produced a marginal return. This book has tried to show that the current situation is somewhat similar and collapsing gracefully is no more than dealing with problems we know how to solve.

References

Bullis, K. (2014). How much will it cost to solve climate change? *MIT Technology Review*, May 15. Retrieved March 18, 2020, from https://www.technologyreview.com/s/527196/how-much-will-it-cost-to-solve-climate-change/.

CNA (Centre for Naval Analyses). (2007). *National security and the threat of climate change Alexandria*. CNA Corporation.

Coady, D., Parry, I., Sears, L., & Shang, B. (2015), *How large are global energy subsidies?* (IMF Working Paper WP15/105). International Monetary Fund. Fiscal Affairs Department May, 2015.

Cozzi, L., & Petropoulos, A. (2019). Growing preference for SUVs challenges emissions reductions in passenger car market. *IEA Commentary*, 15 October Paris. International Energy Agency. Retrieved March 23, 2020, from https://www.iea.org/commentaries/growing-preference-for-suvs-challenges-emissions-reductions-in-passenger-car-market.

References

Delucchi, M. A., & Jacobson, M. Z. (2011a). Providing all global energy with wind, water, and solar power, Part I: Technologies, energy resources, quantities and areas of infrastructure, and materials. *Energy Policy, 39*(3), 1154–1169.

Delucchi, M. A., & Jacobson, M. Z. (2011b). Providing all global energy with wind, water, and solar power, Part II: Reliability, system and transmission costs, and policies. *Energy Policy, 39*(3), 1170–1190.

EEA. (2020). *Average CO_2 emissions from new cars and new vans increased again in 2019.* Copenhagen, European Environment Agency, 29 June. Retrieved July 6, 2020, from https://www.eea.europa.eu/highlights/average-co2-emissions-from-new-cars-vans-2019.

Elvestuen, O. (2018). "Norway's low emissions policy" presented by Minister for climate and environment at *The EU's vision for a modern, clean and competitive economy.* Stakeholder consultation high level public event, Brussels, 11 July. Retrieved July 6, 2020, from https://www.regjeringen.no/en/aktuelt/norways-low-emissions-strategy/id2607245/#:~:text=Electric%20vehicles%20.%2080%20percent%20of%20Norway%27s%20emissions,fulfilment%20with%20the%20EU%20in%20its%202030%20target.

Ghosh, S., & Vale, R. (2006). The potential for solar energy use in a New Zealand residential neighbourhood: A case study considering the effect on CO_2 emissions and the possible benefits of changing roof form. *Australasian Journal of Environmental Management, 13*(4), 216–225.

Grossman, V. (2013). Structural change, urban congestion, and the end of Growth. *Review of Development Economics, 17*(2), 165–181.

Henderson, V. (2003). The urbanisation process and economic growth: The so-what question. *Journal of Economic Growth, 8*, 47–71.

IMF. (2020). *World economic outlook database.* October 2007 Washington DC. International Monetary Fund. Retrieved March 5, 2020, from https://www.imf.org/external/pubs/ft/weo/2007/02/weodata/weorept.aspx?sy=2007&ey=2007&scsm=1&ssd=1&sort=country&ds=.&br=1&c=001&s=NGDPD&grp=1&a=1&pr1.x=57&pr1.y=10.

Liu, Z., Tan, Y., & Chai, Y. (2019). Neighbourhood-scale public spaces, inter-group attitudes and migrant integration in Beijing, China. *Urban Studies, 19* pp.

Lomborg, B. (2001). *The Skeptical environmentalist: Measuring the real state of the world.* Cambridge University Press.

Lomborg, B., & Rubin, O. (2009). *The dustbin of history: Limits to growth* (Foreign Policy Nov 9). Retrieved March 5, 2020, from https://foreignpolicy.com/2009/11/09/the-dustbin-of-history-limits-to-growth/.

NOAA (National Oceanic and Atmospheric Administration). (2020). *Monthly Average Mauna Loa CO_2.* Retrieved March 5, 2020, from https://www.esrl.noaa.gov/gmd/ccgg/trends/.

OECD. (2007). *OECD annual report 2007.* OECD Publications.

Oswald, Y., Owen, A., & Steinberger, J. K. (2020). Large inequality in international and intranational energy footprints between income groups and across consumption categories. *Nature Energy, 5*(3), 231–239.

Otto, I. M., Kim, K. M., & Lucht, W. (2019). Shift the focus from the super-poor to the super-rich. *Nature Climate Change, 9*(2), 82–84.

Oxfam. (2015). *Extreme carbon inequality.* Oxfam Media Briefing 2 December 2015. Retrieved March 7, 2020, from https://oi-files-d8-prod.s3.eu-west-2.amazonaws.com/s3fs-public/file_attachments/mb-extreme-carbon-inequality-021215-en.pdf.

PRB (Population Reference Bureau). (2015) *2015 world population data sheet.* Washington DC. Retrieved July 6, 2020, from https://www.prb.org/2015-world-population-data-sheet/.

PRB. (2020). *World population highlights 2007: Overview of world population.* Washington DC. Retrieved March 5, 2020, from https://www.prb.org/623worldpop/.

Reichgelt, T., D'Andrea, W., Valdivia-McCarthy, A., Fox, B., Bannister, J., Conran, J., Lee, W., & Lee, D. (2020, 20 August) Elevated CO_2, increased leaf-level productivity, and water-use efficiency during the early Miocene. *Climate of the Past, 16*(4), 1509–1521.

Rowlatt, J. (2020). *Greta Thunberg: Climate change 'as urgent' as coronavirus.* London, BBC News. Retrieved July 6, 2020, from https://www.bbc.com/news/science-environment-53100800.

Shahbandeh, M. (2019). *Global value of the cosmetics market 2018–2025*. Nov 27. New York, Statista. Retrieved July 10, 2020, from https://www.statista.com/statistics/585522/global-value-cosmetics-market/.

SIPRI (Stockholm International Peace Research Institute). (2019). *World military expenditure grows to $1.8 trillion in 2018*. 29 April. Stockholm. Retrieved July 14, 2020, from https://www.sipri.org/media/press-release/2019/world-military-expenditure-grows-18-trillion-2018.

Stern, N. H. (2007). *The economics of climate change: The Stern review*. Cambridge University Press.

The World Bank Group. (2019a). *Military expenditure (current USD)*. Retrieved March 5, 2020, from https://data.worldbank.org/indicator/MS.MIL.XPND.CD.

The World Bank Group. (2019b). *GNI per capita PPP (current international $)*. Retrieved March 6, 2020, from https://data.worldbank.org/indicator/NY.GNP.PCAP.PP.CD.

The World Bank Group. (2021). *GDP (current US$)*. Retrieved February 5, 2021, from https://data.worldbank.org/indicator/NY.GDP.MKTP.CD.

Thiers, M. A. (1850). *The history of the French revolution*. Translated, with notes and illustrations from the most authentic sources, by Frederick Shoberl. Philadelphia: A Hart, late Carey and Hart.

UCS (Union of Concerned Scientists). (2014). *The US Military and Oil*. June 1. Cambridge MA. Retrieved July 13, 2020, from https://www.ucsusa.org/resources/us-military-and-oil#.Wlf idBQYRJY.

US EPA (United States Environmental Protection Agency). (2019). *Global greenhouse gas emissions data*. Last updated September 13. Retrieved March 7, 2020, from https://www.epa.gov/ghgemissions/global-greenhouse-gas-emissions-data.

Worldometers. (2020a). *World population by year.* 13 July. Dover, Delaware. Retrieved July 13, 2020, from https://www.worldometers.info/world-population/world-population-by-year/.

Worldometers. (2020b). *Current world population*. Retrieved July 14, 2020, from https://www.worldometers.info/world-population/.

York Civic Trust. (2020). *Shambles*. Retrieved August 10, 2020, from https://yorkcivictrust.co.uk/heritage/civic-trust-plaques/the-shambles/.

GPSR Compliance

The European Union's (EU) General Product Safety Regulation (GPSR) is a set of rules that requires consumer products to be safe and our obligations to ensure this.

If you have any concerns about our products, you can contact us on

ProductSafety@springernature.com

In case Publisher is established outside the EU, the EU authorized representative is:

Springer Nature Customer Service Center GmbH
Europaplatz 3
69115 Heidelberg, Germany